Electrical Workshop

Safety, Commissioning, Maintenance & Testing of Electrical Equipment

Electrical Workshop

Safety, Commissioning, Maintenance & Testing of Electrical Equipment

THIRD EDITION

R.P. SINGH

Director

Bimla Devi Educational Society's
Group of Institutions
JB Knowledge Park
Faridabad

TECHSAR™

NEW DELHI

Published by

Techsar Pvt. Ltd.
4435-36/7, Ansari Road, Daryaganj
New Delhi–110 002 (India)
E-mail: tech_info@tehsarworld.com
Website: www.techsarworld.com

ISBN 978-93-81141-20-5

Reprint 2025

Printed by Shree Maitrey Printech Pvt. Ltd. Noida

Dedicated to

The engineers, the nation builders

Preface

In order to prepare engineering students to apply their technical skill in solving the problems in their day-to-day lives, the subject knowledge of electrical workshop is necessary. Awareness and proper knowledge of this subject is necessary for supervisors, engineers, planners, designers and technicians involved in preparation of project proposal, installation, testing, commissioning, maintenance and operation of electrical plants and machineries.

Though it is essential but electrical safety principles do not find sufficient coverage in the curriculum of the professional courses. There is tremendous industrial growth as well as use of electricity in agricultural and domestic applications. A large number of lives and huge amount of properties are lost due to negligence of safety aspects. Electric shocks, fires, explosions, flashovers and faults in the industrial plants and machineries are the common risk factors. Electrical accidents can be reduced up to a great extent by establishing and following the safety requirements and procedures.

There are very few types of equipment, which fail due to aging effects. Majority of equipment get damaged due to improper supervision and maintenance. In the present competitive and commercial scenario, investment on the proper maintenance of the plants and machineries are paid back rich dividends in the terms of less replacements cost, improved production and supply position and finally a secure operation. Proper installation, commissioning and routine testing of the equipment and their components directly or indirectly expose the nature of the incipient faults. This in turn helps in diagnosing the abnormalities and in taking the remedial measures.

This book presents the subject understanding in two sections. The first section; Section A – Learning by Theoretical Description covers Basics of Maintenance of Electrical Equipment, Details of Electrical Safety, Tools used during Installation, Maintenance and Testing, Earthing and Underground Cables. Details of the following make this book very useful to the field engineers as well as teachers and students.

- Fundamentals of Safety, Installation, Commissioning, Maintenance and Testing of Electrical Equipment
- Maintenance of DC Batteries
- Commissioning, Maintenance & Testing of Power Transformers
- Commissioning, Maintenance & Testing of Electric Motors
- Commissioning, Maintenance & Testing of Alternators

- Maintenance of DG Sets
- Maintenance of Substation Equipment & Transmission Lines
- Automobile Electrification
- Electrical Wiring Systems
- Indian Electricity Rules

To reinforce theoretical instructions to the related practical, instruction sheets are prepared and provided as Section B – Learning by Experiments. The instruction sheets are the solutions of the practical difficulties, correlating the theory and practice. In the instruction sheets, step-by-step instructions in detail for carrying out experiment along with relevant theory in brief is given. A set of relevant questions is given at the end of instructions sheet for each experiment. Students for answering these questions will be required to go for thorough study of the topics.

The author, sincerely, hopes that the book shall earn wide appreciation both from the faculty members and the students. We shall welcome and gratefully acknowledge the advice, suggestions and constructive criticism about the book.

Dr. R.P. Singh

Contents

SECTION B – Experiments

SECTION-A

Theoretical Description

1

Maintenance of Electrical Equipment

1.1 INTRODUCTION

Routine inspection and alertness are two main pillars of the proper upkeepment of any system or apparatus. It is not only visual look after, but proper record entry of the conditions of different components of the system is also essentially required. Logbooks are placed at different sections of the systems, e.g., at substation control room, battery room, switchyard for recording and condition monitoring of the equipment and their parts.

For keeping up-to-date record of the condition of a substation and its associated equipment, protective and measuring devices, a daily log sheet is prepared by the substation authority. Log sheet gives following information on hourly basis.

- Voltage at different feeders
- Power factor of the systems and substation
- Transformer tap position
- SF_6 gas pressure of the circuit breaker
- Battery condition and charging requirement
- Loading and running schedule of the feeders
- Supply frequency and variation if any
- Report regarding major equipment like transformer, circuit breakers & reactors, etc.
- Ambient temperature and weather conditions

1.2 INSPECTION OF THE MACHINES

Examination of the machine and their components to judge their condition or state is known as inspection. Besides routine inspection, a detail inspection is must at the time of receiving the machines. By inspecting the machine properly, one can ensure that machines are received in good condition and as per required specifications. The competent person who has in-depth knowledge of those particular machines should do inspection.

Intimation regarding missing accessories of the machine should be sent to the supplier immediately after unpacking and preliminary inspection of the machine. Marked instructions and advice note on the tag should be strictly followed while unpacking the machine. Inspection of the machine includes following:

- External inspection of the machine
- Opening the terminal lids, inspect terminals
- Removal of the dust with the help of blower
- Ensure free rotation of the rotating parts of the machine by turning with help of the hand
- Caution notices are visible and clear
- Ensure that bearings of the machines are functioning well

Special attention should be given during inspecting the machines stored for long duration. Exposed portions of the machines should be brushed and cleaned nicely.

1.3 HANDLING OF ELECTRICAL APPARATUSES

Proper earthing and earth connections are must before electrical apparatus and connecting conductors are handled. Earthing protects the working personnel by discharging apparatus being charged accidentally. No person should work on any live electric supply line or apparatus. First, he or she should take shutdown of that faulty section and take all the safety measures approved by the safety manager.

1.4 MAINTENANCE

To minimize the downtime of any plant or system, quick restoration of the operation of the machineries is essentially required. Maintenance involves those activities, which keep the machineries in healthy operating conditions. Maintenance can be classified as:

- Corrective maintenance
- Contracted maintenance
- Breakdown maintenance
- Preventive maintenance

1.5 CORRECTIVE MAINTENANCE

Corrective maintenance is required to be carried out when any equipment does not perform its job satisfactorily. This is also required when any equipment fails to perform.

1.6 CONTRACTED MAINTENANCE

Contracted maintenance is carried out according to the terms and conditions agreed upon by both the parties, the owner of the machine and firm authorized to take up the maintenance work. Both the preventive maintenance as well as corrective maintenance may be undertaken under contracted maintenance.

1.7 BREAKDOWN MAINTENANCE

Breakdown of a plant is referred as sudden failure and interruption in the production. The main reasons of sudden failure or breakdown may be one or more of the following:

- Accident
- Negligence
- Overload
- Wear and tear
- Incorrect design
- Mishandling
- Incorrect installation and commissioning
- Improper testing
- Inadequate protection
- Inadequate preventive maintenance

Following are the requirements for carrying the breakdown maintenance:

- Readily available inventory and spare parts
- Availability of the suitable tools, mechanized lifting, infrared heating and jointing the rods of high capacity wire by welding must be preferred
- Proper analysis is recommended if any equipment requires unexpected and frequent repairs
- Overall condition of any equipment is accurately accessed by the proper upkeepment of the record entry of all the defected faults. Proper-recorded logbook helps to plan better maintenance work and identify the faulted section
- External as well as internal inspection should be carried out by an experienced personnel to identify the fault

1.8 PREVENTIVE MAINTENANCE

Due to aging, wear and tear, adverse climatic and working conditions machines may suffer partial or complete breakdown. Sudden and unplanned breakdown creates chaos for supply and production scenario and loss to the companies. Timely repair of the components without neglecting smallest part of the machine is necessary, otherwise even small fault or maloperation of any component may lead to a major breakdown. A well-scheduled preventive maintenance gives trouble free continuous service. To keep the electrical equipment in a fit operating condition, following functions should be observed honestly.

- Regular visual inspection of all parts of the equipment
- Redressal of the incipient faults and minor defects such as tightening loose fastenings
- Routine testing such as periodic measurement of the insulation resistance, insulation testing air gap measurement between rotor and stator, dielectric measurement of the transformer oil

While preparing the preventive maintenance schedule, following aspects are required to be sought.

- When, what and how to inspect?
- How often to inspect?
- Who to inspect?
- What and how to record?

Keeping in the mind above aspects, the planning of the preventive maintenance should be made considering the following factors.

- Check up the surroundings of the machineries
- Check up the working conditions of the equipment
- Listing of the consumable items required
- Check up the stationary parts
- Check up the rotating parts

1.8.1 Advantages

1. Prevents premature failure and unscheduled interruptions.
2. Determine need for minor and major repairs.
3. Reduces chances of sudden breakdown.
4. Provides low cost maintenance.
5. Gives accurate condition monitoring for machineries.
6. Provides better safety and protection to the workers.
7. Increases lifespan of the machines.
8. Less use of standby equipment are needed.
9. Preventive maintenance helps to plan flexibility in operation and makes working conditions better.

1.8.2 Classification

Routine maintenance, periodic maintenance, overhaul and immediate fault maintenance may be the different types of the preventive maintenance.

Routine Maintenance

- Visual inspection should be done daily
- Replacement of small parts and adjustment
- Removing the dirt and dust
- Cleaning of the fittings and covers of the machines
- Cleaning the fixed and moving contacts of equipment
- Replacement of the burnt contacts
- Washing and lubrication of bearings

- Replacement of carbon brushes
- Checking the control equipment

It is done without dismantling the machine. However, before going for repair work, disconnection of supply is necessary.

Periodic Maintenance

It is also known as '*Scheduled Maintenance*'. Periodic maintenance requires correcting:

- Ventilation
- Loose parts
- Oil change
- Insulation checking
- Replacement of worn parts.

Depending upon the machine condition the nature and criticality, periodic maintenance may be planned and carried out daily, weekly, fortnightly, monthly, quarterly, or half yearly basis.

Overhaul

Overhaul of the machine includes:

- Dismantling the machine
- Replacement of the main units
- Renewal of their assemblies and main working parts, i.e. rewinding of the stator and rotor of the rotating machines, complete oil changing, varnishing of the winding of the transformers, dehydrating the oil, replacement of the bushings of high voltage circuit breakers, etc.

Overhaul is required after continuous run of the machine for a year or longer duration.

1.9 COMMON MATERIALS REQUIRED FOR MAINTENANCE WORK

1. Lubricants, i.e. oil, grease
2. Sandpaper for removing the dust, dirt, contaminants, pollutants
3. Carbon brushes
4. Lamp holders
5. Solder wires
6. Embry paper
7. Thimbles
8. Ball bearing
9. Nuts, bolts and washers
10. Rubber gloves
11. Brushes

12. Socket outlet and plugs
13. Varnish
14. Lead terminals
15. Fuses
16. Double pole iron clad switch and triple pole iron clad switch
17. PVC/VIR cables
18. Capacitors
19. Winding wires
20. Contacts of starters
21. Fuse carriers
22. Fuse wires

1.10 LUBRICATION

Rotation of the shaft of machines at high-speed generates substantial amount of heat. Heat adversely affects the life of bearings and their performance. So, it is obvious that mating surfaces of the shaft-journal and the bearing surfaces should not come direct in contact with each other during running of the machine. Hence, lubrication is required to avoid direct contact between two mating surfaces, shaft-journal and surfaces of the bearing. Application of the lubricants develops a thin film between the mating surfaces. Avoiding direct contact enables the shaft to rotate free and smooth without getting overheated. Lubricants also work as coolant.

Liquid, semisolid and solid are three types of lubricants. Graphite is solid lubricant while oils and grease are liquid and semi-solid lubricants respectively. Lubrication is done by adapting any of the following methods.

- Ring lubrication
- Forced lubrication
- Wick lubrication
- Bottle lubrication
- Gun lubrication

Bearings carrying normal loads are lubricated with oil by ring lubrication. Using forced oil lubrication method preferably lubricates shafts running at high speeds and bearings carrying heavy loads.

Greases are used as very common lubricants in low rated machines, especially for domestic appliances. A grease gun is a suitable apparatus for doing grease lubrication.

Types of Lubricants

Lubricants are named under different trade names. Some are stated here under:

- Mineral oils composed of hydrocarbons
- Fatty oils and acids sometime mixed with mineral oils
- Waxes derived from petroleum. These are soft wax, bee's wax, paraffin wax

- Graphite suspensions like amorphous graphite suspensions, colloidal suspensions and dry flake suspensions
- Sulphurous mineral oils and chlorinated compounds

1.11 IMPACT OF CONTAMINATION ON THE ELECTRICAL INSULATION

With the industrial growth, on one side power demand is increasing day by day and on the other side pollution is becoming severe and severe throughout the world. The increase in demand of power necessitates going in for high voltage transmission due to economic reasons. As the transmission voltage level increases, the insulation contamination poses a major problem to maintain the reliability of power supply. Earlier, it was felt that insulation performance in higher voltage system was dictated by its switching surge behaviour. However, it has now been realized that the constraints are mainly imposed by pollution.

Insulation protects against atmospheric as well as switching over voltages and works as a component at first line of defence to protect costly generators, transformers and other installed equipment in substations, and other types of factories using electrical machineries and electricity.

Generally, the pollution on a solid insulation surface, i.e., housing of the surge arresters or body of the insulators do not cause serious problem under fair weather because in this condition, layers are substantially non-conducting. Lightly wet layers start conducting significantly. Due to Joule heating, there will be uneven drying up of the wet layers. Thick layers can have, under wet condition, a very different conductivity than the superficial layers. So, it is obvious that drying of the wet conducting layer will also be non-uniform. Due to non-uniform drying of layers, current gets constricted to a few narrow paths, where there may be intense heating giving rise to hot moisture, formation and growth of dry bands. The voltage across dry bands may rise to a very high value leading to arcing. Continuous arcing sometimes cause complete flashover across the insulation and loss of its insulation property.

Insulations near a cement factory, face a little different types of problems. The cement particles, those come in contact with humid air, start setting in air itself. Before they are permanently set, these cement particles are carried by winds and thrown on insulation surface and moist cement particles deposited get permanently set on the insulators themselves. These deposits are not soluble in water and rain.

1.11.1 Causes of Pollution

The worldwide power generation largely depends on the combustion of fossil fuel-coal, which produces major pollutants during its combustion as fly ash, SO_x, NO_x, CO and CO_2. These are the gases responsible for acid rain.

According to IEC Publication 71-2: Insulation Coordination Application guide, pollution severity of sites has been divided into four levels given below. For each level of pollution, a rough description of some typical corresponding environments is made available. These pollutants and contaminants become conductive in the presence of fog, dew, rain and snow. With concentrating effects of pollutants on electrical transmission system, corona also becomes more effective on the line in the presence of conductive pollutants. These pollutants speed up ionization processes around the conductor resulting into increased loss of energy. Early corrosion on malleable iron caps, forged ball pins and stem of insulators, surge arresters and other equipment take place in the presence of pollutants under adverse atmospheric conditions like rain, fog, ice, etc.

Pollution levels and examples of typical environments

Light

- Areas without industries and with low intensity of houses equipped with heating plants.
- Areas with low density of industries or houses but subjected to frequent winds and /or rainfall.
- Agricultural areas (*)
- Mountainous areas

All these areas must be situated far from the sea (10 to 20 km) and must not, in any case, be exposed to winds from the sea.

Medium

- Areas with industries not producing particularly polluting smokes and/or with average density of houses equipped with heating plants.
- Areas with high density of houses and/or industries but subjected to frequent clean winds and/or rainfall.
- Areas exposed to winds from the sea but not too close to the coast (at least a few kms).(**)

Heavy

- Areas with high density of industries and suburb of large cities with high density of heating plants producing pollution.
- Areas close to sea or, in any case, exposed to relatively strong winds from the sea.

Very Heavy

- Areas generally of moderate extension subjected to conductive ducts and to industrial smoke producing particularly thick conductive deposits.
- Areas generally of moderate extension, very close to the coast and exposed to sea sprays or to very strong and polluting winds from the sea.
- Those desert areas, characterized by no rain for long periods, exposed to strong winds carrying sand and salt and subjected to regular condensation.

(*) Use of fertilizers sprayed by gun or by plane, or where crop burning has taken place can lead to a higher pollution level.

(**) These distances from seacoast depend on the topography of the coastal area and on the wind condition.

1.11.2 Remedial Measures to Avoid Surface Contamination

The main objective of the field engineers and researchers has been to know what economically and technically sound measures should be taken to avoid insulation failure. The common way of determining the behaviour of external insulation is to carry out artificial pollution tests, as representative as possible of the actual pollution. During the pollution test methods based on total flashover need a costly power source and lead to a considerable maintenance cost. So, the most economic method of assessment of pollution is by investigating the physical processes of the pollution flashover, either on model insulators or using computer modelling techniques. At the same time on the principle of "precaution is better than cure" remedial measures should be adopted to prevent flashover/failure.

The extension of industrial areas and cities can contribute to increase in the different types of pollution for sites. Transmission and distribution lines have to pass through areas of this pollution. The following primary remedial measures to minimize accumulation of contaminants and flashover on the surface of insulators, surge arresters and bushing are generally adopted.

Manual Wipe Cleaning

In this method, first a wipe cleaning with a non-fluffy cloth is done, then the same insulator is cleaned with water and finally wipe cleaned with dry cloth at regular intervals. This is recommended for very light pollution only.

Recoating of Silicon Greasing

The silicon greases are composed of a liquid phase, a solid phase and additional chemical agent to protect the coating from ultraviolet rays. The mechanism of action of these compounds has been divided into three steps.

- Oil exudation
- Encapsulation of the particles of contaminant
- Gradual saturation of coating

Ultraviolet rays make the exuded oil hard on the surface and decrease the trapping efficiency, which can stop before the grease is saturated.

Since there is no self-cleaning on coated insulators, so after some time, leakage current can decay the silicon grease. This decay appears as whitish trails on the coated surface. The coating must be renewed before long whitish trails are observed. RTV (Room Temperature Vulcanized Silicon Rubber) coating is easier to apply as compared to silicon grease and can be hot washed.

Fixed Hotline Washing

The direction, intensity and duration of wind influence the rate of built-up of pollutants on the insulator surface to a large extent. If rainfall of sufficient intensity occurs at intervals less than the period in which the built-up of contaminants on the insulators reaches critical values, the rain acts like natural hot-line insulator washing system.

The only way to prevent the building up of the contaminants to the critical limits is to clean the insulators before these limits are reached. Using Automatic Hotline Insulator Washing System can do that. The housing of energized conventional and metal oxide arresters can be cleaned without service interruption by applying a suitable flow of high resistance hot water. The system of high resistive water supply consists of pumps, distribution pipes and nozzles. Pollution sensors, differential anemometers and sequential control systems make it an automatic system. The two commonly used fixed hotline washing systems are:

The Spray System uses fixed nozzles around each insulator. It is efficient only if the wind velocity is low and does not divert the spray water.

The Screen System consists of a number of nozzles arranged in a line producing a vertical water curtain in front of the equipment. The wind blowing from the opposite direction diverts the water jets towards the insulator. At the same time, the water screen dissolves part of the salts brought by wind and reduces the pollution intensity. During long lasting heavy winds, fast building up of contaminants takes place on the insulator surface. For such situations, the screen washing method is more effective. The capital and operational costs of hotline washing installations are very high.

Mobile Manual Hotline Washing

This system comprises a stainless steel tank of required capacity in which demineralized water of conductivity of 0.5 to 1.0 micro mho per cubic centimetre and carbon dioxide gas of commercial quality are mixed. Out of the five nozzles on the top of the tank, one nozzle is connected to the CO_2 cylinder through a flexible hose and a regulator. A pressure gauge, a relieve valve set and a drain valve are each mounted on the separate nozzles. To the remaining nozzle, a long flexible hose is connected and the other end of this is connected to an insulated spray stick. This stick has a controlling lever at the bottom from where the flow of spray on the other end can be controlled.

The CO_2 gas partially dissolves in water under pressure. Due to acidic nature of the fluid ($H_2O + CO_2 = H_2CO_3$), the conductivity of the fluid that passes through the hot spray stick increases. The dissolved CO_2 tries to escape from the water at the outlet of nozzle and hence the spray is in the minute droplets and not as a continuous jet, thus increasing the resistance of the washing column of jet. CO_2 is a very good fire-extinguishing agent also. The conductivity of the fluid at the outlet of the nozzle should not be greater than safe values such that leakage current passing through the operator while doing hotline washing is limited to five microamperes.

Wind Deflectors

These are used near critical equipment. It does not provide effective and adequate protection to the insulators of ground equipment such as isolators, circuit breakers, potential transformers, etc. due to the fact that wind direction keeps on changing continuously. The height of the structure (erected in M.S. angles and asbestos sheets) is normally not increased beyond four metres as it may pose a clearance problem with energized lines. It has not been found feasible and cost effective.

Green Belts

To avoid deposition of contaminants on the electrical insulation near chemical or smoke producing factories or salt deposition intensity partially on the insulators near the seashore, green vegetation of sufficient height along the factories and coastline, diversion results into minimization of the contaminant deposition and salt laden spray on to the switchyard equipment.

Semi-conducting Glazing

In order to attain the necessary power frequency insulation strength during wet contamination, insulator is provided with semi-conducting glaze. Power dissipated in semi-conducting glaze raises the surface temperature several degrees above the ambient. It prevents the fog condensation on the surface. In wet and contaminated conditions, it helps in the formation of very large dry bands or in the process of drying of the insulator surface.

Dielectric glaze in which Barium Titanate is used as a major constituent is also in use. Barium Titanate increases the capacitance of the insulator as a whole, which as a result linearizes the voltage distribution across the leakage path, increases withstand/flashover voltage, suppresses the transients during flashover, reduces corona loss and radio noise. This also improves glaze stability.

1.12 SOLDERING

Soldering is the process of joining the two or more than two pieces of electric wires by using an alloy known as *solder*. Soldering provides protection to the surfaces being corroded. Loose joints of the wires heat up when current passes through it. Some times sparking takes place and finally causes separation of the conductors while loose connected. So soldering serves the purposes like:

- Provides all electrical joints of good mechanical strength and electrical contact.
- It increases electrical conductivity and gives better appearance.
- It protects the joined surfaces from the corrosion.

1.12.1 Types of Solders

Soft Solder

Soft solders are used for tinning electrical and joining them. It is an alloy of tin and lead. Composition of tin and lead depends on the working temperature and application of the wires. The general proportions and uses of the soft solder are shown in Table 1.1.

Table 1.1

Composition	Working temperature	Use
Tin 50%; Lead 50%	200°C	Fine soldering
Tin 60%; Lead 40%	185°C	Tinning electrical wires
Tin 90%; Lead 10%	219°C	Tinning electrical wires
Tin 35%; Lead 65%	237°C	Heavy duty solder

Hard Solder

Hard solders are made to make the joints when the wires are to be applicable for very high temperature environment. These are high zinc brasses, which melt around 850 to 900°C. These are used to recondition electrical contacts tips, etc. by brazing.

Silver Solder

Alloy of silver, copper and zinc makes the silver solders. This has very good corrosion resistance property. They make joints of better conductivity and strength. These are used to join high voltage grade conductors, i.e., armature conductors with commutator risers.

Aluminium Solder

These are used to join aluminium joints. These are alloy of tin, lead, zinc and cadmium.

1.12.2 Flux

This is a substance, which assist the solder during joining of metal surfaces. The main function of flux is:

- To remove oxide layers of joining surfaces.
- To permit easy flow of solder (or joining metal) on joining surfaces.
- To protect the joining surfaces against re-oxidation.

Different types of fluxes are Zinc Chloride, Sal ammoniac rosin, Tallow turpentine, Eyre 7 Para fluxal, etc.

1.12.3 Soldering Equipment

Soldering Iron

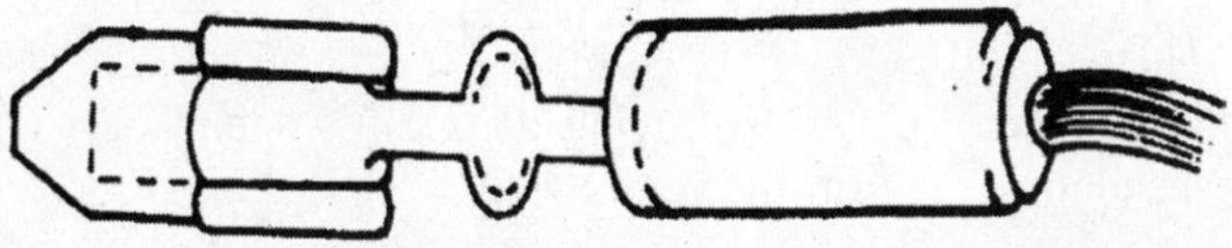

In an electric soldering iron a piece of copper drawn to a point or edge is fastened to an iron rod fixed in a wooden handle. In it, the heat is obtained from a nichrome wire element of power, 10 W for smaller sizes to 2 kW for bigger sizes. These works with the voltages, 10 V to 250 V.

Soldering Pot and Ladle

A cast steel pot to melt the solder is generally used. The flame of a blowlamp heats up the pot. The pouring hot solder is done by means of a ladle.

Blowlamp

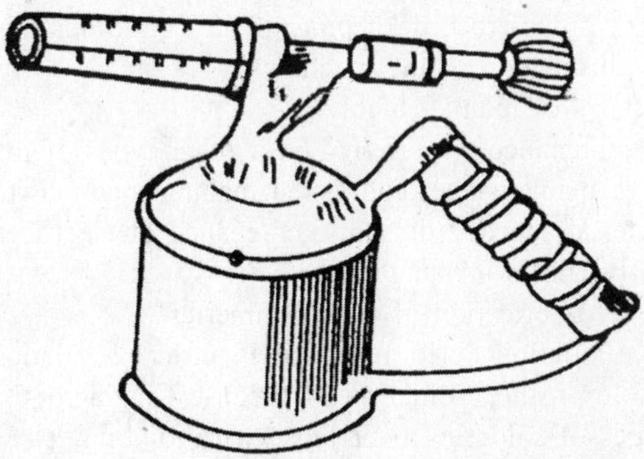

It is used to give open flame to melt the solder. Blowlamp is filled with petrol, kerosene or spirit. The flame obtained from a blowlamp directly heats up the job and solder or a soldering iron. The flame is controlled by the pressure of air of the tank and the wire pin inside the nipple.

1.12.4 Soldering Process

Following are the steps for soldering:

- Cleaning the surfaces to be soldered
- Applying the flux evenly on the wires

- Maintaining the temperature within prescribed limit
- Heating the soldering bit
- Tinning the soldering bit
- Applying flux on the clean surfaces
- Wait till the solder solidifies
- Cleaning of the soldered joint by removing the extra solder
- Wrapping the insulation tape on the joints

1.12.5 Precautions to be taken During Soldering

- Be careful for the flame, hot iron, hot jobs and hot parts during soldering.
- Ensure that solder has melted well before making joints.
- Do not use the solder horridly.
- Avoid excess solder, which gives troubles in tapping.
- Wash and clean the soldering equipment properly.
- Must use the safety goggles and gloves during soldering process.
- Ensure that soldering is being done in a well-ventilated place. Special care must be taken particularly while doing aluminium soldering.

EXERCISES

1. Write importance of proper record keeping of the conditions of the machines.
2. Write note on the 'Inspection of the machines'.
3. Discuss different types of maintenances.
4. Why preventive maintenance is preferred over other types of maintenances?
5. Compare, corrective maintenance, contracted maintenance and preventive maintenance.
6. Make a list of materials needed during maintenance of electric machines.
7. Describe in brief 'Breakdown maintenance'.
8. Write in brief about 'Lubrication of the machineries'.
9. Write in detail on 'Pollution Performance of Electrical Insulation'.
10. Write a detailed note on the 'Soldering of Electric Conductors'.
11. Write the advantages of soldering over loose wire jointing.
12. Describe different remedial measures to avoid contamination on the insulation surface.

2

Electrical Safety

2.1 ELECTRIC SHOCK

2.1.1 Introduction

Though the electricity is very useful in bringing the human life very comfortable. But, it may be fatal, if not handled carefully. The human failure in handling the electrical components or repairing the distribution system without switching 'off' the supply, or not following the rules strictly, may cause severe problems, which may cause threat to the human life.

When a live conductor is touched by hand or any part of the body, current escapes through the body to earth. It may pass through the heart also. It is difficult to specify the exact voltage, which is likely to be fatal voltages below 100 volts may also cause a severe accident.

The result of electric shock on human body may be fatal due to:

(a) Fibrillation of heart, i.e., damaging the heart and causing the stopping of breathing.
(b) Stopping of breathing action caused by blockade in the nervous system causing respiration.
(c) Local over heating or burning of the body due to sparking.

Note: The fibrillation of the heart is the most serious cause of death and there is no cure, although there is a possibility of rescuing a man who has suffered by the causes mentioned above (b) and (c).

2.1.2 Electric Shock and its Effects

Frequency of Current

When the frequency is low, the electric shock is more severe and dangerous. This is the reason that direct current shock is most severe. Direct current does not vary and its full value passes through the body for the entire duration of contact. Alternating current varies and its value comes down to zero in every half cycle. At the time of zero value of the current, there is chance to detach the contact which lessen the damage to the human body.

Current Path through Human Body

If the path of the leakage current does not involve the chest or heart, survival is possible but there are severe burns on the parts of the body involved in the shock depending upon the intensity of the current.

Current

The current flow through the body or parts of the body gives shock although it depends upon the voltage. Leakage current between 20 mA and 50 mA, if passes through chest, it may stop breathing and currents between 100 mA and 200 mA may cause fibrillation of heart.

Table 2.1: The effect of current on the human body system

Current (mA)	Effect
1 to 8	Perceptible but not painful
8 to 15	Painful shock, individual can let go at will, muscular control is not lost
15 to 20	Painful shock muscular control of adjacent muscles lost, cannot let go
20 to 50	Severe muscular contraction, breathing difficulty
50 to 100	Possible death, ventricular fibrillation, a heart condition that results in instant death, no known remedy
200 and above	Certain death, severe burns, muscular contraction extremely severe that chest muscles clamp down the heart and stop it during the duration of shock

The current beyond 200 mA will cause burns and if it passes through heart, even it will not cause fibrillation of heart but may stop breathing temporarily. It is observed that alternating currents of low frequency at low current level between 1 mA and 8 mA are just bearable, but currents between 8 mA and 15 mA give a painful shock, which sometimes contract muscles too. The leakage current is given in the Table 2.1.

Body Resistance

Human body resistance depends on body size, height, weight, bone's conditions, etc. The high voltage causing currents beyond 20 mA punctures the outer skin causing burns. The dry human body resistance varies between 70 kilo-ohms and 100 kilo-ohms per square centimetre and wet body resistance reduces to between 700 ohms and 1000 ohms per square centimetre. The average effective resistance of the body may be taken as 50 kilo-ohms when dry and 1000 ohms when wet.

In wet conditions, the body resistance and contact resistances are low and therefore, even a small leakage current becomes dangerous. Although under such conditions, 50 volts is sufficient to prove fatal if applied for an appreciable time, however, this can only be considered as exceptional. Table 2.2 gives the observations regarding results of shock under different conditions and under different voltage levels.

Table 2.2: Observations regarding results of shock under different conditions and under different voltage levels

Body Condition	Body resistance (Ohms)	100 V Amp	Effect	500 V Amp	Effect	10 kV Amp	Effect
Complete wet	1,000	0.1	Certain death and slight burns	0.5	Burns and probable death	10	Severe burns and may survive
Partially wet	5,000	0.02	No burns or injury but painful shock	0.1	Certain death and slight burns	2	Severe burns and may survive
Dry	1,00,000	0.001	Very light shock but no burns	0.005	Light shock and no burns	0.1	Sure death and slight burns

Duration of Current Flow

For higher duration of the current flow, the situation becomes dangerous due to lack of sensation in the body at the contact. Such electric shock becomes dangerous and sometimes fatal if contact is not broken within the earliest possible time.

Contact Area of the Human Body

In the situations when resistance decreases at the area of contact with the live part and the contact pressure, the accident is nearly fatal. In the event of electric shock and contact near the chest and brain areas, death is likely.

2.1.3 Precautions to be Taken if a Person is Getting an Electric Shock

1. Do not allow any person to come in contact with the victim, physically. Try to disconnect the person from the supply by any means available. Immediate action is necessary and switch 'off' the electric supply at the location of the accident. If the main switch is not nearby, ask some one to contact the substation supplying the area to cut 'off' the supply temporarily. Use the telephone or send some one immediately.
2. Avoid direct contact with the victim when he is in contact with a live line. Switching 'off' the supply is obligatory, if that is high voltage circuit. First protect yourself with the help of dry non-conducting material and if the voltage is 500 volts or below every attempt should be made to free the person from contact with the live wire.
3. Stand on a dry plank, stool, table or any other insulating object and pull the man away from the mains in order to free the person from contact with live wire. If these are not available, pull the victim's dry clothes and use there clothes. You may also use your own coat, dry cloth, paper, bamboo stick, wooden pole, etc., either to pull or push the person without touching his body directly. Sometimes the live wire itself may be pulled or pushed away using a dry bamboo or any other dry woods make stick such as walking stick.

4. Send for a doctor.
5. Observe the victim whether he is unconscious and breathing normally. If the victim is not breathing normally, start giving artificial respiration without any delay. Proper training and practice is needed to give artificial respiration. After achieving the consciousness, stimulants should be avoided unless advised by the doctor.

2.1.4 Cure of Electric Shock

"Think twice; act once, correctly and quickly." In case, the victim becomes unconscious and stops breathing and his or her heart still beats, the most urgent and immediate cure for the victim is that he should be given artificial respiration in the proper manner. The step-by-step methods are described below and it should be continued until the victim starts breathing normally. It should be borne in mind that if the artificial respiration is stopped just after the victim recovers, he or she is liable to become unconscious again. In some cases, the artificial respiration is required for one or more hours. Since fresh air drawn into the lungs prevents the blood from becoming impure and will thus keep alive the sensitive brain cells.

2.1.5 Artificial Respiration

First Method

1. Lay the patient in a comfortable position on his stomach with his face to one side. Remove any obstruction from the patient's mouth by freeing the neck from clothing and see that passage of his throat is clear. Remove false teeth, tobacco, if any from the mouth.
2. Kneel over him or at his side as may be convenient and place the palms of your hands flat on his back, with the thumbs nearly touching and the fingers spread out on each side of the body under the lowest side of the ribs.
3. With arms straight, lean forward gradually over the patient bringing the weight of your body to bear on the patient for about a second. Next release all pressure for about three seconds by swinging your body backward but without lifting your hands from the patient.
4. Repeat this application and relaxation of pressure as described above, without any marked pause between the movements at the rate of about 12 to 15 times per minute until natural breathing is restored.
5. Do not give up efforts to restore natural breathing, until the victim is pronounced to be safe by a doctor. There is every possibility that the victim may survive even after more than an hour of unconsciousness.
6. The patient should be kept warm with blankets or coats and hot water bottles, if available, should be applied to the feet.
7. Do not give any liquid until the patient is conscious.

Figures 2.1 A & B describe the steps.

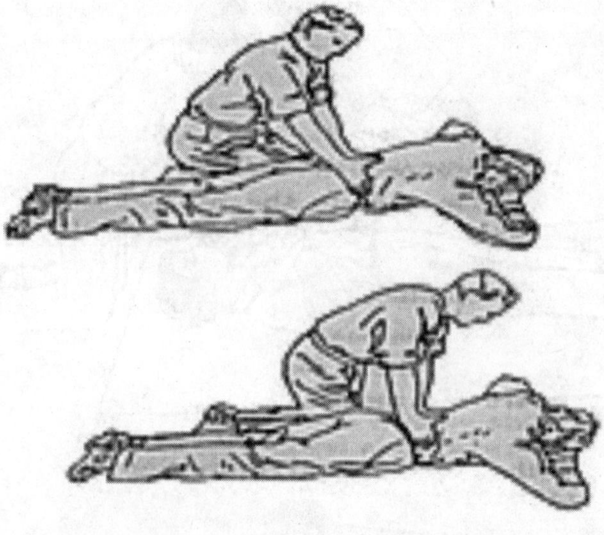

Figs. 2.1 A & B. First method artificial respiration.

Second Method [Silvester's Method]

When the patient has burns on his chest or anywhere on front side, then the patient must definitely not be laid on his front downward. The main process of artificial respiration is to expand and contract his chest to draw the air in and expel it out alternately. In this method:

1. Immediately place the patient on his back on a flat surface in a comfortable position.
2. Loosen the clothes around the chest and stomach and place a rolled up coat or improvised pillow beneath the shoulders so that the head falls backwards.
3. Remove all the tight clothing from the patient.
4. Raise his shoulder on a cushion or folded clothing so that his head hangs forward. Maintain a free entrance of air into the patient's windpipe.
5. With the help of someone, draw the tongue forward as far as possible with a handkerchief.
6. Draw the patient's arm upward, outwards and towards you with a sweeping movement, pressing his elbows towards the ground. This causes inspiration.
7. Bring his arms back along the same route.
8. Press them firmly against the front and the ribs of his chest. This causes expiration.

Repeat these cycles of movement smoothly and rhythmically about twelve times per minute till the patient begins to breathe by himself. See the pictorial demonstration in the Figure 2.2 A & B.

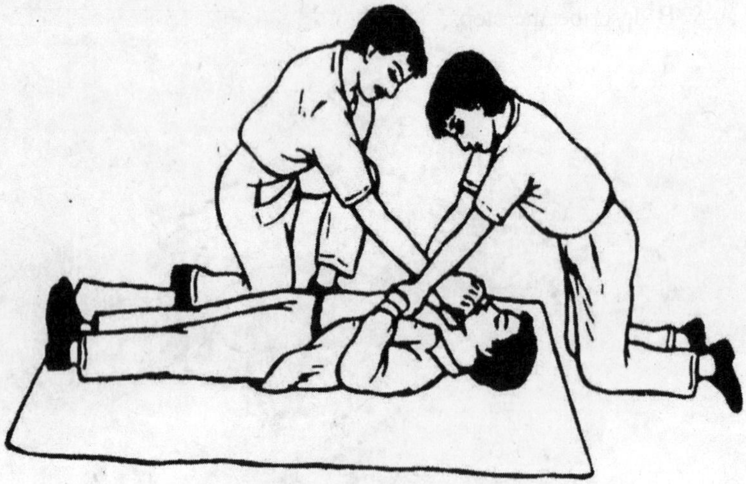

Fig. 2.2 A. Artificial respiration (Step-I).

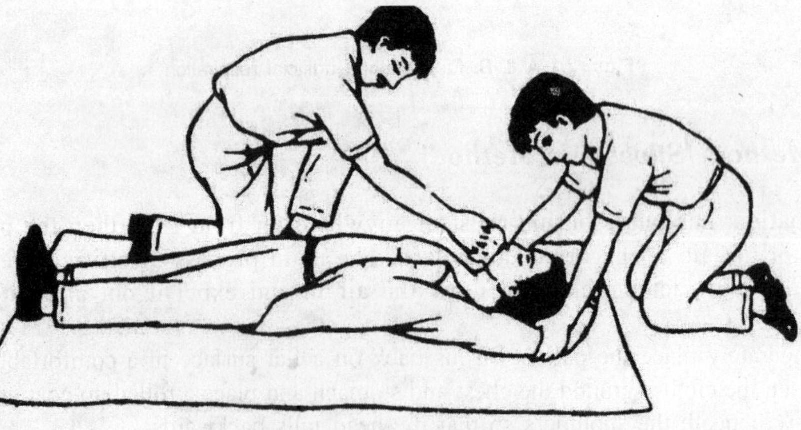

Fig. 2.2 B. Artificial respiration (Step-II).

Modern Mouth-to-Mouth Respiration Method

It is easily applied and particularly suited where the patient has suffered chest injuries.

Insert thumb of your left hand between victim's teeth. Hold the jaw upward so that the head is tilted backwards. Close victim's nostrils with your right hand. Take a deep breath and place your mouth tightly over victim's mouth and your own thumb. Blow forcefully enough to make victim's chest rise. Repeat inflation every three or four seconds. The main precaution to be adopted is that a handkerchief or some other cloth should be placed over the victim's mouth to prevent direct transmission of germs from one mouth to the other. Pictorial demonstration is presented in the Figure 2.3 A & B.

The mouth-to-mouth method has been found to supply ten to twelve times more volume of air into the lungs than any of the other methods. There is danger of sagging of the victim's

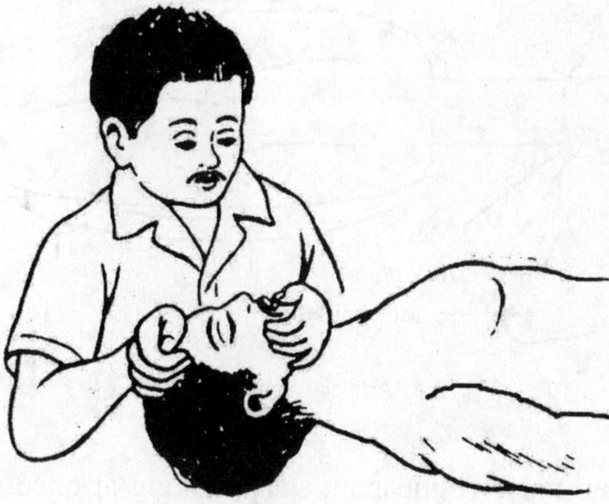

Fig. 2.3 A. Modern mouth-to-mouth respiration (Step-I).

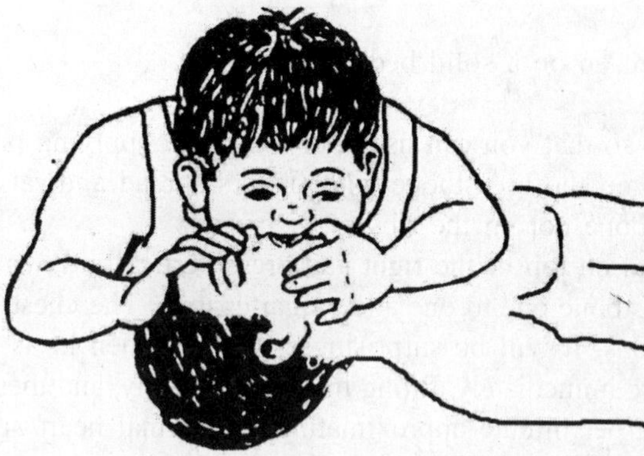

Fig. 2.3 B. Modern mouth-to-mouth respiration (Step-II).

tongue into the throat and blocking the air passage. But this can be prevented in the new technique if the victim's head is tilted back.

Artificial Resuscitator

It is recommended to have artificial resuscitator in all the electrical installations and maintenance and repair personnel. Its main components are rubber balloon, a valve, a mouthpiece and tubing. During artificial respiration, the mouthpiece is cupped on the mouth of the patient. Air is pumped into the chest of the affected person by deflating the balloon. In the next step, the balloon is allowed to be released so that it gets inflated by entering the atmospheric air through valve. The balloon is deflated again. The process is required to be repeated till the affected person restores normal breathing. Artificial resuscitator is shown in the Figure 2.4.

What to do if the heart stops beating?

1. The following main points may be of great help in reviving heart stoppage due to drowning, choking or electric shock:

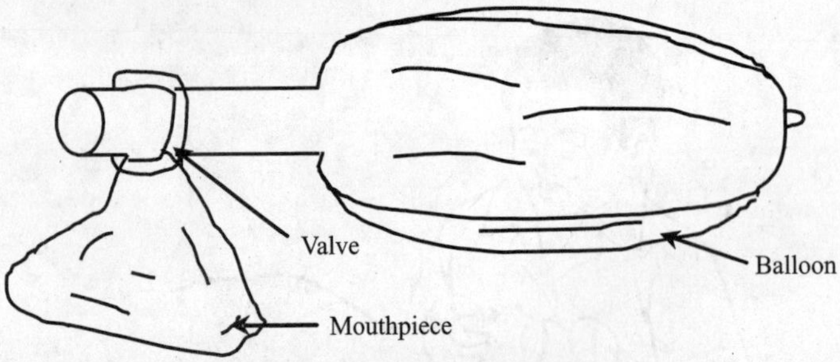

Fig. 2.4. Artificial resuscitator.

2. Check for pulse, if no pulse is apparent, start working at once without waiting seconds in asking for equipment or help. The lack of sufficient oxygen carried in the blood to feed the brain may result in peril of any heart. The brain is the most sensitive tissue of the body.
3. Lay the patient face up on a solid bed or floor.
4. Tilt the head back.
5. Kneel on the body so that you can use your weight in applying pressure. Place the heel of your right hand on the breastbone with fingers spread and raised so that pressure is only on the breastbone not on the ribs.
6. Place your left hand on top of the right and press vertically down wards, firmly enough to depress the breastbone one to one and a quarter inch. The chest of an adult is resistant when he is conscious. It will be surprisingly flexible when he is unconscious.
7. Release the pressure immediately, lifting the hands slightly, and then repeat in a frequency of 60 to 80 thrusts per minute approximating the normal heart action.
8. If normal heartbeat and respiration has resumed, even then the patient should be shifted to hospital as soon as possible.
9. Continue the massage until you get professional medical aid for the patient. Also continue if possible, the mouth to mouth breathing until the patient is put on oxygen.

2.2 ELECTRIC SAFETY

2.2.1 Introduction

"*Electricity respects those who obey safety laws. Risk is not always apparent; be careful; take no chance*". Electricity has revolutionized our lives and level of livings but human errors are the causes of most electrical accidents. It comes with risk and becomes lethal if not handled properly. One should not forget that electricity and machines have no 'feelings' for men's 'failings'. Safety rules and codes must be religiously followed. Electrical faults occur due to various non-electrical causes such as mechanical, thermal, chemical and magnetic.

Electrical safety essentially embraces four aspects, namely:

- Safety of people and other living beings
- Safety of materials and premises

- Safety of devices and machinery
- Protection of electrical equipment and circuitry

2.2.2 Classification

The electrical safety is categorized into two broad areas:

- Indoor and
- Outdoor

Above each of both are again subdivided into:

- Low voltage and
- High voltage

The areas can further be classified as follows:

- Domestic: Electric fittings, appliances, toys, showpieces, etc.
- Industrial: Administrative building, plant, campus, and common amenities, etc.
- Public utility: Streetlights, gardens and playgrounds, railway-substations, etc.
- Public services: Water supply, firefighting, sanitations, etc.
- Business organizations: Offices, banks, etc.
- Educational institutions
- Commercial establishments: Shops, service centres, etc.
- Workshops
- Substations
- Generating stations
- Distribution lines
- Hospitals-Medical centres
- Theatres, Auditoriums
- Religious places: Temples, Churches, and Mosques, etc.
- Temporary connections: Festivals, functions, pilgrimages, trade fair, exhibitions, circus, etc.
- Electric traction

2.2.3 Safety Precautions for Indoor Installations

Human body is a good conductor of electricity. Following safety precautions must be followed to avoid the risk of injury, or even death:

1. Do not put on the switches with wet hands especially during monsoon season.
2. Do not build houses/slums below the high-tension lines.
3. Frayed wires should be replaced at once.
4. Repair any appliance immediately that sparks, emit smokes, or shocks you.
5. Train children not to put things into electrical outlets. Plastic outlet guards are a good idea.

6. Keep work areas clean. Oily rags, newspapers, and sawdust can catch fire from electric spark.
7. Use the cutout fuses of appropriate ratings. Always switch off the main switch before replacing a blown fuse.
8 The main distribution board should have separate MCB for power (heating) and lighting besides a common DP switch. Any minor job should be done only after main DP switch and MCB are switched off.
9. Always maintain the earth connection in a satisfactory condition. Safety depends upon good earthing.
10. Be aware of live wires or conductors.
11. Always ensure that single pole switch is connected to the live line and not to the neutral.
12. Never disconnect a plug point by pulling the flexible wire.

2.2.4 Safety Precautions for Outdoor Installations

1. Keep the items like ladders, antennas, kites, balloons, model aeroplanes, and trees away from overhead power lines.
2. Never use electric power tools or appliances in the rain or while standing in water.
3. Do not ever climb utility poles or transmission towers when they are charged.
4. Do not let anyone shoot or through anything at insulators.
5. The transformers are inside sturdy metal cabinets, which are locked for safety. Never pry them open.
6. If you are caught in lightning storm, stay away from trees and whenever possible, stay dry. Go, indoor and keep clear of windows.
7. Before digging, learn the location of underground power lines. Call your electric utility company or assistant.
8. The main control panel must be located at ground floor, out of reach of general public.
9. Periodic inspection, measurement of insulation level of equipment and of installations, resistance of earth system must be made and documented. Remedial measures must be confirmed to relevant standards.
10. 'First aid' and 'Firefighting' facility should be located at the centre of gravity of loads with easy access.
11. Even low-tension overhead line should not pass just over the terrace of any building. Otherwise human being may get electrocuted which may cause severe shock via sensual nerves leading to death.
12. Whenever overhead lines or even busbars are under repair, first job is to earth them.
13. Never pull out the porcelain HV fuse cutout without switching off the main switch or all the connected loads. Otherwise the eventual arcs may chase the person to burn him to death.
14. The stored energy in the capacitance of HV motors, transformers, U/G cables and busbars gets discharged while switching off for repair if these are not properly earthed.
15. Never attempt to remove a person from an energized circuit with your bare hands.

16. Outlets near water sources, i.e., bathrooms, kitchen sinks, garage, outdoors should be GFCI (ground fault circuit interrupter) and ELCB (earth leakage circuit breaker) protected.
17. Always wear a safety belt before starting work above ground level, or on a pole.
18. In case of fire, disconnect the supply immediately. Do not throw water on equipment.
19. Electrical equipment should be regularly inspected and tested.
20. Rubber mats should be placed in front of electrical panels or switchboards. Always use electric rubber gloves while working above 650 volts. Always seek permission to the work, while servicing live switchgear.

The electric safety is a very elaborate subject and every element such as generation, transmission, distribution and load end users are responsible. No one can be exception. Selection of quality product, proper use and observing adequate care are essentially required and no compromise should be done.

Electricity gives us light but do not take electricity lightly. Otherwise you may not get scope to say sorry.

2.2.5 First Aid Practice

First aid is necessary to meet out the preliminary treatment after getting electrical accidents. These are required for:

- Burns and scalds received by a person due to electric shock.
- Bleeding because of wounds sustained by the victim due to electric shock.
- The victims of chemical burns.

The aim and objectives of the first aid are:

- To reduce the effects of the heat and to relieve pain.
- To lessen contamination and the risk of contamination.
- To reduce discomfort and swelling.
- To ensure that the patient takes sufficient fluids.

Burns

The contact burns are generally found on the body of the victim at the points at which the current has entered and left. The deeper tissues of the bodies through which the current has passed are also affected. If the victim is severely burnt, do not waste time. Take him to the hospital for treatment immediately.

Burns are very painful, if the large areas of the body are affected. For the burns caused by chemicals, the chemicals must be washed off very carefully by clean water before starting the treatment. Any wound bleeding profusely specially on the wrist, hand or fingers must receive professional attention. For immediate relief, first aid by pressing the wound is the best means of stopping bleeding and avoiding infections. However, trained personnel or doctors should be preferred.

Equipment and First Aid Materials

The ready stock of the first aid for use in the emergency is listed below.

S/N	Items	Quantity
1.	First aid kit	One set
2.	Adhesive wound dressings of different sizes	One set each
3.	Safety pins	One dozen
4.	Sterilized eye pad with bandage in a sealed packet	One packet
5.	Sterilized un-medicated dressings (small, medium and large)	One set each
6.	Adhesive plaster 25 mm x 5 metres	One spool
7.	100 gm packet of absorbent sterilized cotton wool	Two packets
8.	Triangular bandage of unbleached calico, the largest side of which measures not less than 125 cm and each of the other sides not less than 90 cm.	One set

Procedure of Initial Treatment

S/N	Electric burns and scalds	Chemical burns	Severe bleedings
1.	Dip the parts affected in the cold water	Flush the areas of injury with running water	Make the patient lie down and rest.
2.	Make the parts dry and clean	Remove the contaminated clothing from the victim	If possible, raise the injured part above the level of the body.
3.	Remove the items restricting blood circulation, i.e., belts, rings, bangles, etc	Cover the affected areas with clean dressings till the medical assistance is received.	Apply pressure to the wound.
4.	Give small quantity of cold drinks at frequent intervals		Summon medical assistance immediately.
5.	In case burn is liable to get dirty, it should be covered by clean dressings till the medical assistance is received.		

2.3 ELECTRIC FIRE

Fire generated danger to life of the working personnel in any plant, electrical generation, transmission or distribution system is most important issue of safety division. Safety to persons is always given higher weightage on property losses. Working personnel of the construction, operation and maintenance must be properly trained to prevent themselves during event of outbreak of the fire.

Fire gets ignited at a hot spot and if heat energy generated is more than the heat energy absorbed by the immediate atmosphere or radiated to the nearby medium is higher, it spreads along the combustible material to neighbouring area. However, expansion of the fire depends on the availability/presence of combustible material, air and local temperature rise.

High-risk zones must be equipped with automatic fire sensing and alarm system along with firefighting system. Automatic firefighting system minimizes the losses to the life as well as property in the event of fire. This also helps in minimizing the spreadover of the smoke and poisonous gases.

2.3.1 Initiation of Electric Fire

In general, the main causes of the occurrence of fire are:

- Lapse in safety management
- Negligence of working personnel
- Poor housekeeping
- Poor quality of wiring
- Old and worn out equipment
- Loose connections
- Ignorance of gas leakage
- Inadequate testing and maintenance
- Lack of awareness and training
- Sometimes fires takes place due to accidents beyond human control

2.3.2 Fires in Electrical Power Plants and Substations

Following may be the reasons of fires in electrical generating stations, substations and plants.

(i) Flashover due to failure of air insulation. Arc produced by the flashover may continue to spread over if there are flammable materials around the flashover point.

(ii) Short circuit may take place between phases, phase to ground or inter-turn in electrical machines.

(iii) Joints of the conducting path and connections at terminals sometimes found loosely connected and cause sparks. The repetitive sparks may overheat the parts of the equipment/ machine resulting to fire and melting.

(iv) There is repetitive charging and discharging of the capacitance between phases to ground if required clearances are not maintained properly. This is called Corona Discharging and initiates from sharp points. Corona is prominent in moist weather. In the presence of flammable materials nearby the discharge may cause arcing ground and fire hazard.

(v) If due to any reason, temperature of windings or oil of the oil filled equipment exceeds permissible limits, electrical fires and explosion takes place.

(vi) Sometimes explosion takes place due to bursting of capacitors. Capacitor cans burst due to overheating by over voltages.

(vii) Flammable materials catch fire if static charges accumulated on the metallic surfaces of transformer tank and metallic parts placed below high voltage transmission line get discharged in the absence of earthing.

(viii) Underrated current resistor elements in the electrical equipment become red-hot and release heat leading to catch fire in insulation and other flammable materials.

(ix) Over-voltage caused by switching operations or lightning generates external/internal flashovers in the electrical equipment. Capacitors and circuit breakers are prone to such flashover.

2.3.3 Classes of Fires

For practical purposes fires are classified into Class A, Class B, Class C, Class D and Class E. These classes are adapted by European and United States Standards. The classes confirm the combustion materials involved and required extinguishing medium as given in the Table 2.3.

Table 2.3: Classes of fires

Fire class	Involved combustion medium	Fire extinguishing medium
Class-A	Wood, Coal, Plastics, Cloth, Paper, Rags, Rubbish, Construction and Packing materials, Rubber	Cooling and wetting of materials with the help of water or high water content solution helps in extinguishing the fire
Class-B	Transformer oil, Diesel oil, Solvents, Liquid chemicals, Lubricating oils, Paints, Varnishes, Thinners, Greases	(Water is not suitable) Limiting the air or oxygen supply, Inhibiting fire dry chemicals, Foam, Halon
Class-C	Fires involving in live electrical equipment and/or in energized state. If equipment is dead Class is A or B	(Water is not suitable) CO_2 gas, Dry chemicals
Class-D	Fires involving in Metals like Magnesium, Titanium	(Normal extinguishing media not suitable) Special chemicals and techniques are used
Class-E	Fires involving Flammable Gases and Fuels like Hydrogen, Ammonia, Acetylene, LPG, Petrol, Furnace oil	Starvation of fire is most useful, Special methods, Necessary inlet valve closed

2.3.4 Removal of Fire

Heat, oxygen and flammable materials are the basic needs for initiation of fire and its sustenance. For essentially removal of fire, following processes are helpful:

- Cooling
- Smothering
- Starving
- Interrupting chain reaction of fire

Cooling means removal of heat. Temperature of the affected area should not increase beyond the ignition temperature. So, by cooling overall temperature of the vicinity of the affected area is lowered down. By maintaining the higher rate of cooling than the rate of heat generation by combustion extinguish the fire.

First disconnect the electrical supply and then spray the easily available coolant (that may be water) on the affected area of the fire. The other important method is to provide sufficient ventilation that is release of heat. Slabs above fire zone should be ruptured to release the heat to the atmosphere.

Smothering also known as blanketing is removal of oxygen supply and it is done by supplying the CO_2, foam, halon, etc.

Starvation means removal of flammable or combustible materials. If combustible materials like oil or gases are there, its supply must be stopped by closing the inlet valve from the supply end. Dry nitrogen gas should be supplied if possible. Oil tanks should be immediately drained and evacuated.

Interrupting chain reaction of fire is separation of combustible materials from ongoing fire. Dry chemicals and hydrogenated hydrocarbons are commonly used for breaking chain reaction of fire.

2.3.5 Prevention of Fire

Electrical power stations and substations have combustible materials like fuel oils, transformer oils, insulating materials, cooling gas, chemical gases. It is better to prevent occurrence of fire. Several preventive actions starting from the planning, designing, installation and testing are essential. Preventive actions for fire are also equally important during commissioning, operation and maintenance of electrical equipment and plants.

Preventing the fire means avoiding the simultaneous presence of flammable materials, oxygen (air), local temperature rise and ignition or spark.

Following suggestions are advised to be incorporated for prevention of fire in electrical plants and apparatuses:

- Necessary precautions should be taken during storage and handling of combustible materials. Suitable guard should be provided to isolate the flammable materials from flashovers, arcs, sparks, hot spots, etc. that causes the temperature rise.
- Welding and carrying of open flames must be carefully handled.
- Explosive materials, cigarette buts and open resistance heaters should not be allowed in the control room.
- Prescribed moderate temperature must be maintained surrounding insulating materials.
- During filling up the oils in transformers, circuit breakers and furnaces, special care should be taken.
- Overloading of the electrical equipment must be avoided at any cast. Proper protective schemes for overloads, over-current and temperature rise must be commissioned and maintenance.
- Leakage must be attended without delay. Scrap, waste oil, waste wood and waste papers should be disposed off as early as possible.
- Correct ratings of fuses and MCBs should be installed along with good quality of wiring and wiring materials.
- Scheduled and regular periodic inspection and corrective measures are inevitable for fire prevention.
- Metallic bodies of the equipment and pipes, cabinets, etc. should be properly earthed. Earthing helps in discharging the static charges.

Subsystems must not be energized unless the fire protection systems are commissioned and are in alert workable state.

Gas cylinders and chemicals must be stored separately as per safety recommendations.

Following fire-extinguishing materials should be kept in ready stock.

Water (H_2O) is mostly used as fire extinguishing medium. Water jet is required to be injected on the fire-affected area. Water evaporates absorbing the heat of fire and cools down the combustible materials below ignition temperature. Water is suitable for extinguishing Class-A fires.

Carbon dioxide (CO_2) is an inert gas having property of not supporting the combustion. Carbon dioxide is suitable for extinguishing Class-C fires and not proper for extinguishing Class-A fires.

Foam is prepared by mixing aluminium sulphate and sodium bicarbonate in water. This is not recommended for electrical equipment. However, foam is suitable for extinguishing the fire in oils, varnishes, paints and flammable solvents.

Dry chemical powder is prepared by mixing sodium bicarbonate (90%), magnesium sterate (1.5%), magnesium carbonate (1%), and tricalcium phosphate (1%). This is recommended for extinguishing Class-C fires, Class-B fires and Class-E fires.

Halon (Halogenated Hydrocarbon) is prepared by composing Carbon and Halogen (a combination of chlorine, bromine, and fluorine). Halon has capability to break down the chain reaction of fire and effectively extinguishes the fire. These are recommended for extinguishing the fire of electrical equipment, flammable liquids and gases in open space only. Firefighters must use mask while using it, because with fire, halogen decomposes and releases corrosive poisonous gas like hydrogen fluoride (HF).

2.3.6 Fire Extinguishers

The prime jobs of a fire extinguisher are:

- Extinguishing fire
- Stopping spreading of fire
- Minimizing the smoke and poisonous gases
- Saving life
- Saving property
- Minimizing the risks to adjacent buildings and premises

Actions of Defence for Fire Fighting

- First act of defence must be to use Portable Fire Extinguishers immediately.
- Second act of defence must be to use Fire Hydrant, Hose and Nozzle.
- Third act of defence must be to use Automatic Fire Extinguishing System installed in the plant.
- Inform fire brigade.
- Vacate and leave the affected location.

Note: While operating the extinguishers, first remove the guard cap and then strike the plunger against a hard surface (may be the floor). Finally, direct the jet emerging from the nozzle on the fire.

Types of Fire Extinguishers

1. Water Expelling Fire Extinguisher

These extinguishers use water as fire extinguishing agent. Water is released in the form of jet while gas pressure is applied from the upper part of the container of the extinguisher. Applied water on the burning materials is directly converted into steam. This steam helps either way to reduce the percentage of available oxygen.

These are not used on fire involving electrical equipment. However, if there is no other option first de-energize the electrical equipment, then apply the water expelling fire extinguishers.

Following three types of water expelling fire extinguishers are generally in use but soda acid type is most common.

(i) Constant Air Pressure Type Water Expelling Fire Extinguisher

The container of this type of fire extinguisher is filled with water and dry air, which develops desired pressure. When it is required to use, safety pin is withdrawn and pressure level inside gets depressed which results a jet of water through the hose under internal air pressure.

(ii) Gas Pressure Type Water Expelling Fire Extinguisher

There are two compartments in such types of water expelling fire extinguishers. The inner compartment is filled up with CO_2 carrying cartridge under pressure. The outer container carries water. While using such type of water expelling fire extinguishers, the CO_2 carrying cartridge is first pierced open. Now, release of CO_2 under pressure takes place which drives water out through the discharge tube.

(iii) Soda Acid Type Water Expelling Fire Extinguisher

In this type of fire extinguishers, sodium bicarbonate solution is filled up in the main container. A bottle having Sulphuric acid is fitted at the top of the container fixed with a plunger. While using this fire extinguisher, the plunger is getting struck first to rupture the bottle of the Sulphuric acid. Now, the Sulphuric acid and the sodium bicarbonate solution react together and release CO_2. CO_2 gas works as propellant of the water and water extinguishes the fire by cooling effect. Figure 2.5 shows the sketch of a Soda Acid Type Water Expelling Portable Fire Extinguisher.

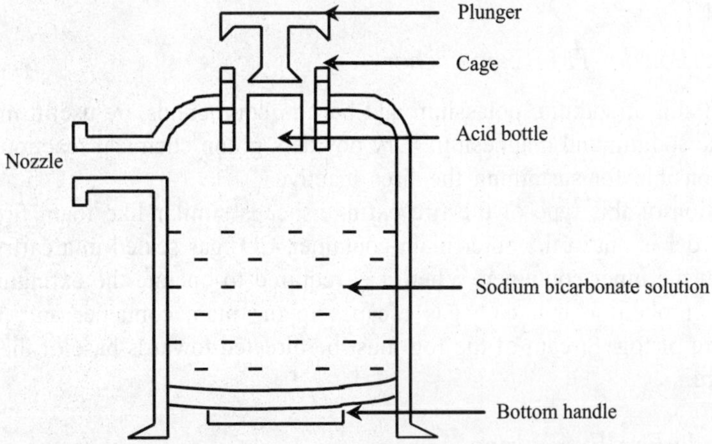

Fig. 2.5. Soda acid type water expelling portable fire extinguisher.

2. Foam Fire Extinguisher

This type of fire extinguisher expels foam and is very useful to extinguish fire in a liquid filled in any container. So, these are recommended for class-B fires involving flammable liquids such as oils, solvents, petroleum products, varnishes, paints, etc. Foam jet from the extinguisher is directed either inside edge of the container or at adjoining vertical surfaces above the level of the burning liquid. The foam forms a blanket over the surface of the burning liquid.

Foam fire extinguisher has two compartments in the container. The inner compartment carries Aluminium sulphate. The outer compartment carries sodium bicarbonate with foam stabilizer. When valve gets open, Aluminium sulphate and sodium bicarbonate get mixed up and forms foam, which comes out through nozzle. Figure 2.6 shows the drawing of a foam type fire extinguisher.

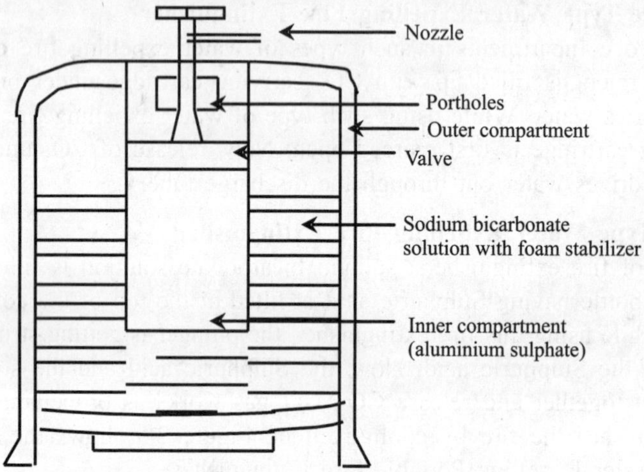

Fig. 2.6. Foam fire extinguisher.

3. Dry Chemical Powder Fire Extinguisher

Dry powders made up of sodium, potassium and barium compounds are useful in extinguishing fires in metals like sodium and magnesium. Dry powders go on chemical reaction and make the free radicals responsible for sustaining the fires inactive.

The construction of this type of the fire extinguisher is similar like foam fire extinguisher. The chemical powder is put in the outer main container. CO_2 gas sealed in a cartridge with high pressure is held in the inner container. When it is required to operate the extinguisher, CO_2 gas cartridge is to be broken so that CO_2 gas escapes to the main container and pushes out the powder in the form of fog. Stream of the fog must be directed towards base of the flame or near the seat of the fire.

4. Carbon Dioxide Fire Extinguisher

A non-combustible agent CO_2 does not react with most of the substances and penetrates to all areas affected by fire. Liquid CO_2 under pressure is kept in a cylinder. Cylinder is kept vacant about one-third from the top. Carbon dioxide fire extinguisher is shown in the Figure 2.7. After actuating the extinguisher, CO_2 comes out from the cylinder with velocity into the atmosphere. This gas coming out from the hose is 1.5 times heavier than the air which form vapour blanket across the fire and reduces the oxygen supply in order to stop the combustion.

This extinguisher should not be used for extinguishing the fire involving reactive metals like sodium, potassium and magnesium. This extinguisher should also not be used to extinguish fires involving their own oxygen supply like cellulose nitrate.

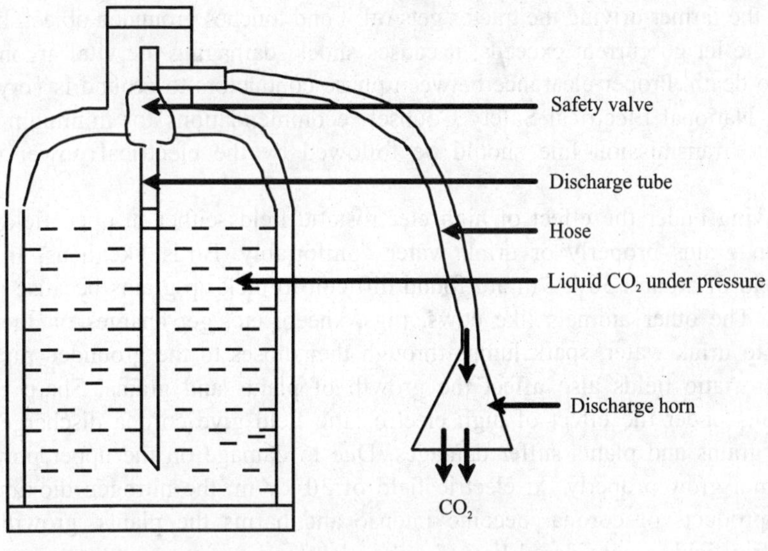

Fig. 2.7. Carbon dioxide fire extinguisher.

Carbon dioxide fire extinguisher is recommended to extinguish the fires in oils, petroleum products, electronic apparatus and gaseous substances under pressure.

Maintenance of Fire Extinguisher

Since fire extinguishers are very important equipment for saving the life as well as property, its maintenance must be done with utmost care. Following are the recommended schedule:

Weekly checkup and maintenance	Clean and polish the exterior of the extinguisher. Check nozzle outlet and vent holes. Ensure there is no clogging. Ensure that plunger is clean and is in fully extended position.
Yearly checkup and maintenance	Do the simulated test practically to ensure their satisfactory performance.
3-yearly checkup and maintenance	Discharge the extinguisher once in 3 years. Check the pressure hydraulically. Check the mechanical damage and correct if any.

2.4 HIGH ELECTROSTATIC FIELD AND LIVING BEINGS

There is danger of the high electrostatic fields near and below the HV and UHV transmission lines. The presence of human beings, animals, plants, vehicles get affected due to static charge accumulation and then get discharged through living beings.

High electrostatic field on human beings has potential of shocking effects, which may lead to loss of life. Vehicles get charged when either stopped or parked under a transmission line. Umbrella of a tractor running in the field through which EHV/UHV transmission lines pass gets charged. When the farmer driving the tractor gets off it and touches grounded object, he discharges current and if the let go current exceeds, it causes shock, damaging the vital argons like brain, heart leading to death. Proper clearance between phase conductors to ground is very essential to be maintained. National Electrical Safety Counsel recommendations for minimum clearance at mid-span of the transmission line should be followed by the electrical power transmission authority.

Animals living under the effect of high electrostatic fields either in open field or in cages do not pick up grains properly or drink water comfortably. Birds like hens, if affected by electrostatic field of around 30 kV/m are found difficult to pick up grains because of chattering of their beaks. The other animals like cows, pigs, sheep, etc. get charges on their body and when they try to drink water, spark jumps through their noses to the grounded pipe or trough.

High electrostatic fields also affect the growth of plants and grains. Sharp edges of the plants and grains under the effect of high electrostatic field give corona discharges and upper portion of the grains and plants suffer damages. Due to damage on the upper portions, grains and plants do not grow properly. At electric field of 30 kV/m, the nitrogen dioxide and ozone gases, the by-products of corona, become intense and harms the plants' growth. Generally, 20 kV/m, electric field is considered limit for the plant and grains.

EXERCISES

1. Why should we not touch the victim, when he is in contact with a live line at the time of an electric shock?
2. What should be the rate of artificial respiration that should be administered?
3. Why the precautions be listed and stick on the walls near the electrical installations?
4. What are the main causes of electrical accidents, and means to avoid them?
5. What precautions, the every householder, using domestic appliances, must take?
6. What do you understand by the term 'Let-go current'?
7. Describe about the electric resistance of human body.
8. How characteristic of the electric shock differs from female to male and why?
9. Is any difference in chances of getting death in humans and animals having four legs?
10. What will you do if downed line is touching a guardrail?
11. How do you handle an electric shock?
12. How do you prevent an electric 'dig in' when doing excavation work on your property?
13. What especial electricity safety arrangement should be done for the zones prone to severe lightning?

14. List the main IE rules pertaining to electrical safety.
15. What first aid is recommended to stop severe bleeding?
16. List the precautions you take in attending to a victim affected by burns.
17. What cares will you take in attending the victims of chemical burns?
18. What should be the safety precautions observed while handling portable equipment?
19. What are the precautions to be taken before starting work on a live line?
20. Write a note on 'initiation of electric fire and extinguishing it'.
21. Describe 'Fires in Electrical Power Plants and Substations'.
22. Write notes on:
    (a) Effect of high electrostatic field on living beings
    (b) Different types of fire extinguishers
    (c) Fire extinguishing mediums
    (d) Classes of fires

3

Tools and Specifications of Electrical Materials and Appliances

3.1 INTRODUCTION

Selection of proper hand tools is very necessary for getting perfection in jobs. There are number of hand tools of different sizes and specifications, which are used in electrical works. Some of the common tools are described below.

3.2 COMMON TOOLS

Pullers

Pulley pullers are used to fix/fit and remove the pulleys. Overstraining should be avoided.

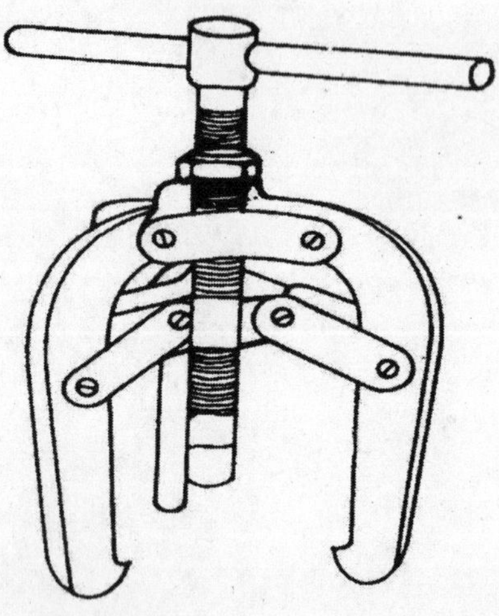

Bearing pullers are used to fix/fit and remove the bearings. Overstraining should be avoided. And bearing must be set properly.

Pipe Bender, Pipe Cutter, Pipe Wrench

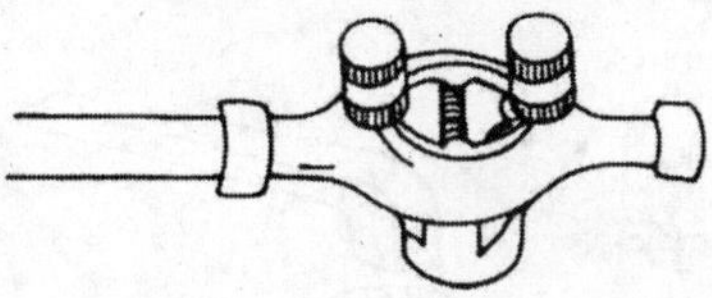

Stock & Die

Guide

Die

These are used in conduit wiring for cutting, setting and shaping of the conduits.

Chisel

This is most common metal cutting tool and is of four types:

Flat chisel is used for cutting and surfacing of the metals.

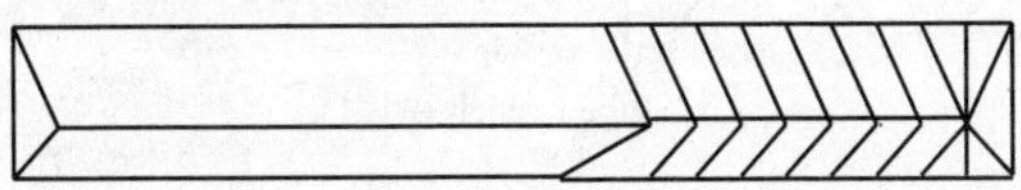

Round nose chisel is used for cutting grooves and channels.

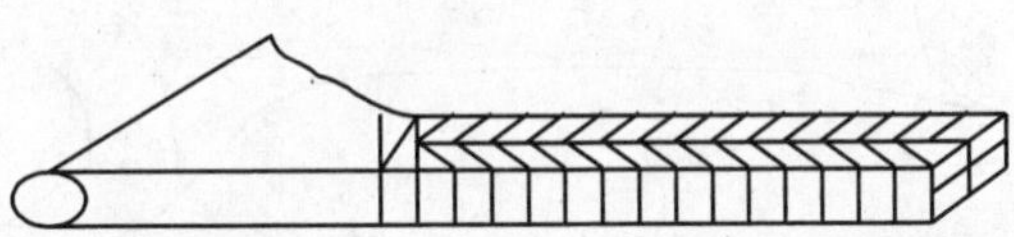

Crosscut chisel is used for roughing purpose.

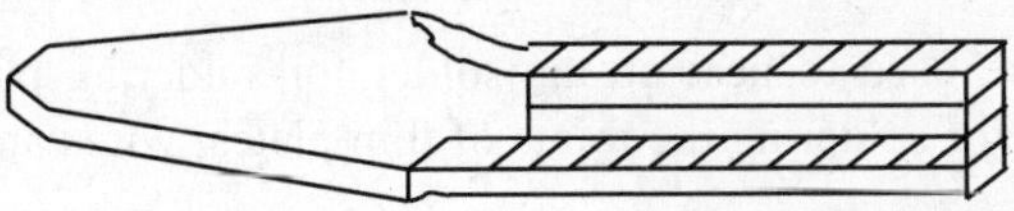

Diamond point chisel is used to clean corners and cut grooves for small holes.

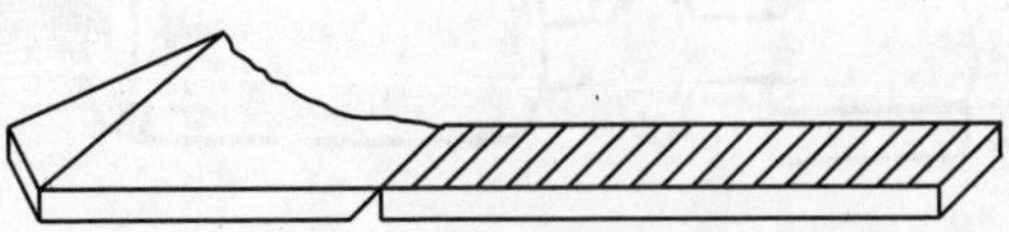

Spanners and Wrenches

These are used to tighten and loosen bolts, nuts and screws, etc. These are of different sizes to suit the different types of works. Examples are ring spanners, double ended spanners, torque wrenches, slide wrenches, etc. These are also used for assembling and disconnecting the conduits.

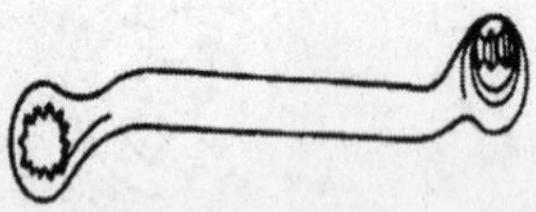

Ring spanner

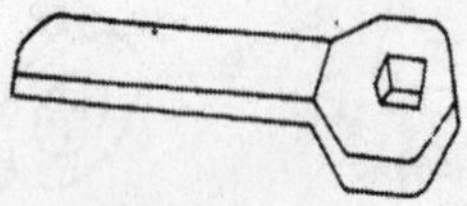

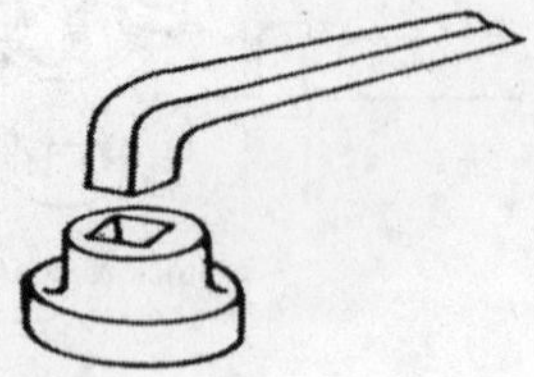

Socket wrenches

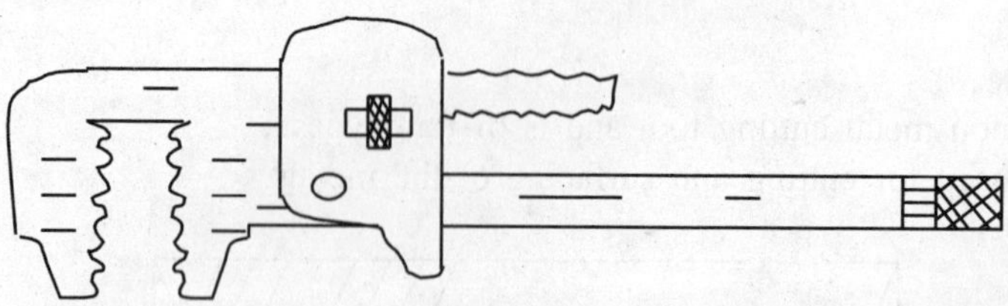

Pipe wrench

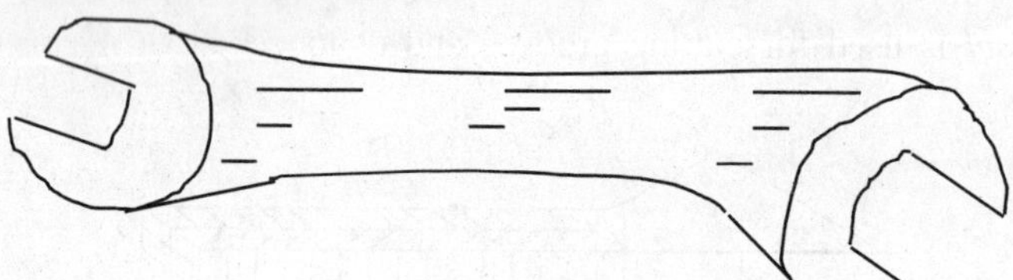

Double end spanner

Soldering Instruments

Soldering pot and ladle are used to heat up the solder for soldering purpose.

Soldering iron is used for soldering the joints of thin plates, wires and cables. The blowlamp heats up its bit.

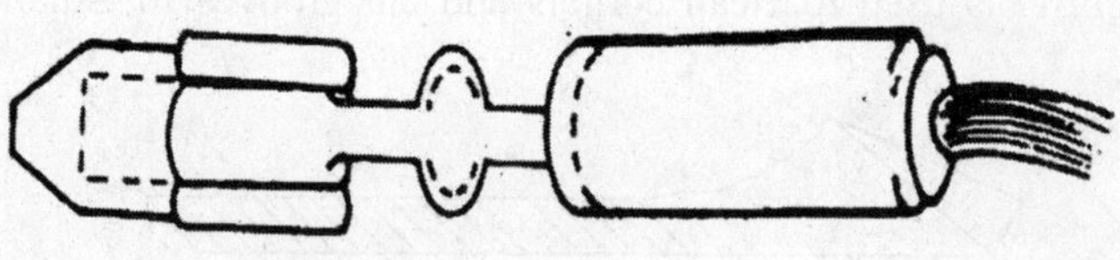

Blowlamp is used to produce heat to heat up solder pot in which solder is kept to melt.

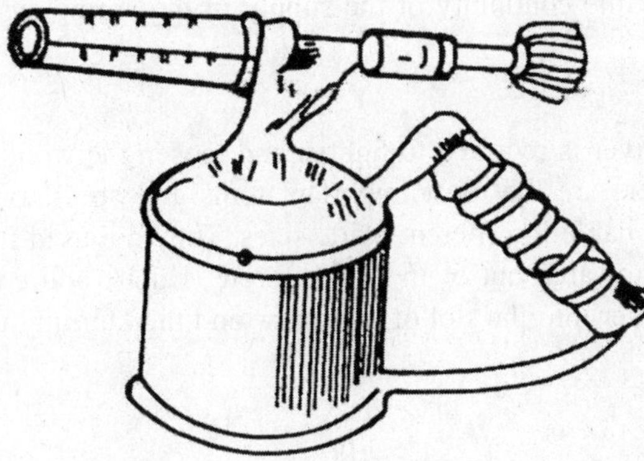

Pocker and Scratch Awl

In order to make holes by punching the wood for fixing the nails and screws, these are used.

Plumb Bob with Twin Thread

For marking vertical lines in the walls, plumb bob with twin thread is used.

Electric Hand Drills

These are used to drill the small holes on wooden boards or walls for doing small jobs.

Saw

Tenon, keyway and handsaws are used to cut batten, metals such as conduits, cables, round blocks, boards, etc.

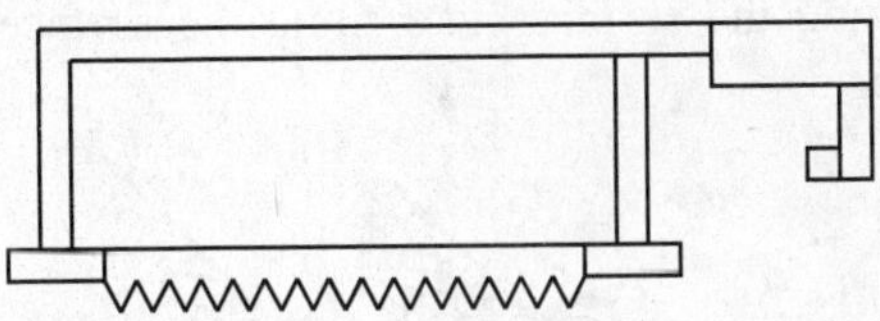

Mallet

It is used to strike firmer chisels and to bend metal plates.

Hand Tongs

Hot joints and hot substances are held by these tongs.

Test Lamp

Test lamp is used to check the continuity of the supply in the wiring systems, cables and fuses.

Screwdriver

A complete set of screwdriver is required to tighten and loosen the wood and metallic screws of different sizes. Screwdrivers are the most important tools and are among the most often used tools. Screwdrivers are available in different blade sizes. They are used for taking out or driving in the screws. Movement towards out or in of the screw is done and controlled by putting the pitted head of the screwdriver into the slot of the screw and then turning the screwdriver forward or backward directions.

Electrician's Knife

These are used to remove insulation covers from wires and cables. These are also required to cut the insulating materials.

Files

These are used for smoothening the surfaces like filing the surfaces of the metallic ends like conduits, bus bars, holes, etc.

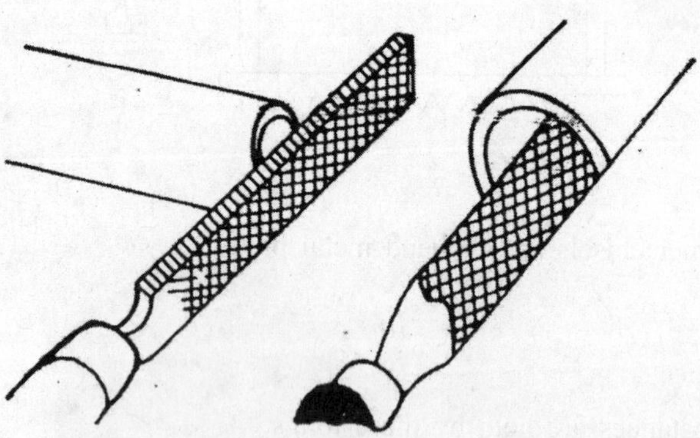

Pliers

Combination pliers are common and essential tools for gripping, cutting, bending, drawing and twisting the wires. These pliers are useful in our day-to-day works. Cutting and twisting the wires are done with the help of the combination pliers. For electrical work, always use insulated pliers.

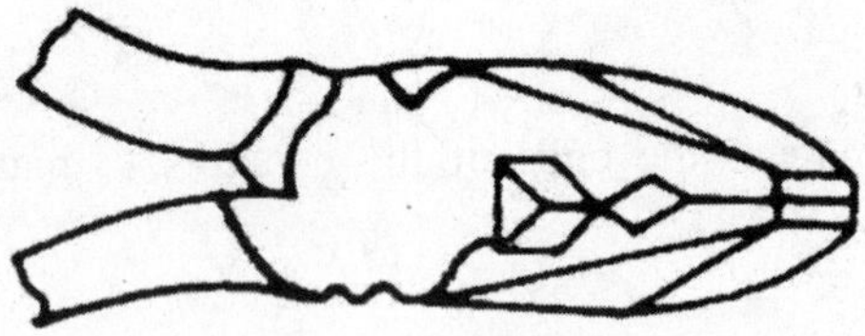

End cutting pliers are exclusively for cutting the wires.

Diagonal cutting pliers are for cutting the wires in confined places and panel wiring. Its overall length is generally 150 mm.

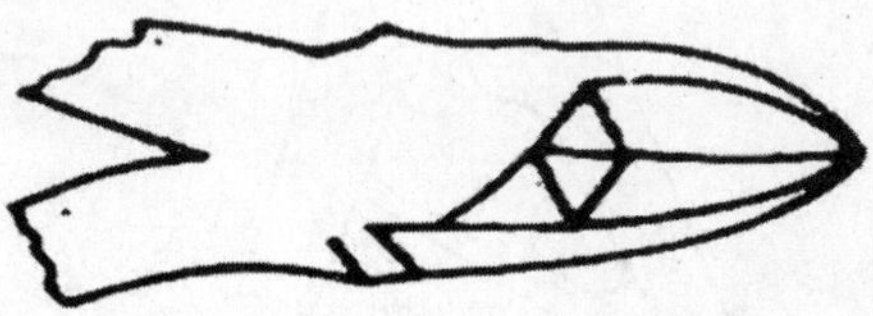

Slip joint pliers are used as a pipe wrench on small jobs and to bend and grip steel tape of armoured cables. The joint of these pliers is made to slip and hold jobs of conduits and cables of different sizes.

Flat nose pliers are for holding the terminals and lugs for soldering.

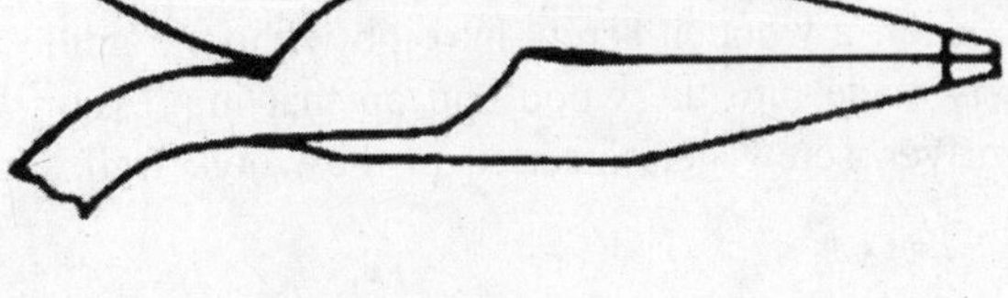

Round nose pliers are helpful for making the loops of conductors to be connected to the terminals.

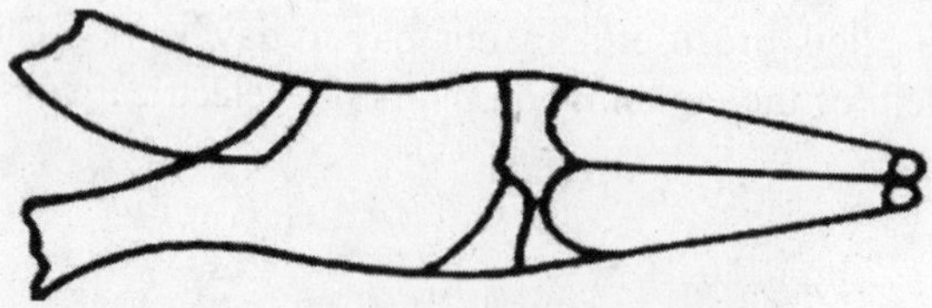

Reamers are required to ream the ends of the conduits to make the end surface of the conduits smooth.

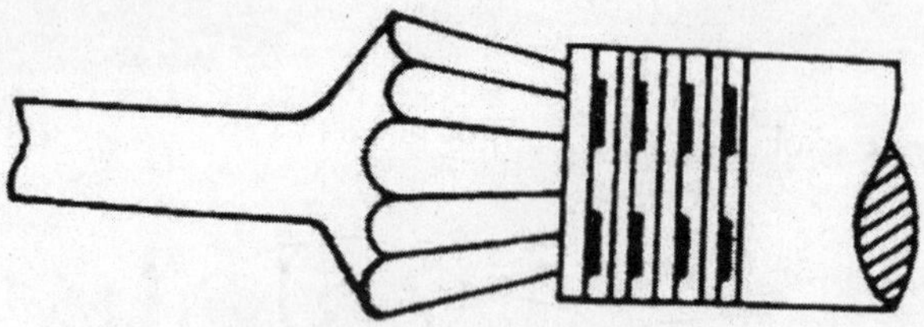

Crimping tool is required for crimping purposes.

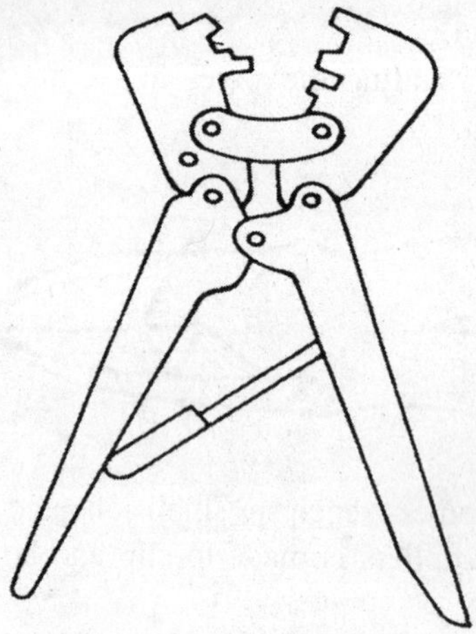

Poker

It is generally difficult to insert a wooden screw into job without a guide-hole. With the help of the poker, an impression is made into the wood and on that impression, screw is set and then with the help of a screwdriver, screw gets inserted in the wood, wall, etc.

Hammer

There are several types of hammers, but following types are frequently used.

Ball-pin hammer: It is used for riveting purposes.

Cross-pin hammer: It is used for riveting purpose in awkward places.

Hand Drill

This is used for drilling holes in wooden blocks and wooden boards to facilitate the passage of insulated wires.

Drilling Machine

It is a belt driven consisting variable speed transmission tool used to make small size holes on walls, boards, etc.

Spirit Level

Spirit level is used to check and adjust the horizontal surfaces.

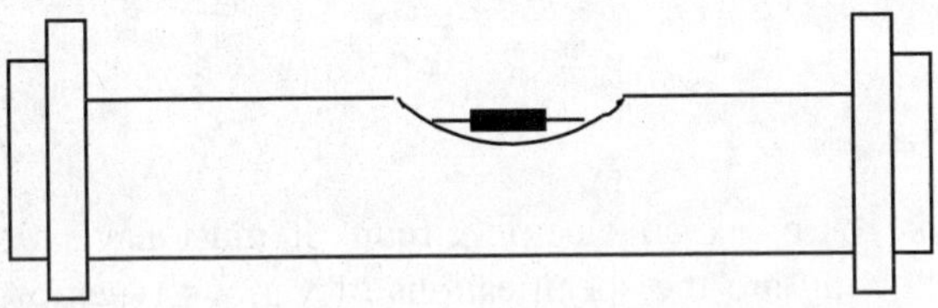

Vice

To hold the work while being operation, vice is used. Bench vice is generally useful in electrical works. Another type of vice is Machine Vice which is used for holding the items while they are being drilled.

3.3 SPECIFICATIONS OF THE ELECTRICIAN HAND TOOLS

S/N	Name of the Tool	Specification
1.	Conduit stock and die	20 mm and 25 mm dia
2.	Blow lamp	1 pint and 2 pint (1 pint = 473 ml)
3.	Test lamp	40 W, 60 W, 100 W
4.	Poker	10 cm
5.	Rawl tool	No. 8, 9, 10
6.	Wood saw	15, 25 and 35 cm
7.	Hack saw	15, 20, 25 and 30 cm
8.	Measuring tape	1, 2, 5, 10, 25 and 50 m
9.	Hammer ball pin	250 gm, 500 gm, etc.
10.	Electric hand drill	Drill of dia 2, 3, 4, 5 mm, etc. (for both twist and flat drill)
11.	Chisel	10, 15, 20, 25 mm wide (for both types firmer and cold)

Table Contd.

S/N	Name of the Tool	Specification
12.	File 8, 10, 12 inch, etc	
13.	Outside Micrometer	0-25 mm, 25-50 mm
14.	Screwdriver	10, 15, 20, 25 and 30 cm
15.	Electrician knife	15, 20 cm
16.	Pliers	
	(i) Combination pliers	14, 20, 23 and 25 cm
	(ii) End cutting pliers	13 and 16 cm
	(iii) Diagonal cutting pliers	15 and 20 cm
	(iv) Flat nose pliers	18, 20 and 22 cm
	(v) Round nose pliers	18, 20 and 22 cm
	(vi) Slip joint pliers	20 and 25 cm

Note: It is always advisable to use specific tools for specific purposes. A good finished job can only be executed out of good tools.

3.4 SPECIFICATIONS OF ELECTRICAL MATERIALS AND APPLIANCES

Conception of any project first requires estimation. Estimation means an assessment of the quantity of material and its cost required for the work. Estimation includes the listing of the materials stating complete specifications and approximate cost.

A specification is a detailed description of a material or equipment to be procured for a given work.

Switches

A switch is used to make or break an electric circuit. It must have a strong base and high insulation resistance. Following are the specifications of various types of switches.

S/N	Type of switch	Specification
1.	Tumbler switch	Type : Single way or Two way Material : Complete bakelite or porcelain base and bakelite base Capacity : 5 A or 15 A; Working Voltage: 250 V
2.	Flush switch	Type : Single way or Two way Capacity : 5 A or 15 A; Working Voltage: 250 V
3.	Pull switch	These are normally used in bedrooms. Capacity : 5 A or 15 A; Working Voltage: 250 V
4.	Grid switch	These are normally used in portable machines i.e. hand drills. Capacity : 5 A or 15 A; Working Voltage: 250 V
5.	Architrave switch	Type : Single way or Two way Capacity : 5 A or 15 A; Working Voltage: 250 V
6.	Main switch*	Type : ICDP or ICTP single phase or three phase Capacity : 15 A, 30 A, 60 A, 100 A, 200 A
7.	Rotary snap switch**	Capacity : 5 A or 15 A Working Voltage: 250 V

**A rotary snap switch is different from an ordinary tumbler switch. It consists of an insulated handle to which the blades are fixed. The movement of the handle makes these blades move in a stepwise fashion and make contact with the terminals to which are connected the wires in electric circuits. The handle movement is controlled by a cam or spring.

These switches are used for the reversal of small motors, speed control, control circuits or electric ranges of heaters.

*The main switch is provided with neutral links and an operating handle of sufficient strength, and controlled by a spring. These are made of cast iron and have a very robust construction. They are filled with requisite fuse units.

Push-Button Switch

It consists of one blade given a rocking motion by push button. Its movement is controlled by a cam and a spring. These switches are normally used to control lighting circuits.

Capacity	5 A or 15 A;
Working Voltage	250 V

Plugs

For tapping power from socket outlets, two-pin or three-pin plugs are used. The plug must have an appropriate finger grip for inserting and withdrawing the plug.

Capacity	5 A or 15 A;
Working Voltage	250 V

Sockets

Two-pin sockets and three-pin sockets are the two types of sockets in use. In the three-pin sockets, third terminal is connected to the earth.

Capacity	5 A or 15 A;
Working Voltage	250 V

DG Sets

Diesel Engine specification	Alternator Specification	Control Panel Specification
Capacity in hp Speed in rpm	Normal full load capacity: (kVA) No. of phases: (Single/Three-phase–3 wire/4 wire)	Type: (Cubicle or Open) Mounting: (Pedestal/Wall)
No. of strokes: (2/4)	Frequency: (50 Hz/60 Hz)	Type of busbar: (Copper/Aluminium)
No. of cylinders: (1/4/6)	Terminal voltage: (230 V/400 V)	Current rating in amperes
Type of starting: (Battery, Pneumatic)	No. of poles: (6/8)	Working voltage: (250 V/400 V)
Arrangement of cylinder: (Vertical/Horizontal)	Speed: (1000 rpm/750 rpm)	No. of phases: (According to the type of circuit breaker and capacity)
Type of cooling: (Oil cooled/water cooled)	Type of coupling: (Rigid/Flexible)	Meters: Voltmeters and Ammeters in alternator output circuit and in excitation circuit. Energy meter, Frequency meter, Power Factor meters, etc.
Type of bearing: (Ball/Roller)	Type of bearing: (Ball/Roller)	
Type of coupling: (Rigid/Flexible)	Excitation voltage: (110 V)	
	Exciter type: (Bridge rectifier/pilot generator) Connection: (Star/Delta)	

Wires/Cables

S/N	Type of wires/cables	Specification
1.	PVC cables	Insulation (PVC); Voltage grade (250 V or 660 V); Conductor (Copper or Aluminium); Single or Twin cores; Size (1.5 mm^2, 2 mm^2, 3 mm^2, 4 mm^2, 6 mm^2, 10 mm^2,15 mm^2, 20 mm^2, 25 mm^2); Generally used in all types of wiring.
2.	TRS wires	Insulation (Tough Rubber Sheath); Voltage grade (250 V or 660 V); Conductor (Copper or Aluminium); Single or Twin cores; Size (1.5 mm^2, 2 mm^2, 3 mm^2, 4 mm^2, 6 mm^2, 10 mm^2, 15 mm^2, 20 mm^2, 25 mm^2); Especially suited to places where acids and alkalines are likely to be present.
3.	Underground cables	Insulation (Paper insulated with lead alloy sheath); Conductor (Aluminium); Voltage grade (600 V or 1100 V); No of cores (2,3 and 4); size (25 mm^2, 35 mm^2, 50 mm^2, 185 mm^2, 300 mm^2, 500 mm^2); Generally used in underground power distribution.
4.	VIR wires	Insulation (Vulcanized India Rubber); Voltage grade (250 V or 660 V); Conductor (Copper or Aluminium); Single or Twin cores; Size (6 mm^2, 10 mm^2, 15 mm^2, 20 mm^2, 25 mm^2, 10 mm^2, 15 mm^2, 20 mm^2, 25 mm^2); Generally used in cleat and wooden casing capping wiring).
5.	Weatherproof cables	Insulation (Vulcanized India Rubber); Voltage grade (250 V or 660 V); Conductor (Aluminium); Single or Twin cores; Size (6 mm^2, 10 mm^2, 1.5 mm^2, 2 mm^2, 4 mm^2, 6 mm^2, 10 mm^2, 15 mm^2, 20 mm^2); Generally used in service connections)..
6.	Flexible cord	These consists of two or three separately insulated flexible stranded conductors. Voltage grade (250 V); Conductor (Aluminium or Copper); Size (0.4 mm^2, 0.6 mm^2, 1 mm^2, 2 mm^2, 3 mm^2, 4 mm^2); Especially used for connecting equipment.

Fuse Units

Types: Semi-enclosed and totally enclosed. The two-piece construction containing a fuse base and fuse carrier is known as semi-enclosed type. HRC cartridge fuse is known as totally enclosed type.

Capacity	5 A, 10 A, 20 A, 50 A, 100 A
Working Voltage	250 V

Lamp-holders

Types (i) Pendant (ii) Bracket (iii) Screw type

Material	Brass, Bakelite or Porcelain
Capacity	5 A, 15 A
Working Voltage	250 V

Ceiling Rose

Types	Two way or Three way
Material	Bakelite

Terminals	Made up of brass (Contacts should be separated in porcelain or bakelite)
Capacity	5 A, 15 A
Working Voltage	250 V

Ceiling Fan

Current Capacity	5 A, 15 A
Working Voltage	250 V
Wattage	80 W
Sweep	900 mm, 1050 mm, 1200 mm, 1400 mm, 1500 mm
Speed	300 rpm, 200 rpm, 155 rpm
Type of Supply	Single phase, 50 Hz
Type of Bearing	Both ball bearing, or one ball bearing and one bush, or both bush

Teak Wood Batten

In practice, generally following size of teak wood battens is used.

12 mm × 12 mm
19 mm × 12 mm
25 mm × 12 mm

Wiring Clips or Link Clips

Material	Brass or Tinned brass
Length	30 mm, 40 mm, 50 mm

Screws

Material	Brass or Iron
Length	4 mm, 6 mm, 10 mm, 12 mm, 15 mm, 20 mm, 25 mm, 35 mm, 50 mm, 100 mm
Size	No. 4, No. 6, No. 8, No 10
Type	Wooden screw and metallic

Conduits

Types	Metallic (Mild steel, Aluminium, and Copper) Non-metallic (PVC)

Coating		Either black enamelled or galvanized.
Size		16 mm, 19 mm, 25 mm, 30 mm, 35 mm
Gauge		SWG 16
Conduit boxes	(a)	Size suitable for conduit size, i.e., 16 mm, 19 mm, 25 mm, 30 mm, 35 mm
	(b)	Material: Malleable cast iron

Incandescent Lamps

Working voltage	230/250 V
Wattage	25 W, 40 W, 60 W, 100 W, 200 W, etc
Type	Vacuum gas filled, clear/milky white
Filament	Coiled coil or not
Cap	Bayonet cap/screw cap

Fluorescent Lamps

Wattage	20 W, 40 W, 80 W or 120 W
Rated voltage	230/250 V
Length of tube	6m, 1.3 m or 1.6 m
Fixture	Single strip fixture or single batten fixture
Tube	Opal or colored tube
Starter	Normal or bimetallic
Choke	Copper wound or aluminium wound

Motors and Starters

DC motor (Specification)	DC motor starter (Specification)	AC motor (Specification)	AC motor starter (Specification)
1. Wattage	1. kW rating and type of motor	1. Wattage	1. kW rating and type of motor
2. Voltage	2. Hand operated/ automatic face plate/drum type	2. Voltage	2. Hand-operated/ automatic/semi-automatic
3. Full-load current	3. Voltage rating	3. Full-load current	3. Voltage
4. Speed	4. Current rating	4. Frequency	4. Current (normal, thermal ratings)
5. Continuous or intermittent rating	5. Three point or four point	5. Speed	5. Single phase or three-phase
6. Series, shunt or compound	6. Open or enclosed	6. Power factor	6. Frequency

Table contd.

DC Motor (Specification)	DC motor starter (Specification)	AC motor (Specification)	AC motor starter (Specification)
7. Types of frame and its size	7. With or without no voltage relay	7. Single phase or three-phase	7. With or without voltage relay
		8. Type of motor: series, universal, shaded pole, split phase repulsion, capacitor start-run, squirrel cage, wound rotor, synchronous and commutator motors	8. Open or enclosed
		9. Continuous or intermittent rated	
		10. Type of frame and its size	

Mercury Vapour Lamps

Wattage	80 W, 125 W, 250 W, 400 W
Rated voltage	250 V
Lamp voltage	15 V
Electrode heating current	2 A
Lamp heating current	2.2 A

Discharge Lamps

Type of lamp	Sodium/Cadmium/Mercury or Cadmium Mercury Lamp
Wattage	40 W, 100 W, etc.
Rated voltage	100-250 V for Sodium or Mercury Lamp 200-250 V for Cadmium or Cadmium Mercury Lamp
Lamp voltage	20 V for Sodium Vapour Lamp 15 V for Mercury Vapour Lamp 30 V for Cadmium or Cadmium Mercury Vapour Lamp
Electrode heating current	1.3 A for Sodium Vapour Lamp 2 A for Cadmium or Cadmium

Mercury Vapour Lamp

Frequency of supply	50 Hz

Earth Electrodes

Material: GI or copper plate, GI pipe, copper pipe, GI or copper rod, Size 0.6 m × 0.6 m and 3.15 thick for copper plate; 0.6 m × 0.6 m and 6.35 mm thick for GI plate; 2 m long and 38 mm dia or 1.37 m long and 51 mm dia for GI pipe for dry ordinary soil; 2.75 m long and 38 mm dia or 1.85 m long and 51 mm dia for GI pipe for dry rocky soil; 16 mm dia and 2.5 m length for GI rod and 12.5 mm dia and 2.5 m length for copper rod.

Maximum resistance 10 Ω

Earth Wire

SWG 8 copper wire or GI wire

Switchboards

Following sizes are normally in use.

100 mm × 100 mm
150 mm × 100 mm
150 mm × 100 mm

EXERCISES

1. Write down the complete specification of a 60 kW DG set.
2. List out the specifications of the following.
   - Ceiling fan
   - Ceiling rose
   - Fan regulator
3. Write a note on the different types of lamp holders.
4. What are the different types of switches? Describe their specifications.
5. What are the different types of lamps? Describe their specifications.
6. What is the use of earth electrode? Describe its specifications.
7. Where the main board in a large wiring installation is located? Write down the specification of the switchboards.
8. Write down the advantages of underground cables over overhead conductors. Specify them properly.
9. How weatherproof cables are specified?
10. Specify the details of a starter required for a 12 hp three-phase squirrel cage motor.
11. When do we use cross-pin hammers?
12. How is blowlamp used?
13. Why and when crimping tools are required?
14. Describe different types of pliers.
15. List out the instruments and tools used mainly for soldering purpose.
16. Write down the complete specification of a 60 kW DG set.

4

Earthing

4.1 NECESSITY OF EARTHING

According to Indian Electricity Rules, earthing is defined as "the earthed or connected with earth means connection with the general mass of the earth in such a manner so as to ensure at all times an immediate discharge of energy without danger". Earthing or grounding refers to connection to the general mass of the earth. It is of two types: Neutral earthing refers to the connection of neutral point of electrical power network, i.e., star point of transformer or alternator to earth. This is to ensure that neutral point remains at zero potential. Equipment earthing is the connection of non-current carrying part of an equipment, e.g., outer body of transformer, alternator or motor, etc. to the general mass of the earth. This provides low resistance path to any leakage current to pass to the ground and hence electric shock to the personnel is prevented.

Earthing provides the body of the equipment to the earth potential means zero potential and thus avoids the shock to the operator or the person needs to touch the body of the equipment or appliance.

So, the term earthing means connecting the neutral point of a supply system to the non-current carrying metal parts used in electrical distribution system to the general mass of earth by wire of negligible resistance in such a manner so that at all times an immediate and safe discharge of electrical energy takes place due to leakage, faults, etc. without danger. The neutral of the three-phase four-wire supply system is also solidly earthed to ensure its potential equal to zero. This neutral commences from the star point of the 11 kV/415 V transformer.

Necessity of Earthing

The earthing is necessary to:

- Avoid electric shock to human body.
- Avoid risk of fire due to earth leakage current through unwanted path.
- Ensure that no current carrying conductor rises to a potential with respect to earth than its designed insulation.

- Maintain a constant voltage in healthy phases in the event of a fault in one phase.
- Protect equipment and buildings from lightning.
- Serve as a return conductor in communication and traction works.

If the insulation of the motor winding fails, the current is liable to leak into the metal body of the motor which would get charged with the supply voltage and become alive and dangerous, unless the charges accumulated are drained away to earth by an earth connection.

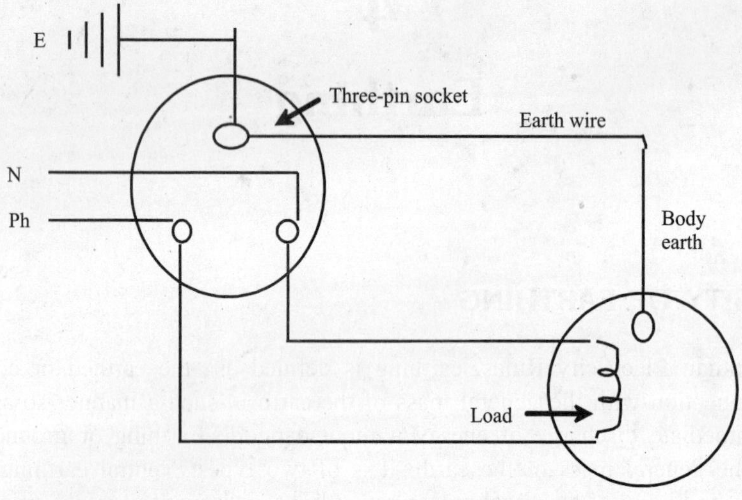

Fig. 4.1. Current flow path through earth.

The earthing helps to automatically clearing the fault in electric equipment and appliances if metallic frame of each non-current carrying electrical equipment is efficiently connected to earth point for draining out leakage of current to earth. The Figure 4.1 shows an a.c. single-phase, 230 volts appliance connected to three-pin plug and single-phase supply. The neutral point is effectively connected to earth. Since the impedance of path of the current is low, a large amount of current can flow to earth. The faulty currents would flow into the earth, which would then blow out the fuse on the phase line as the current exceeds the limiting value of the fuse installed in the circuit. Blown off fuse will result in disconnecting the supply to the motor.

4.2 SPECIFICATIONS OF EARTHING ELECTRICAL INSTALLATION

The various important specifications regarding earthing as recommended by I.S.I. are given below:

- An earthing electrode shall not be situated within a distance of 1.5 metres from the building whose installation system is being earthed.
- The conductor by which the metal body of an equipment or appliance connected to the earth is known as '*earth continuity conductor*'. The earth continuity is generally ensured through metal conduit, metal sheathing of cables or by a special earth continuity conductor.

The cross-section of conductor is normally selected not less than 2.9 mm^2 or half of the installation conductor size.

- The earth resistance should be low enough to cause flow of current sufficient to operate the protective relay or blow fuses in the event of an earth fault. The value of earth resistance does not remain constant but changes with the weather, as it depends upon the moisture content of the soil and becomes maximum during dry season. Following values of earth resistances shown in Table 4.1 are generally recommended:

Table 4.1

Type of installation	Maximum combined resistance of earth in the station
Large power station	0.5 ohm
Major power station	1 ohm
Small substation	2 ohm

Note: For other cases 8 ohms.

- The earth wire and earth electrode must be of same material.
- The earth wire shall be taken through G.I. pipe of 12.7 mm of diameter for at least 30.5 cm length above and below ground surface to the earth electrode to protect it against mechanical damage.
- It is not necessary that earth wire connected to an earth electrode runs along the whole wiring system. All earth wires run along the various sub-circuits shall be terminated and looped firmly at the main board to the main earth wire finally be taken to the earth electrode. The loop earth wires used shall not be either less than 2.9 mm square or half of the size of the sub-circuit conductor.
- The earthing electrode shall always be placed in vertical position inside the earth or pit so that it may be in contact with all the different earth layers.

4.3 EQUIPMENT TO BE CONNECTED TO EARTH

The following equipment shall be connected to earth through earth wire as per Indian Electricity Rule No 61:

- The neutral conductor of three-phase, four-wire electrical distribution system and middle conductor of two-phase, three-wire DC supply system should be earthed at generating station and at the substation by two separate and distinct earth connections. It may also be earthed at one or more points along the distribution system in addition to any connection with earth, which may be at the consumer's premises.
- Metal frame of generators, motors and other metallic parts of the equipment used for regulating or controlling energy and all medium voltage energy consuming apparatus should be earthed by two separate and distinct connections with earth.
- All metal parts of electrical installation such as metal conduits enclosing VIR or PVC cables, light fittings, iron clad main switches, iron clad distribution boards, cable sheathes, etc.
- Earth terminal of the three-pin lighting and power plug socket should be permanently and efficiently earthed.

- Fabricated steel transmission towers, tubular poles, rail poles, concrete poles, etc. used in overhead transmission line should be earthed. For this purpose, a continuous, earth wire is provided and connected with earth at four points in every mile, the spacing between points being nearly equidistant.
- The metal casing of portable apparatus should be earthed. This covers electric drill, soldering iron, hair-driers, heaters, refrigerators and similar electrically operated equipment. If any one of them is installed in a fixed position, a separate direct connection to the earth should be provided in addition to the earth wire in the connecting cable.
- In the case of a system comprising electric supply lines having concentric cables, the external conductor of such cables should be earthed by two separate and distinct connections with earth.

4.4 SELECTION OF THE EARTHING CONDUCTOR

Electrical potential should be kept into the mind while selecting the earthing conductor. That is the voltage level, which will cause the flow of short-circuit current during fault. The conductor must have high conductivity preferably copper. The conductor used should have sufficient diameter in order to carry the fault current safely. The other wire in place of copper if used should be galvanized iron wire so that it does not get rusty in adverse conditions, as it has to be buried into earth.

4.5 DEFINITIONS

Earth electrode: Any wire, rod or metal plate embedded in earth for the purpose of making an effective electrical connection with the general mass of earth is known as earth electrode.

Earthing lead or the main earthing conductor: The wire, which connects overhead earth wire or any other apparatus to be earthed to the earth electrode, is known as earthing lead or the main earthing conductor.

Sub-main earthing conductor: The earth wire which runs between the main switchboard to the distribution board or along the sub-mains is known as sub-main earthing conductor.

Earth continuity conductor: The conductor running between the distribution board and various plugs and appliances is known as earth continuity conductor.

The size of the plate earth electrode for various loads and insulation should be in accordance with the details given in the following Table 4.2.

4.6 FACTORS AFFECTING THE VALUE OF EARTH ELECTRODE RESISTANCE

Earth resistance value depends on the following parameters:

- Condition of soil
- Moisture content of soil
- Soil temperature

- Depth of electrode
- Size and spacing of earth plates
- Size of conductor
- Metal of earth plate and earth wire
- Quality of coal dust and charcoal in the earth electrode pit

4.7 METHODS OF REDUCING THE EARTH RESISTANCE

Maximum value (0.5 ohm), beyond which the earth electrode resistance should not increase. However, for better performance at powerhouse/substation or for high horsepower motors beyond 100 hp, the earth resistance should not be more than 0.5 ohms and 1 ohm respectively. The resistance of earth is mainly dependent upon contact area of the electrode with the general mass of earth, the quantity of soil, coke and salt. However, low earth resistance can be achieved by adopting the following arrangements:

- Using a number of earth electrodes interconnected in parallel can reduce earth resistance. Parallel connection of resistances reduces the resultant resistance.
- By replacing the existing electrode with larger one having more cross-sectional area. Cross-sectional area is inversely proportional to the resistance of the electrode.
- By increasing the depth of pit for earthing system to get more moisture/water content in the soil.

Table 4.2

Type of loads and installation	Size of earth lead				Size of earth electrode	
	Copper (SWG)	Area in mm²	GI (SWG)	Area in mm²	Copper	GI
Consumers residential premises	8	12.97	8	12.97	60 cm × 60 cm × 3.18 mm	60 cm × 60 cm × 635 mm
Pole earthing of transmission or distribution lines	do	Do	do	do	do	do
Industrial loads upto 10 hp	do	Do	do	do	do	do
Industrial loads between 10 hp and 15 hp	do	Do	6	18.68	do	do
Industrial loads between 15 hp and 30 hp	6	18.68	2	38.60	Do	90 cm × 90 cm × 6.35 mm
Industrial loads between 30 hp and 50 hp	4	27.27	Not	Used	90 cm × 90 cm × 6.35 mm	Not used
Industrial loads between 50 hp and 100 hp	(i) 2 (ii) Copper strip	38.60 12.7 mm × 2.54 mm	Not	Used	90 cm × 90 cm × 6.35 mm	Not used
Industrial load above 100 hp	Copper strip	25.4 mm × 2.54 mm	Not	Used	90 cm × 90 cm × 6.35 mm	Not used

Contd.

Table 4.2: Contd.

Powerhouses and substations. Detail as under: Capacity of Transformer/Generating set:	Copper strip	25.4 mm × 25.4 mm	Not	Used	90 cm × 90 cm × 6.35 mm	Not used
(a) Upto 300 kVA	Copper strip	20 mm × 4 mm	—	—	do	—
(b) Above 300 kVA but not exceeding 500 kVA	do	32 mm × 5 mm or 40 mm × 4 mm 40 mm × 6.3 mm				
(c) Above 500 kVA but not exceeding 800 kVA	do	50 mm × 5 mm 50 mm × 6.3 mm	—	—	do	—
(d) Above 800 kVA but not exceeding 1000 kVA	do	32 mm × 5 mm or two- 40 mm × 4 mm	—	—	do	—

- The methods mentioned above have their own limitations, but the common method is that the area around the earthing pipe is increased and charcoal soaked in salt solution is filled in the pit. In dry seasons, the fresh salt water is poured through the pipe over the coal layers to create moisture.
- By excavating a hemispherical hole around the top of the electrode, the rain by its action assists the salt to percolate to the lower level so reducing the resistivity of the soil immediately around the electrode. The effect is however not permanent, lasting approximately 12 to 18 months, the salt then needing regular pouring time to time.

4.8 METHODS OF EARTHING

1. Earthing through a water mains
2. Wire or strip earthing
3. Rod or pipe earthing
4. Plate earthing
5. MAT earthing

4.8.1 Earthing through a Water Main

People should not use the water pipes in the place of earth conductor.

4.8.2 Wire or Strip Earthing

At the places having rocky soils, wire or strip type earthing is preferable. Generally, four SWG copper wire or a copper strip of cross-section not less than 25 mm wide and 1.6 mm thick is used as an earth wire. All the wires or strips are embedded and placed 0.5 m below the soil in the form of a network. If round conductors are used, its cross-sectional area shall not be smaller than 3 mm^2. Wires or strips made of GI of size 25 mm × 4 mm are also used. The length should not be less than 1.5 m as per ISI specifications. Alternate layers of charcoal and salt for reducing the soil resistivity is arranged around the embedded strips or wires. The detail of dimensions and drawing recommended for wire or strip earthing are shown in Figure 4.2.

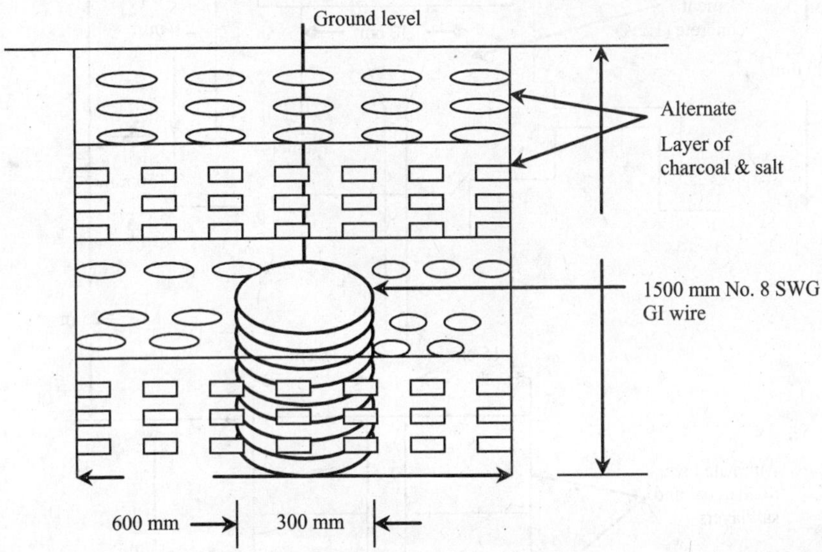

Fig. 4.2. Strip earthing.

4.8.3 Rod or Pipe Earthing

GI solid rod of 16 mm diameter with clean surface or of copper rod having 12.5 mm diameter should be placed in the ground below moisture level. Minimum 2.5 m rod is driven vertically into the earth, which may be up to depth of 5 to 6 metres depending upon depth of moisture in the earth. This system of earthing is suitable for sandy earth. In order to reduce the earth resistance to the desired value, it is necessary to increase the embedded length of the electrodes under the ground. After making proper joints, more than one rod sections are hammered, one above the other. The earth wire is carried in the pipe to the ground and in the case of solid rod the wire is to be tied to it with small sturdy clamps. Since excavation is not required, it is therefore cheaper system of earthing.

Pipe earthing as recommended by IS 3043/1966, GI pipes of 38 mm diameter and 2 m lengths may be used for normal soil. This is a common and best system of earthing as compared to other systems suitable for the same earth and moisture conditions. The pipe having a tapered casting at the bottom is placed up right in that pit. The charcoal and salt are filled in that pit alternately in layers up to about two metres from bottom and for a distance of about 1.5 cm around the pipe to increase dampness and moisture around the each pipe. The pipe placed has 12 mm diameter holes drilled in it so that water poured from top is made to spread in the charcoal layers through the holes to decrease earth resistance accordingly. The pit of about 40 square cm is dug in the soil. At the top, cement concrete work is made for the protection of earth pipe from mechanical damage and also to facilitate water-pouring arrangements for provision

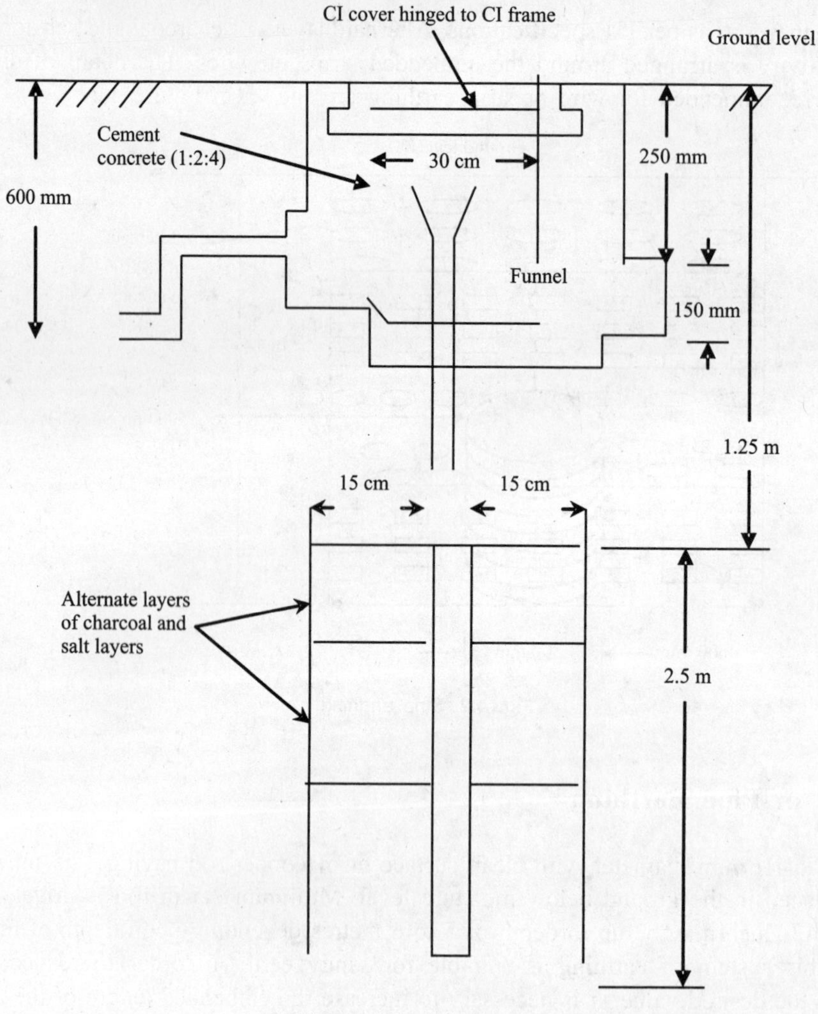

Fig. 4.3. Pipe earthing.

of dampness. In the dry summer season three to four buckets of water should be poured regularly through a funnel. The detail drawing with dimension is shown in the Figure 4.3.

4.8.4 Plate Earthing

This is most preferred earthing for many installations and houses. In this earthing, embedded copper or GI plate is placed in the earth, 5 to 6 metres deep. The size of plates is 60 cm × 60 cm × 6.3 mm for GI plate and 60 cm × 60 cm × 3.15 mm for copper plate.

A pit about 5 to 6 meters deep is dug and the earth electrode is placed in the ground as shown in the Figure 4.4. The plate is kept vertical and is arranged in alternate layers of charcoal and salt for a minimum thickness of 15 cm. The electrode or plate is connected to GI pipe of

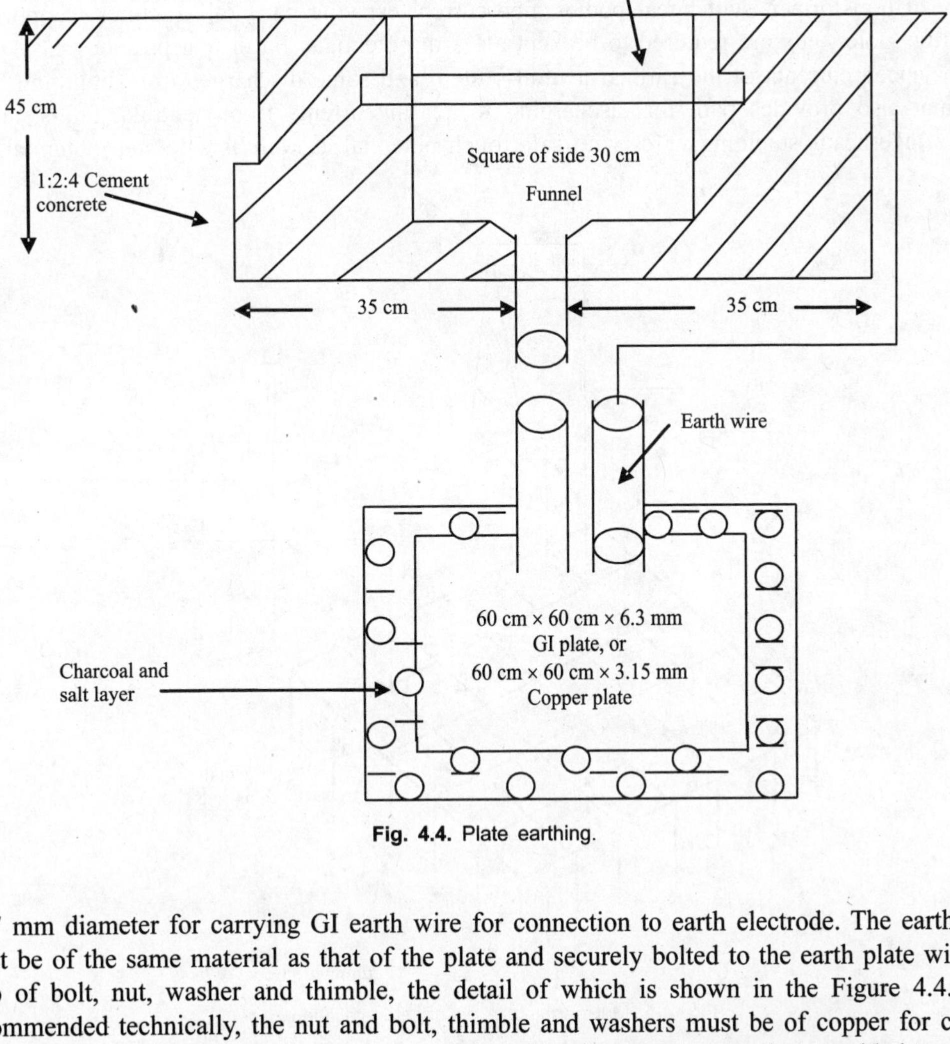

Fig. 4.4. Plate earthing.

12.7 mm diameter for carrying GI earth wire for connection to earth electrode. The earth wire must be of the same material as that of the plate and securely bolted to the earth plate with the help of bolt, nut, washer and thimble, the detail of which is shown in the Figure 4.4. It is recommended technically, the nut and bolt, thimble and washers must be of copper for copper plate and GI for galvanized earth plate or electrode. A water pipe connection provided to the pit filled with charcoal and salt carries water from top to the bottom of the pit for the purpose of increasing dampness and moisture, in order to reduce earth resistance. The cement work is covered with iron plate for periodic opening and pouring the water.

For the large substations and generating stations, the size of the earth plate is kept 1.2 m × 1.2 m × 12.5 mm for cast iron and 1.2 m × 1.2 m × 3 mm for copper plate.

4.8.5 MAT Earthing

Substation earthing is done by MAT earthing and is essential in account of keeping the entire substation switchyard at equipotential level to avoid electrical accidents. Earth mat is required for connecting neutral points, equipment body, and support structures to earth. Metal parts such as body of transformer, switchgear bodies, non-current carrying parts of structures, supporting cross-arms, poles, etc. are required to be kept at earth potentials. Earth mat provides easy path for unbalanced currents of the earthed neutral system and surge discharges from lightning, etc. Earth mat also provides path for discharging the earth currents from neutrals, faults, surge arresters, overheads shielding wires with safe touch potential as well as safe step potential.

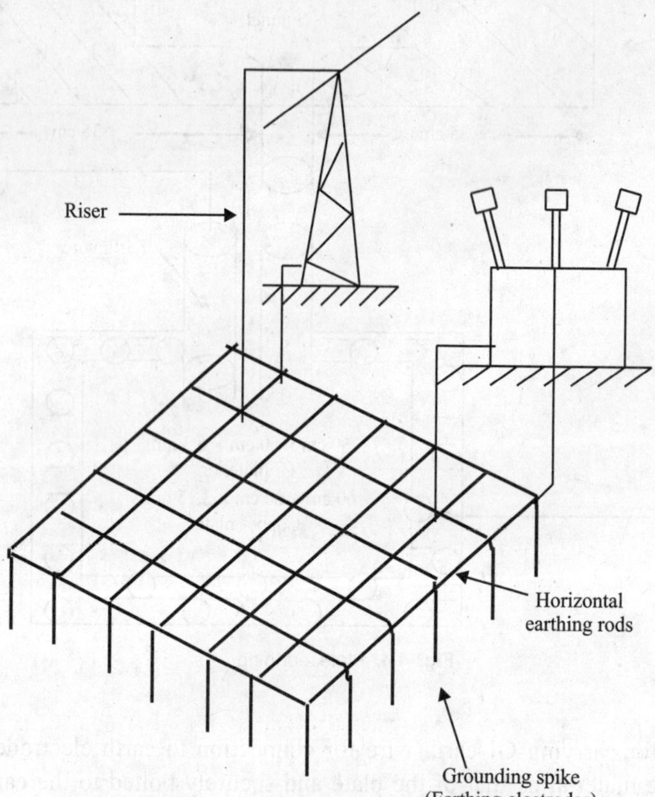

Fig. 4.5. Mat earthing in electrical installation.

While doing the earthing for the smallest distribution substation, the connections to the minimum four earth electrodes should be done. One direct connection from three 11 kV lighting arrester placed at top of the poles supporting busbars. One earth electrode should be connected with each lightning arrester meant for each phase separately. One direct connection from the neutral of the distribution transformer, one direct connection from the body of the 11 kV transformer and one common earth for neutral, body of transformer, non-current carrying parts of structures, supporting cross-arms, poles, etc.

By earthing the equipment, the potential of the earthed body remains within safe limit and in the event of earth fault, current flows through the earth plates, rods, pipes, etc. and may cause the operation of earth fault relay and circuit breakers. Horizontal earthing mat (grid) is prepared by using 75 × 10 mm mild steel flat bars spacing 3 to 4 metres in mesh form. Steel bars of 25 mm diameter and 2 to 3 metres long may also be used. While making the mat, distance between the parallel strips are maintained 2 m. Joints are made by electric arc welding at each crossing and should be covered by 2 mm thick bitumen paint. Entire mat is put 0.5 m below surface of the ground.

The earth mat (grid) covers the entire substation area and sometimes a few metres beyond the fencing of the substation. Earthing electrodes also known as spikes are placed vertically and welded with earth mat. Spikes are made up of 25 mm or 40 mm diameter, 2 to 3 m long. Figure 4.5 depicts the earthing system in an electrical installation.

Steps for Laying Earthing Mat

1. Prepare suitable layout and properly indicate the location of trenches and equipment's foundation. Trenches should be at least one metre deep.
2. Stones should be removed from the base of the trenches. Base should be filled by soft soil and rammed properly.
3. Now lay the earthing rods.
4. Grounding spikes are driven at different places in the earth. Separate spikes should be provided for transformers neutral, surge arresters and coupling capacitors.
5. Risers are used to connect the mat and the neutral of the equipment. These are led between the earthing rods and structure's foundations.
6. Joints must be clamped properly and then welded. Welded joints should be painted with bitumen paint.
7. Trenches should be filled with the earth covered by crushed stone of 5 to 8 cm size.
8. Earth resistance should be measured to ensure that it is within prescribed limit.

4.9 PRECAUTIONS TAKEN WHILE INSTALLING THE EARTH CONDUCTOR

- There should not be chances of being damaged the earthing conductor.
- The earth conductor is fixed with the system clamps saddles, nipples.
- Soldering, or welding of the earth wires must make the joints. Clamps give tight joints.
- Every care should be taken in order to make the earthing electrically continuous.
- Non-conducting materials, i.e., paint, enamel, corrosion materials should be removed from the connecting terminals of the equipment to which earthing thimble is to be connected.

4.10 EARTHING IN INDUSTRIAL PREMISES

In workshops and factories all metal conductors, cable sheath, switchgear, distribution fuse boards, starters, motors and all the other metal parts must be bonded together and connected to an efficient earth system. In large installations, it is recommended to connect earth continuity rods or pipes in parallel.

Earth system should be designed in such a way, so that interlinking of the earth electrodes is easily done. Recommended value of any earth system is always less than 5 ohms unless otherwise specified. All medium voltage equipment, apparatus and appliances shall be earthed separately. Distinct point connection from the electrodes through a busbar arrangement may be done.

In designing a *plate electrode*, the resistance may be calculated from the following formula:

$$R = \frac{1}{400} P \sqrt{(\pi/A)} \quad \text{Ohms}$$

where; P = resistance of soil in Ohms
A = area of the plate in m^2

And for designing pipe earthing:

$$R = \frac{100\,P}{2\pi L} \log(4L/d) \text{ Ohms}$$

where; L = length of rod or pipe in m
d = dia of rod or pipe in m

4.11 SUBSTATION EARTHING

Substation earthing is essential in account of keeping the entire substation switchyard at equipotential level to avoid electrical accidents. Metal parts such as body of transformer, switchgear bodies, non-current carrying parts of structures, supporting cross-arms, poles, etc. are required to be kept at earth potentials. It provides easy path for unbalanced currents of the earthed neutral system and surge discharges from lightning, etc.

The size and number of pipes depend on the size of the substation and the nature of soil.

4.12 EARTHING IN 11 KV/440 V DISTRIBUTION SUBSTATION

While doing the earthing for the smallest distribution substation, the connections to the minimum four earth electrodes should be as shown in Figure 4.6.

- One direct connection from three 11 kV lighting arrester placed at top of the poles supporting busbars. One lightning arrester is for each phase separately.
- One direct connection from the neutral of the distribution transformer.
- One direct connection from the body of the 11 kV transformer.
- Common earth for neutral, body of transformer, non-current carrying parts of structures, supporting cross-arms, poles, etc.

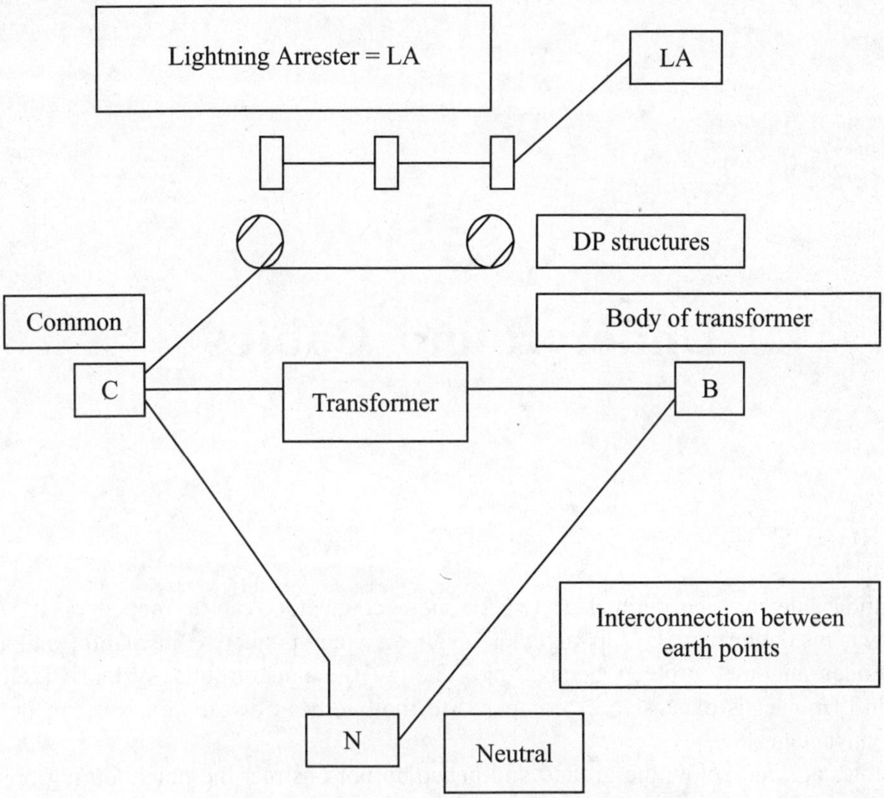

Fig. 4.6. Earthing in 11 kV/440 V distribution substation.

EXERCISES

1. On what basis, earthing conductor is selected?
2. What happens in case of failure of insulation at any one place in any system?
3. Write any five IS-specifications pertaining to earthing of domestic and factory installations.
4. State the main factors to be considered for a substation earthing.
5. Write the factors on which earth resistance value depends.
6. What precautions should be taken while installing the earth conductor?
7. Explain thc location where the earth pit should not be located.
8. Why are the metal frames of electrical apparatus earthed?
9. What are the points to be checked while inspecting the earthing arrangements?
10. The overhead lines are always supported on insulators, then why is it necessary to have a pole-to-pole resistance?
11. Describe Mat Earthing.

5

Underground Cables

5.1 INTRODUCTION

Underground cable may be defined as two or more conductors each wrapped or surrounded with its own insulation and lay up together under an outer protective insulation and laid up together under an outer protective covering. Though the underground system of electrical distribution of power is expensive as compared to the overhead distribution system, but it has following advantages.

1. Cables are laid below the ground and hence do not obstruct the neighbouring objects as overhead lines conductors do. Overhead lines conductors do not look good in the locality.
2. Faults due to lightning flashover, birdage, breaking of conductors due to fall of trees or other objects, etc., do not occur in the underground cables. Faults created due to these reasons cause interruptions in the continuity of power supply in the overhead distribution system and hit the reliability of the system.
3. Underground cables need less maintenance as compared to overhead distribution system.
4. It has longer life as compared to overhead lines.
5. Conductors of the underground cables are covered with the metallic sheath and electrical insulations; this is the reason that underground distribution system does not interfere with the telecommunication circuit.

   Note: Underground distribution lines are used in the densely populated localities, wide road crossings, wide river crossings, around gas plants and refineries, etc.

An underground cable must have the following qualities.

1. Materials used for making conductors as well other components of the cable should be such as to give complete chemical and physical stability throughout its life.
2. Insulation of the cable must be of high quality, so as to provide safe and reliable working voltage.
3. Conductors should be flexible and stranded. It should cause least voltage drop and heating.

4. Proper mechanical strength should be provided to the cables so that cable may survive the rough usage specifically during its laying.

5.2 CLASSIFICATION

Cables are classified as given in the Table 5.1. These cables are of single core or multi-core. In order to keep the cost of single core cables reasonable, these are not armoured. There are low voltage cables. Lead sheath is provided to protect the cable from moisture and mechanical damage. To avoid corrosion sheath is wrapped by an overall serving of fibrous material. Single circular core of stranded aluminium is used. Core is surrounded by insulating material like impregnated paper.

Table 5.1: Classifications of the cables

Cables	Operating voltage limit
L.T. (Low Tension) cables	1,000 V
H.T. (High Tension) cables	11,000 V
S.T. (Super Tension) cables	33,000 V
E.H.T. (Extra High Tension) cable	1,32,000 V

Multi-core cables are used for higher voltage applications, 11000 volts and above. Cores of these cables are made up of stranded circular shape and sometimes oval or sector shape. Cores are wrapped around by impregnated paper. Space within the conductors is filled with insulating material and the cable is made up of circular cross-sections. To avoid moisture entry inside, lead sheath that encloses the entire sheath (sheath of each conductor) is provided. Metallic armouring is done to save the cable from the mechanical damage.

Serving is provided over the armour. Serving consists of a compound and one layer of impregnated hessian tape over it. Lime wash coating on the outer surface of the cable is done to protect it from chemical action of ingredients present in the soil. Different types of high voltage cables used for three-phase applications are described hereunder.

Belted cables can be used for voltage applications up to 22 kV. Three cores of this cable are generally insulated from each other by impregnated paper. Impregnated papers are arranged in layers. A belt prepared by the impregnated paper tape is wrapped round the grouped insulated cores. Jute like fibrous insulating material is used to fill up the gap between the insulated cores. Filling up the fibrous insulating material gives circular shape to the cable. In order to protect the cable from mechanical injury and moisture ingress, a lead sheath covering the belt is used. An outer serving carrying a number of layers of steel armouring is provided to give mechanical strength to the cable otherwise cable may get broken while loading, deloading, laying and handling.

The main limitation of the belted cable is that insulation resistance as well as mechanical strength of the papers is quite small. Leakage current raises the temperature of the insulating paper, which may damage the dielectric strength of the insulation.

Screened cables can be used for voltage applications up to 66 kV. Leakage current, which raises the temperature of the insulation paper in belted cable, is overcome in the screened cable. In the screened cables, leakage current is diverted to the earth through metallic screen. Screened cables are of two types the S. L. types cables and H-type cables.

Pressure cables

These are high capacity cables and used for voltage applications above 66 kV.

5.3 CABLE SELECTION

System voltage current carrying capacity, mode of installation, permissible voltage drop, short-circuit rating and economic consideration are the factors, which should be taken into account at the time of selection of the size and type of cables.

Whether the system voltage is single-phase, three-phase, AC or DC must be stated clearly, while recording specifications of the cable. The most important factor is the current carrying capacity of the cable and continuous current carrying rating should be known.

There are generally following ways which cable carries out installation.

(i) Directly in the ground
(ii) In a duct or pipe.
(iii) In air.
(iv) Special installations like under the sea, inside mine, river crossing, vertical suspensions, etc.

Mode of installation is necessary to be taken into consideration, while deciding the selection of the cable. Recommended permissible voltage drop is $\pm$ 6% of the declared voltage.

Double line fault and three-phase fault generate very high short-circuit current. Cable must be capable to withstand electric stress as well as temperature rise due to the maximum fault current. Economic consideration is also an important factor, which governs cable selection. Generally, basis of minimum running cost is preferred while selecting cable size. But if cables are installed in an important locality or in high reliability application area, higher cost is acceptable.

5.4 CURRENT RATING OF THE CABLE

Temperature rise up to withstand value is determined by the maximum current flow in the cable. So, current rating of the cable is indirectly governed by the amount of heat produced by the load current and temperature around the cable. Besides these, temperature rise is also dependent on the depth of the cable, spacing between conductors, number of cores, methods of cable laying, thermal conductivity of the soil, etc.

5.5 CABLE HANDLING

Cable packing is done in wood make drums. During opening the drum, precautionary measures should be taken so that cable should not get damaged.

While rolling the heavy drums, direction of the arrow mark on the drum should be followed otherwise the cable will get loose and damaged. While pulling the cable from the drum, the drum should be kept fixed on a stand for free movement. Starting of the pulling the cable must be from the top position. If it is pulled from the bottom position, cables will rub each other and may get damaged.

Sometimes, drum gets damaged during transportation. So, to unwind the cables from damaged drum, help of the rotating turntable should be taken. Without taking help of the rotating turntable, there always remain chances of breakage of stranded wires and tearing of insulation.

Drums loaded with cable should never be dropped from the truck. Drums should be rolled on small slopes with same supporting arrangement. Strong rope support is necessary otherwise loss of control is possible.

Cable should be tested for continuity before laying it. Meggers are generally used to test the continuity. Cable jointing is required, if longer space is to be laid. After completion of the cable jointing, high voltage cables are pressure tested and low voltage cables are Megger tested. The cable testing includes continuity test, insulation resistance to earth and insulation resistance between conductors determination.

5.6 CABLE JOINTING

Each steps and operations of the cable jointing requires great attention and care. A faulty joint leads to short circuit and leakage current. Maintenance cost becomes high if cable joint is not proper. Following preparations are required while performing the cable jointing.

1. In order to provide protection against dust, rain and moisture, plastic cover should be kept ready.
2. Cable compound heating should be suitably placed.
3. In the event of collapsing pit, the planks should shore the sides of it upwards.
4. Rubber gloves are must for the cable jointer.
5. Cutting pliers, Metal pots and ladle, Calipers, Straight scissors, Hacksaw with blades, Lead scraper, knife, Screwdriver, Ball-pin hammer, Spanner set, Cold chisel, Blow lamp, file, brush, Foot rule are the tools which are required during cable jointing.

There are two types of cable joints, namely, Straight-through joint and T-joint.

5.6.1 Straight-through Joint

Standard accessories are joined by using straight-through joints. Ninety per cent tin and ten per cent zinc constitute the solder for jointing the aluminium conductors. For sealing the joints standard lead or copper sleeves are used. For sealing the cables into terminating boxes, the ordinary brass-wiping gland is used.

The various steps required to carry out straight-through joint are described below:

Steps

1. Making the seven and half centimetre overlapping on each end from the centre of joint, cable is laid end to end.

2. Make the marks, the portions on the cables to be joined. Remove the serving and armouring from the marked portion.
3. Push the splayed out strands close together to fit in the strands of one conductor into the other. Cut the surplus length of the exposed cores of the cables so that they will properly butt each other. Grip firmly the centre of the joint with a plier and wound round the one after the other.
4. Prepare the lead sleeve and slide it over the insulation wrapped portion. Ensure that inside of the sleeve is properly cleaned to slip it over.
5. Place a piece of thick paper behind the conductor to be joined to protect the other cores of the cables from molten solder splashes.
6. Pour the molten solder using flux over the end gap. Let the solder cool and form a perfect joint. Remove the excess solder with the help of impregnated paper tape or sharp knife.
7. Plumbing of the lead sleeve should be done now and free the ends of the lead sleeve. Heat the taped joint to remove the moisture and tin the end of the lead sleeve.
8. Tight the armour clamps firmly. Fill the compound in the groove of joint in the bottom half of the box. Place top half of the box in the position and bolt it up.
9. Warm up the box with glow lamp and fill the compound. Finally, tight all the bolts.

These steps are depicted in the Figure 5.1 to Figure 5.4.

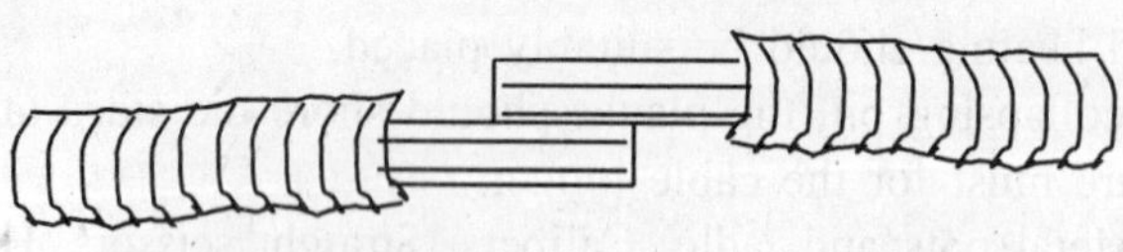

Fig. 5.1. Placing the conductors for making joint.

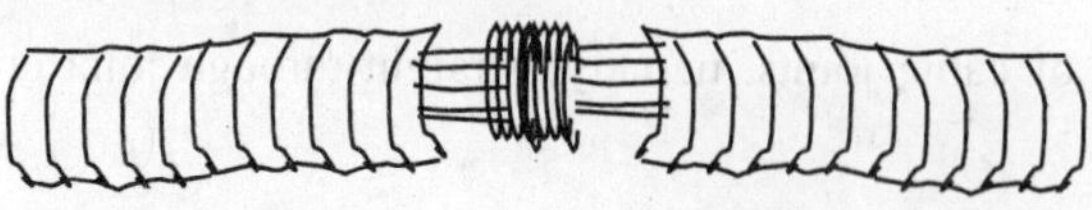

Fig. 5.2. Wrapping of the conductors for making joint.

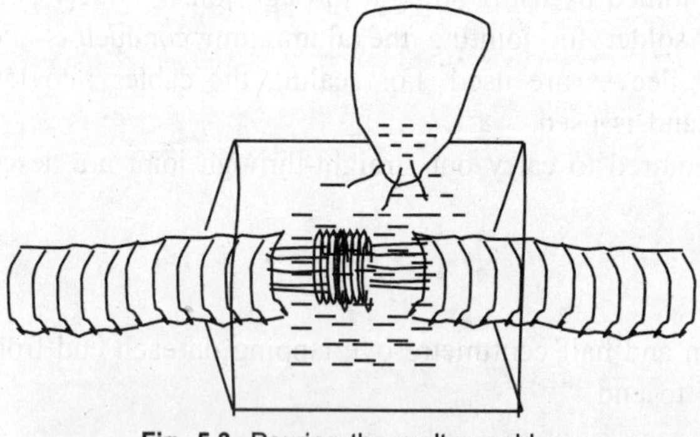

Fig. 5.3. Pouring the molten solder.

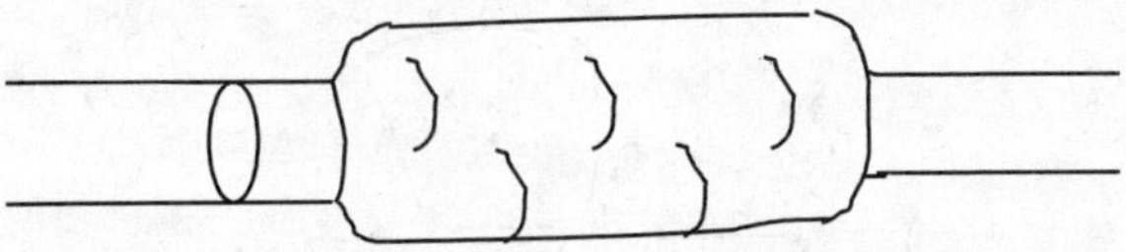

Fig. 5.4. Ready and complete straight-through joint.

5.6.2 T-joint

This joint is required for taping and serving connections. Tools required are the same as for straight-through joint. The various steps required to carry out T-joint are described below:

Steps

1. Open/remove the top half of the T-box.
2. The branch cables are laid over the main cable. Now, mark off the cable, by placing lower half of the box around serving of the service cable strip.
3. Wire binders should be placed around serving of the service cable. It should be at 6 mm outside the armour clamps of the box.
4. Remove the serving and armour of the main cable.
5. Also remove the serving, armour and sheath of the branch cable.
6. Apply cotton tape binder round the belt insulation up to a distance of 2.5 cm from the sheath ends of both the cables. Strip off the belt paper to expose the cores of the cable. Now, cut the filters of the cables.

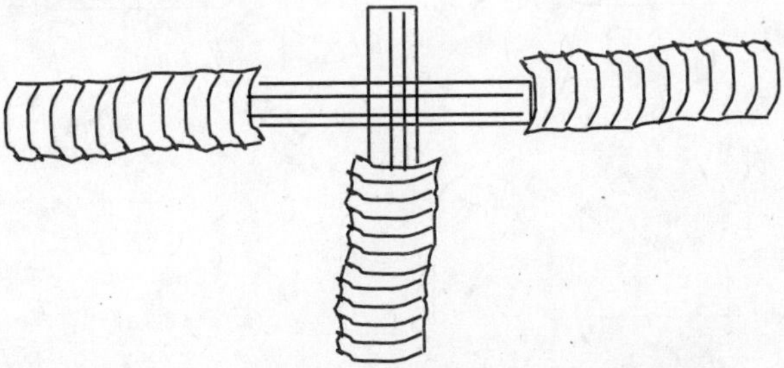

Fig. 5.5. Placing the conductors for making joint.

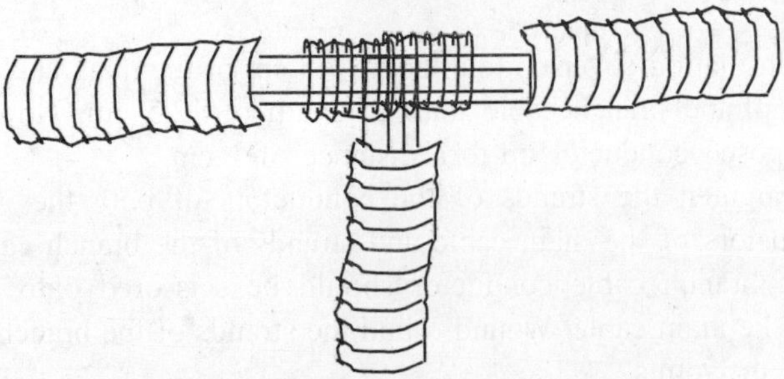

Fig. 5.6. Wrapping the conductors for making joint.

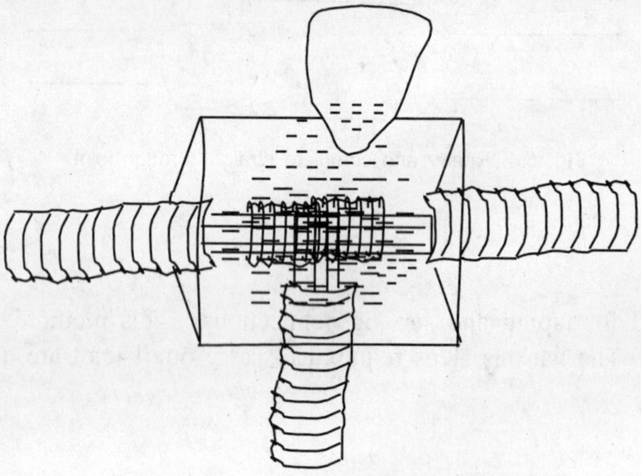

Fig. 5.7. Pouring the molten solder.

7. Cores of the branch cable should be displayed out. Now tap the core with impregnated cotton insulating tape.
8. Wood wedge should be inserted between the cores of the main cable, which makes easier to connect cores of the main branch cable.

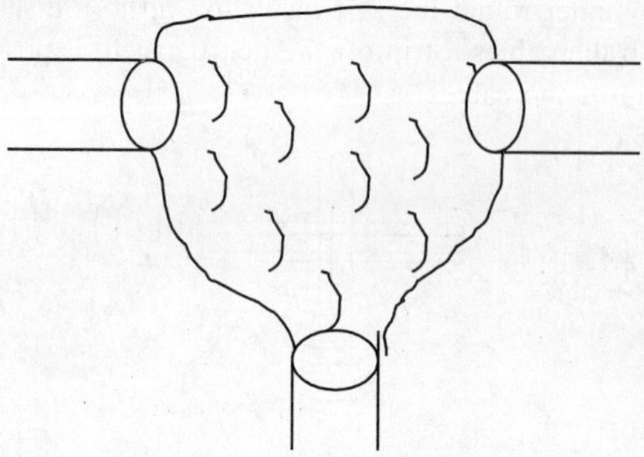

Fig. 5.8. Ready and complete T-joint.

9. Remove core insulation cover up to a length 7.5 cm of the main cable. Similarly, remove the insulation of the branch cable unto a length of 12.5 cm. Now apply binder wire around the exposed conductor up to a distance of 1 cm.
10. Clean and straighten the strands of the conductors of both the cables. Now tin the exposed conductors of the main cable and strands of the branch cable. The spilled out strands of the branch cable conductor should be scissored onto the exposed tinned conductors of the main cable. Wound round the strands of the branch cables on the main cable. Tighten the joints.

11. Pour the molten solder on the joint assembly. Apply the flux on the hot surface of the jointed portion. Before application of the flux, wipe the excess solder.
12. Similarly, join the other cores following the above steps.
13. When all the cores are joined the joint assembly must be taped up with insulating cotton tape and place the joint in a cast iron box.
14. Warm the lower half of the box and fill compound in the grooves of the box.
15. Place top half on the joint assembly in position and tight the belts and fill the gland with compound. Seal the box. The T-joint is ready for service.

These steps are depicted in the Figure 5.5 to Figure 5.8.

5.6.3 Precautions

Proper cable jointing may give uninterrupted service for twenty years or more. Following precautions should be taken during cable jointing.

1. Well-qualified and experienced skilled cable jointer should be employed for cable jointing.
2. Joint should be kept clean and dry.
3. After cutting the cable for jointing, joint should be completed without delay or lapse of time.
4. The main paper belt should be reinforced for a length of 5 cm, after tripping the lead sheathing from the cable. The paper-insulated cores should be lapped with dry tape immediately after the cable is stripped. It protects cable from dirt.
5. Surface of the core should be made smooth without any sharp joints after soldering the joints.

5.7 MATERIALS FOR CABLE JOINTING

Ferrule

It is tinned copper make slotted metal tube. Size of the cable conductor decides it size.

Sleeve

Sleeves are made up of copper, lead or brass. These are required to be employed during cable jointing.

(i) to prevent moisture.
(ii) to bond the sheath of the cables.
(iii) to protect the taped cores against mechanical damage.

Joint-Box

Cast iron joint boxes are used for mechanical protection of taped cores of cables. Sizes of the joint boxes are decided by the working voltage and type of joint. These are made up of two halves. Both the halves of the boxes are fixed together by nut and bolt. Filling holes are provided in the boxes from where compound is poured.

Solder

The main components of solder are tin and lead. Using the solder made up by mixing equal composition of tin and lead joints copper conductors. The composition of solder with organic flux suitable for using for aluminium conductors jointing is:

Lead = 51%;
Tin = 31%;
Zinc = 9%;
Cadmium = 9%.

This composition works temperature in the range of 280°C to 290°C.

Organic Flux

An admixture of Triethanolamine and Fluorine compound is used to make organic flux. It decomposes at temperature about 316°C a little below the jointing temperatures.

Other Materials

Impregnated paper, tape binder, lead strip, sealing sleeves and insulating tape are some other materials which are used during cable jointing.

Filling Compound

Bitumen base compound should be having following essential qualities are used as filling compound of the cable joints.

(i) Do not crack on cooling.
(ii) Should have high softening point.
(iii) Should have very good adhesive property.
(iv) Should have high flash point. (above 250°C)
(v) It should contract less on cooling.

Up to rated voltage 11 kV, the same filling compound is used for filling the sleeve and the boxes. But above 11 kV applications, the compound used in filling the sleeve is oil resin compound.

5.8 CABLE TERMINATION

End of the conductors should be prepared in such a way so that these are easily and suitably be connected with the terminals to which these are to be connected by mechanical means. So, all the bare conductors and terminals for cable connections should be easily accessible. Mechanical or electrical stresses should not be imposed on the terminals. Similar metallic conductors such as copper-copper or aluminium-aluminium should be joined. In the case if dissimilar metallic conductors are joined it should be ensured that there is no corrosion.

5.9 CABLE LAYING

Preparations

Before going for cable laying, following preparations are required to be made:

1. Planning and selecting the suitable route for cable laying.
2. Marking of the proper size trenches. Trench width depends on the number of cables to be buried and depth depends on the operating voltage level.
3. Unwinding of the cable from the cable drum.
4. Laying of the cable includes pulling of the cable parallel to the trench and placing it inside the trench. Cable jointing wherever required.
5. Filling of the trench.

Methods of Cable Laying

There are following two methods employed for cable laying.

1. Direct cable laying in the ground
2. Drawn in system cable laying
   (i) Cable laying on brackets, racks and sleets
   (ii) Cable laying at river crossing or submarine
   (iii) Cable laying on bridges
   (iv) Cable laying in tunnel
   (v) Cable laying under railway track
   (vi) Cable laying under the road
   (vii) Cable laying in open ducts

5.9.1 Direct Cable Laying in the Ground

Cable drawing into the duct is costly as compared to laying the cable directly in the ground. But repair and maintenance is costly for direct laying in the ground. Repair or alteration can be easily done without reopening the ground for cable laid or drawn in ducts or pipes so cable drawing into the duct is most economical system for cable laying.

Cable directly laid in the ground undergoes the process of chemical corrosion and mechanical damage due to soil. Cables drawn into ducts or pipes are not affected by soil.

Good quality and galvanized steel tape armouring is the basic requirement for cables to be laid into the ground directly. Cost of the armoured cables is high.

Once the route of the cable is decided, the next step is to dig the trench of suitable dimension according to the cable rating and size. Following two factors should be taken into consideration while deciding the depth of the trench.

- Cable should not be easily accessible to the children, animals, farmers, etc.
- Sufficient depth ensures that cable does not expose to the temperature fluctuation.

The width of the cable trench is decided by the number of cables to be buried. In general 0.5 m width for a single cable is found suitable. Before laying the cable, riddle and level the bed of the trench. Chemically inert materials like sand or clay should be filled upon the bed of the trench, where soil contains potentially corrosive elements such as clinker, slug or ash. Sand should also be used surrounding the cable if stones or rock pieces are present as shown in the Figure 5.9.

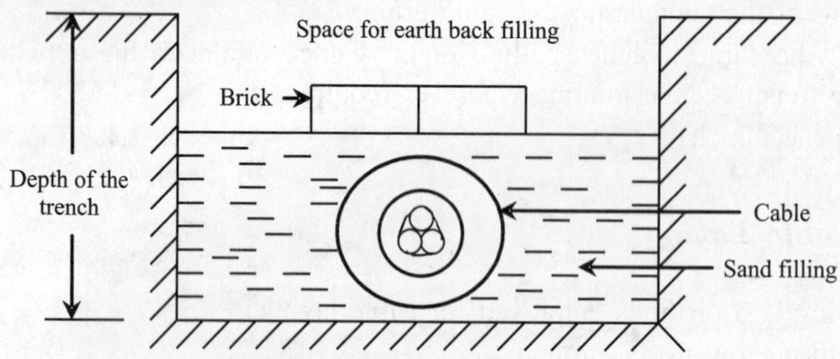

Fig. 5.9. Direct in ground cable laying.

After setting the cable in the correct position in the bed of the trench, proper cover to the cable is provided and then trench is back filled with excavated soil in layers. For better setting up of soil in the trench, watering is preferred. Five to ten centimetre high crowning of the sand at the centre of the trench should be done for future setting of soil.

5.9.2 Drawn in System Cable Laying

This system is preferred when excavation is inconvenient or expensive. In this system of laying, pipes or ducts are used to be laid in the trench. In regular interval manholes are made. Choice between ducts (made up of stone, cement or concrete) and pipes (made up of stone ware) depends on the availability of the materials, economic factor and degree of protection required. Figure 5.10 depicts this system of laying.

The depth of trench should be such as the top pipe remains two feet below the ground level. The internal diameter of the duct should be chosen so that cables are pulled obstruction free. 25 cm to 75 cm space would be maintained between two adjacent pipes. It should be ensured that while back filling is done, alignment shifting of the pipes should not take place.

Manholes are made at the terminal points of the ducts or pipes. These manholes are used to pull in cables through ducts. The size of the manholes should be enough to allow entering

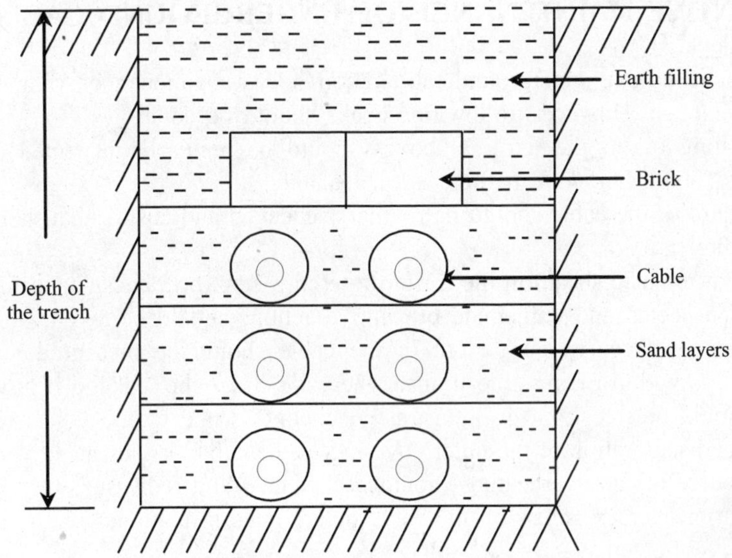

Fig. 5.10. Drawn in system cable laying.

working personnel without any difficulty. Its size also depends on the number of ducts terminating from different directions.

In industries and workshops, small rating cables are also laid on brackets, racks and cleats. These are preferred inside the buildings.

5.10 CABLE FAULTS

Overloading and short-circuiting are the common cause of cable faults. However, non-electrical causes also cause cable faults. Causes of cable faults are described hereunder.

1. Corrosion and pitting on the cable. This may happen due to chemicals presents in the soil.
2. Vibrations of the lead sheathed cable cause cracks in sheath of the cable.
3. Sometimes oil leaks from cable boxes, which reduces the insulation strength and generates fault.
4. Waterlogging near the cable laying creates dampness in the soil. Dampness damages the sheath of the cable.
5. Overloading raises temperature around the cable conductors and insulation gets damaged.
6. During excavation by pickaxe or crowbar, puncturing of cables takes place. Care should be taken to avoid such damage.
7. Mechanical damage occurs if the cable is mishandled at the time of lying of the cable. This may also happen due to poor joints.
8. Improper protection of the cable is also cause of the cable fault.

5.11 PREVENTIVE MAINTENANCE OF UNDERGROUND CABLES

Generally, it is believed that underground cable requires least maintenance. An annual inspection is proved to be enough. However, following checks should be done:

1. Impregnating oil of the terminal boxes should be regularly inspected and corrective measures should be taken, if there is oil leakage.
2. Temperature of the cable should be regularly checked and ensure that there is no excess temperature rise.
3. Damage of the lead sheath in the form of the corrosion, brittleness or insulation swelling must be inspected and immediate preventive actions should be taken.
4. Insulation resistance as per length of the cable should be measured and ensure that resistance is within permissible limits, otherwise check the leakage if any.
5. Cable should be free from dripping of oil, chemicals and water.
6. If there is discoloration of the outer covering of the cable, it indicates overheating of the cable. Source of the overheating should be detected and redressed.
7. Excessive mechanical stress on the cable deteriorates insulation, sheath and servings. It should be corrected.
8. Water content should not be present in the cable ducts. If water is present in the ducts, it should be evacuated.

5.12 MURRAY LOOP TEST FOR CABLE

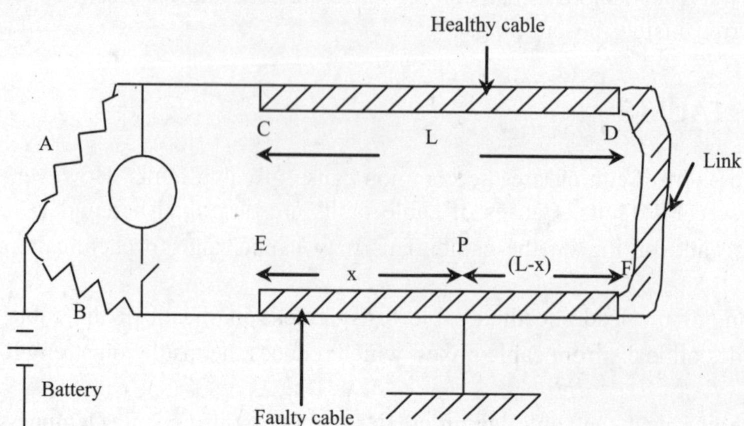

Fig. 5.11. Murray loop test circuit.

Cable fault location is detected with the help of Murray loop test. This test is performed with the help of Bridge Megger. Basically, Murray loop works on the principle of Wheat-stone Bridge. This is also useful in finding the fault location in telephone lines. This test detects the fault in the form of leakage path between the core of the cable and its covering.

In this test, one healthy cable of same length is put parallel with the faulty cable along the same route as shown in the Figure 5.11. A battery is used to supply the test current. Two known resistances say A and B make two arms of the bridge and other two arms are made by the resistance of healthy cable say CD and faulty cable say EF. The ends of the two cables say F and D should be joined by a mechanically strong wire of negligible resistance.

Let;

L = totals length of the each cable, separately.

r = resistance of the both cables in ohmmeter length.

x = length of the faulty cable from one end to the fault point as shown in the Figure 5.12.

(2L-x)r = resistance between fault point P and end of the healthy cable C in ohms

xr = resistance between fault point P and faulty cable end E.

Now, the following relationship will be applicable under the balanced condition.

$$A/B = (2L-x)r/xr$$

or $$x = 2LB/(A+B) \text{ metres}$$

Here, is the fault point 'P' at a distance 'x' metre from the end E.

EXERCISE

1. Describe and compare various systems of cable laying in the underground distribution system.
2. Explain cable testing and commissioning.
3. Describe different steps of cable jointing.
4. Where T-joint is required to be done? Describe how is T-joint done for cables.
5. Describe procedure of straight-through joint for cables.
6. What precautions should be taken during cable jointing?
7. Write short note on:
   (a) Classification of cables
   (b) Cable handling
   (c) Cable selection
   (d) Causes of cable faults
8. What preventive checks should be performed for underground cables?
9. Describe different materials required for jointing the cables.

6
Maintenance of DC Batteries

DC battery is a device in which energy in the form of chemicals is stored and when it comes into a close electric circuit, transformation of chemical energy into electrical energy takes place.

6.1 CLASSIFICATION

The batteries can be classified as:

Primary Battery

The stored energy finishes and ends after its use in this type of battery or cell. These batteries cannot be recharged after use. Dry cells used in torches, radio transistors, etc. are the primary cells. The other primary cells are Mercury cell and Weston cadmium cell.

Secondary Battery

These are the batteries, which can be recharged after getting discharged. Recharging is done by flow of electric current through it. These are lead acid storage battery, alkaline type storage battery, etc.

The advantages of the secondary cells over primary cells are:

- Internal resistance of the secondary cell is low and hence can carry higher current.
- Its life is much higher than primary cells.
- It gives higher voltage than primary cells.

Comparison between primary and secondary batteries is given in the Table 6.1.

Storage Batteries

These batteries come under the category of secondary cells. Energy can be stored in these. These are of the following types.

- Lead acid storage battery and
- Alkaline type storage battery

Different types of alkaline type storage batteries are:
- Nickel cadmium alkaline battery
- Nickel iron alkaline battery
- Silver oxide storage battery

These alkaline storage batteries last long as compared to the lead acid storage batteries. Mechanical efficiency, over charging and discharging capacities of the alkaline storage batteries are higher than that of lead acid batteries. But its use is limited because the terminal voltage is low and cost is higher. But due to lightweight, silver oxide storage batteries are used in the defence sector but these are very costly and their life is very short.

DC supply is required essentially for control, annunciation and signalling, protection and emergency lighting in the event of failure of AC supply. Also, in the electrical high voltage substations, availability of the DC supply is of utmost important for the safe operation of the costly equipment installed in the substations and any failure of DC supply may lead to serious outages and damage to the costly equipment like power transformers, circuit breakers, reactors, etc. Lead acid storage batteries are usually provided at the electrical substations to act as a reliable source of DC supply even when AC supply fails and rectifiers are rendered inoperative.

Table 6.1: Comparison between primary and secondary batteries

S/N	Primary batteries	Secondary batteries
1.	Electrical energy is directly taken from chemical energy	Electrical energy is first stored into the cell and thereafter chemical energy is converted into electrical energy.
2.	Chemical and electrical actions are irreversible	Chemical and electrical actions are reversible.
3.	Cell is replaced after use	Passing current in opposite direction charges cells.
4.	Low efficient and low capacity	High efficient and higher capacity.
5.	Less cost and use & throw	High initial and maintenance cost.
6.	Presence of polarization	Practically no local action and polarization.

Lead acid storage batteries are of the following types:
- Stationery lead acid storage battery (CS-Type)
- High rate discharge type stationery lead acid storage battery (HS-Type)
- Portable lead acid storagc battery (PS-Type)
- Traction lead acid storage battery (V-Type)

Lead acid storage batteries are of the following types:
- Stationery lead acid storage battery (CS-Type)
- High rate discharge type stationery lead acid storage battery (HS-Type)
- Portable lead acid storage battery (PS-Type)
- Traction lead acid storage battery (V-Type)

6.2 ALKALINE STORAGE BATTERY

Alkaline batteries require no maintenance and have very long life. Its cost is high and terminal voltage is low. This is the reason that these batteries are very useful for mine locomotives, miner's lamp, satellites and wireless stations.

Nickel-plated perforated steel pockets or tubes are used to hold the active chemical materials of the positive and negative plates. Positive plates are made up of nickel hydroxide enclosed in a perforated steel tube. In the nickel iron alkaline battery, the negative plate is made up of iron oxide with a little mercuric oxide that reduces the resistance. In the nickel cadmium alkaline battery, cadmium mixed with little iron is used as active chemical material. Mixing iron prevents the cadmium from caking and losing its porosity. Separators between the positive and negative plates are insulated rods.

Operation of Alkaline Storage Battery

In a charged alkaline battery, nickel hydroxide, $Ni(OH)_3$ appears on the positive plate and iron on the negative plate. Iron on the negative plate gets oxidized to ferrous hydroxide during discharging the battery. At the same time, $Ni(OH)_3$ is converted into $Ni(OH)_2$, a lower oxide. The chemical reactions are reversed during charging the battery. The chemical reactions may be written as below:

$$2Ni(OH)_3 + Fe \underset{\text{Charge}}{\overset{\text{Discharge}}{\rightleftarrows}} 2Ni(OH)_2 + Fe(OH)_2$$

Exactly similar chemical reaction takes place in the nickel cadmium battery only cadmium replaces iron.

$$2Ni(OH)_3 + Fe \underset{\text{Charge}}{\overset{\text{Discharge}}{\rightleftarrows}} 2Ni(OH)_2 + Fe(OH)_2$$

6.3 LEAD ACID STORAGE BATTERY

The parts of lead acid batteries are:

- Positive and negative plates
- Separators
- Electrolyte
- Container
- Bottom blocks
- Plate connectors
- Cell connectors
- Terminal posts
- Vent plugs

Inside view of a battery is shown in Figure 6.1. *Positive plates* are the clad type plates which consist of lead dioxide filled in glass fibre tubes having tiny holes and lead alloy rod. Lead alloy rod is placed in the middle of the tube. This plate has longer life and capable to withstand overcharging and over discharging. Pasted type plate is another type of plate in which lead dioxide is filled in the grid framework of the plate. The grid holds the active material and also works as a conductor for current. These plates are widely used due to simple structure and easy availability of varying thickness. Active materials do not fall on the bottom and battery can live longer, because the plates are fixed in fibreglass mats.

Lead is the active material for *negative plates*. Preferably pasted type plates are used. These consist of grids with ribbed structure over which active material is pasted.

PVC separators made up of high quality microporous PVC materials are used to separate the positive and negative elements. These elements are clubbed together and are interspersed. A

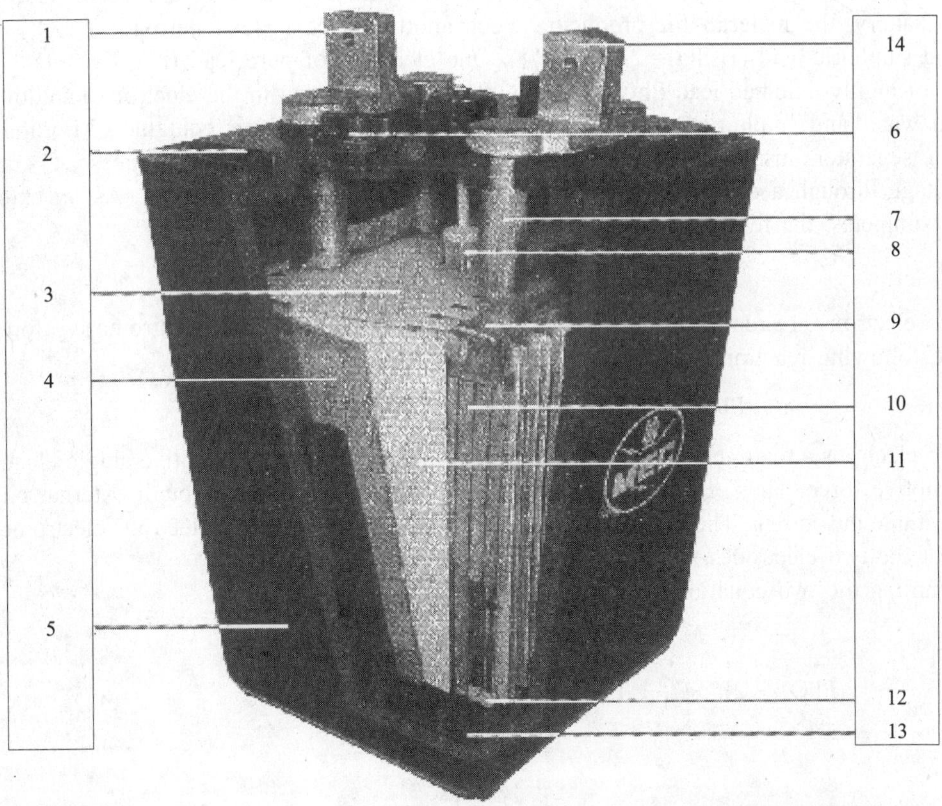

1. Negative terminal
2. Cell cover
3. Separator guard
4. Negative plate
5. Hard rubber container
6. Vent plug
7. Pole
8. Float
9. Busbar
10. Positive plate
11. Microporous PVC separator
12. Bottom bar
13. Prism
14. Positive terminal

Fig. 6.1. Inside view of a battery.

separator guard is fitted around the plate lugs so there is no displacement during the operation of the cell. The separator guard is necessary to prevent the damage to separators and plates while checking the cells with hydrometer or thermometer. The elements with the separator guards are housed in the hard rubber or plastic container. It has high impact resistance and high insulation value, which can withstand temperature up to 60°C.

For indicating the level of electrolyte an acid resistant *float* with level markings is located at top of the cell. Markings indicate the level of the electrolyte and serves as a visual aid for periodical topping up of the cell with distilled water.

Operation of Lead Acid Storage Battery

With the electrolysis of dilute sulphuric acid between lead plates, the anode becomes coated with lead peroxide, PbO_2 while the cathode remains unaltered. Now current in the reverse direction can be achieved on breaking the circuit and connecting the terminals by a wire. In lead acid storage battery, the material for producing electromotive force is lead peroxide (PbO_2), lead (Pb) and sulphuric acid ($H_2SO_4 = 2H^+ + SO_4^-$). One electrode of pure lead (negative plate) and another of highly oxidized lead PbO_2 (positive plate) are immersed in the electrolyte solution of distilled water and sulphuric acid. Solution is taken in a hard rubber container. Microporous synthetic separators insulate the plates from each other. The lead acid cell produces DC current and voltage through a chemical reaction (oxidation reduction) within the cell. As an external circuit completes, the following reaction takes place at lead plate

$$Pb + SO_4^- = PbSO_4 + 2e \qquad (1)$$

Two electrons are repelled due to electron negative single electrode electromotive force. The following reaction takes place at lead dioxide plate

$$PbO_2 + 4\,H^+ + SO_4^- + 2e = PbSO_4 + 2H_2O \qquad (2)$$

Two electrons are attracted through external circuit, due to electron positive single electrode electromotive force. Thus, electrons flow from Pb plate to PbO_2 plate through external circuit and constitute the current. The electromotive force is the sum of the single electrode electromotive forces of the two electrodes.

Summing the two equations, we get:

$$PbO_2 + 2H_2SO_4 + Pb \underset{\text{Charge}}{\overset{\text{Discharge}}{\rightleftharpoons}} 2PbSO_4 + 2H_2O \qquad (3)$$

Equation number (3) above depicts the chemical reaction that, as the battery discharges the active material, i.e., lead dioxide in positive plate and sponge lead in the negative plate react with the sulphuric acid in the electrolyte to form lead sulphate and water. The lead sulphate deposits on both the plates. With the battery in use, the specific gravity of electrolyte gradually decreases because the proportion of acid decreases and the water increases.

When the battery is placed on charge, reverse action takes place. The acid in the updated active material of both the plates is driven back into the electrolyte reducing the sulphate on the plates and increasing the specific gravity of the electrolyte. The specific gravity rises until all the acid is driven out of the plates and back into the electrolyte. Then there will be no sulphate in the plates. All the acid gets returned to the electrolyte and additional charge does not increase the specific gravity.

6.4 ELECTROLYTE AND ITS PREPARATION

Battery grade sulphuric acid is usually available in 1.840 or 1.400 specific gravity concentration. Mixing up distilled water with battery grade sulphuric acid changes the specific gravity to another value. Acid mixing chart is given as Table 6.2, which may be used as a guide to determine the volume ratio of acid to water to obtain the desired specific gravity. During preparation of the electrolyte, following precautions are recommended:

- Always wear goggles, rubber gloves and apron while handling acid.
- Store sulphuric acid only in plastic or lead lined containers.
- Add acid slowly to water while mixing. Never add water to acid. Take measured quantity of water in a vessel and add measured quantity of acid slowly. Stir the solution with a long glass rod.
- Be careful not to spill or splash acid.
- Specific gravity of acid is filled into new cells should be 1.180 or 1.190 at 27°C.
- After initial filling, the cells should be at rest for a minimum of 12 hours and not more than 24 hours before charging is commenced.

The other cares required to be taken are:

- The acid should be battery grade sulphuric acid confirming to IS: 266-1961.
- The distilled water should be confirming to IS: 1069-1964
- The specific gravity of electrolyte varies with temperature. Any reading on the hydrometer should therefore be corrected to read to 27°C.
- For every 1°C above 27°C, add 0.0007 to the specific gravity as read on the hydrometer. Similarly, for every 1°C, below 27°C, subtract 0.0007 from the specific gravity as read on the hydrometer.

Table 6.2: Acid mixing chart

When mixing by volume using 1.840 acid distilled water required to one part acid	When mixing by volume using 1.400 acid distilled water required to one part acid	Desired specific gravity
2.80	0.42	1.280
2.90	0.53	1.270
3.00	0.60	1.260
3.09	0.64	1.255
3.17	0.67	1.250
3.35	0.75	1.240
3.55	0.82	1.230
3.68	0.86	1.225
3.80	0.90	1.220
4.39	1.15	1.200
5.00	1.41	1.180

6.5 INITIAL CHARGING

The procedure of the initial or first charging affects the battery performance to an incredible extent. So, special cares should be followed during initial charging. These are given below.

- First record the open circuit voltage of each cell. After this, batteries can be charged at the specified rates. As soon as the battery is put on charging, take another set of voltage readings to check for any reverse connections.
- The readings of the individual cell voltage, specific gravity and temperature should be recorded every hour.
- During charging, the temperature should not be allowed to exceed 50°C. If the temperature tends to rise beyond 45°C, the current is reduced to one-third the value so that the temperature ceases to rise. If at any time, the current is reduced, the time of charging shall be correspondingly increased. So at the end of charging, the battery is loaded with specified ampere-hour.
- The charging may be suspended for sometime to allow the cells to cool down, provided 50% of the total ampere-hour is fed. The charging should not stop prior to 50% of the total input is given.
- The electrolyte level should be maintained by topping up with electrolyte of specific gravity as originally filled in. Towards the end of charge, all the cells start gassing freely. The cells are supposed to be fully charged if voltage reaches from 2.55 to 2.65 V per cell and specific gravity to 1.210 ± 0.005 at 27°C. Battery may be declared charged if cell voltage and specific gravity values remain constant for three consecutive hourly readings towards the end of the charging period.

  Note: Irrespective of constant specific gravity and voltage readings, the charging is not to be terminated until the specified ampere-hour input is fed into the battery. At the end of the charging, the specific gravity is adjusted to 1.210 ± 0.005 at 27°C.

6.6 BATTERY CAPACITY

Theoretically, Ah (Ampere-hour) capacity of the battery is determined by the amount of active material converted into sulphate for each ampere-hour of discharge (3.83 gm) of sponge lead and (4.42 gm) of lead dioxide independent of the rate of discharge. The amount of active material present in the plates is kept some 5 to 6 times in excess. Part of this excess is present to give long life.

Normally the ampere-hour capacity denotes the 10 hours discharge rate. For example, 450 ampere-hour capacity means that the DC battery can deliver 45 Amps continuously for 10 hours.

Capacity Test

The capacity test is carried out by the way of charge discharge cycle in per year interval. The battery is allowed to discharge through a variable resistance or water load at constant current equivalent to 10 hours rate of discharge. The discharge is stopped when closed circuit voltage across the battery falls to (1.85 × n) volts, where 'n' is the number of cells in series in the battery or when voltage across any one cell falls to 1.70 volts, whichever is earlier.

$$\text{Capacity of battery} = \text{Amperes} \times \text{Discharge Time}$$

A constant normal recharge current of 10% of the 10-hour rated capacity of the battery recharges it till the cells start gassing at a voltage of around 2.35 to 2.4 volts per cell. Thereafter, the charging is reduced to the finishing rate of 50% of the 10-hour rated capacity of the battery and charging is continued at this current till the voltage of each cell reaches 2.55 to 2.65 volts and the specific gravity to 1.210 ± 0.005 at 27°C. Both remain constant for three consecutive hourly readings.

6.7 EFFICIENCY OF LEAD ACID BATTERY

There are two methods to evaluate the efficiency:

Ampere-hour (Ah) Efficiency

This is evaluated using the following formula and the desired value of Ah efficiency of lead acid battery is 90 to 95%.

$$\text{Ah efficiency} = \frac{\text{Amp-hour discharge} \times 100}{\text{Amp-hour charge}}$$

Watt-hour (Wh) Efficiency

During discharging of the battery, the voltage of cells comes down from 2.16 V to 1.85 V, whereas during charging, it increases from 1.85 V to 2.6 V. The Wh efficiency considers the charge in voltage at the terminals and is evaluated as;

$$\text{Wh efficiency} = \frac{\text{Average voltage on discharge}}{\text{Average voltage on charge}} \times \text{Ah efficiency} \times 100$$

The desired value of Wh efficiency of lead acid battery is 72 to 80%.

6.8 METHODS OF RECHARGING

Normal Charging

The normal charging carries out at a charging current rate of 10% of rated 10-hour capacity of cell/battery till the voltage reaches to 2.4 V per cell. After this, charging at a rate 5% of rated 10-hour capacity of the cell/battery till full level of charging is achieved.

Equalizing Charging

In order to normalize the lagging cells equalizing charges are fed to the cells with the other cells. Indication of this is obtained, when specific gravity readings vary by 20 points than specified. Equalizing charge is given at a rate of 7.5% of the 10-hour rated capacity of the cell till full level of charging is attained. Equalizing charges are required to be fed once in every three months.

6.9 OPERATING MODES

Trickle Charging

This charging of the battery is required only during emergency situations. The battery is kept under trickle charging at 2.25 to 2.30 V per cell trickle charge rate may be required to be adjusted depending upon ambient, operating temperature, age of battery. The few examples are given as below.

(a) If the specific gravity of pilot cell show declining tendency, it calls for increase in trickle charging rate.
(b) If the topping up is required very frequently then trickle charging rate is to be reduced.

Float Charging

This charging of the battery comes into operation during standby periods when the electrical load is supplied with power from charger. Charger gives continuous charge at float voltage to the cells. The float voltage is regulated between 2.16 and 2.20 volts per cell depending upon age of the cells. High floating voltage should be avoided because it results in excess consumption of water. Low floating voltage results in reduction of battery's discharge capacity.

6.10 COMMON FAULTS AND REMEDIES

Sulphation

It is the process of accumulation of a film of lead sulphate ($PbSO_4$) on the plates, terminals, nut and bolts of connecting strips. It happens due to continuous discharging of battery. If battery is kept in discharge condition for longer duration the lead sulphate layer becomes hard and difficult to remove. To avoid sulphation, the battery should be charged at rated current and discharge at current of ten-hour rate of discharge. Again charge the battery at rated current. This voids the sulphation. Clean the terminals, nut and bolts of connecting strips.

Shedding

Due to the ageing, excessive charging, vibrations, buckling of plate's active material of the plates accumulate at the bottom of the cell. This causes shedding. Washing the cells with weak electrolyte and removing the sediments are the remedial measures of shedding.

Buckling

Buckling means tilting of the plates and coming in contact with the other plates. It occurs due to failure of the separator and heavy charging. The separator may get deteriorated and develop holes. To avoid buckling take out the plate assembly and wash with distilled water. Tilted plates should be straightened with the help of wooden stick.

Internal Short-circuiting

Sedimentation may be the major reason of internal short-circuiting between the plates. It is indicated by low or no voltage builds up between the cell terminals or failure of gassing action. It may occur due to buckled plates, sulphation, damaged plates, heavy charging or discharging and damage of separator or development of holes in the separator. Washing the cell with weak electrolyte as done for shedding may avoid the short-circuiting. If the cell's performance does not improve, take measures as mentioned for buckling.

Treeing

Treeing is the growth of the active material on the sides of the plates caused by the impurities or shedding of active material. Active materials may accumulate due to overheating of the plates during heavy charging or discharging and by using impure active material. This causes treeing. Its remedy is to wash the cell with distilled water.

Leakage of Container

Leakage of the container must be repaired on priority without delay. The cracked cell may be repaired with sealing compound, Bitulac NOC-161. Mix up sealing compound with petrol or Carbon tetrachloride in such a way that it may flow in the cracks. Leave the cell for 48 hours to allow the compound to dry and set properly.

Hydration

It is the white deposit on the plates caused by moisture being observed from the air by plates. It usually happens in new batteries if the seals are broken and the batteries are fitted with dry wood separators.

Stratification

It is uneven density of the electrolyte caused by the restriction to the free circulation of the electrolyte.

6.11 MAINTENANCE CHECKLIST

Following list helps the routine maintenance check of a battery.

Check	Ensure	To prevent
Cell surface for dust or spilt electrolyte	Surface is kept clean, dry and tidy	Self-discharge
Condition of container and sealing components	No cracks or spillages occur	Increase in the rate of self-discharge
All connections in terminal charging circuit	Connections are tight and firm and terminals are clean	Corrosion or loose clamps
Vent holes in vent plugs	Vent holes and plugs are in clean condition	Deterioration of the cell

6.12 CONSUMABLES AND SPARES FOR BATTERIES

Following materials and spare parts must always be kept in ready stock for maintenance purpose.

1. Sealing compound
2. Distilled water confirming to IS: 1069–1964
3. Battery clips of adequate capacity
4. 10% of soda
5. Copper strips, nut and bolts
6. Electrolyte float as per model of battery
7. Vent plugs
8. Petroleum jelly
9. Cotton waste
10. Emery cloth
11. Clips
12. Positive plates
13. Negative plates
14. Wooden separators
15. Gaskets for sealing

6.13 MAINTENANCE EQUIPMENT

(A) Following equipment should be kept ready in working conditions. These are the equipment needed during battery maintenance:

1. Hydrometer to read specific gravity with 0.005 divisions.
2. Voltmeter of sufficient accuracy and scale length to read cell voltage.
3. Thermometer for cell electrolyte temperature and room temperature having a range of 0 to 60°C with graduation of 1°C.
4. Thermometer for room temperature
5. Rubber gloves
6. Rubber apron
7. Rubber goggles

8. Porcelain, glass or vinyl plastic vessel
9. Glass or ebonite rod
10. Pliers
11. Knife
12. Screw drivers set
13. Hand lamp
14. Vice
15. Steel scoop

(B) The following neutralizing agents and other material should also be kept ready to use:

1. Saturated solution of sodium bicarbonate
2. Solution of 5% sodium bicarbonate
3. Dilute solution of ammonia
4. Olive oil
5. First aid box
6. Wooden dust
7. Protective clothing

6.14 MAINTENANCE SCHEDULE FOR BATTERIES

S/N	Description of work	Period
1.	Measuring specific gravity and temperature of electrolyte of the pilot cells	Daily
2.	Measuring of total voltage across the battery	Daily
3.	Measuring of float voltage and float charging current at battery charger	Weekly
4.	Measuring of specific gravity of all cells and measuring of voltage of all cells after switching off the battery charger	Fortnightly
5.	Topping up of electrolyte up to correct level	Fortnightly
6.	Cleaning of all cell terminals and applying petroleum jelly	Fortnightly
7.	Tightening of all nuts and bolts	Monthly
8.	Checking the earth leakage with the help of meter provided on charger	Monthly
9.	Checking the voltage fall after switching off the battery charger for some time (Say half an hour).	Monthly
10.	Give equalizing charge if fall in specific gravity is indicated	Quarterly
11.	Remove connecting strips, nuts and bolts and clean with boiling water, wooden stick and cloth. Do not scratch any connector or nuts and bolt with sandpaper or metallic tools. Apply petroleum jelly after refitting and tightening the connectors.	Half yearly
12.	Give discharge-charge cycle. Capacity test	Yearly

6.15 CARE AND MAINTENANCE OF BATTERIES

- Battery room should be well ventilated. The battery and surroundings area should be clean and dry to prevent self-discharging.
- All electrical connections should be tight and checked on regular basis for heating. Clean corroded parts with warm water and coat with petroleum jelly.

- If any of the cell jars is broken, the grooves should be kept immersed in distilled water till container is replaced.
- If the battery is not regularly undergoing the discharge/charge cycle but maintained in charged condition by trickle/float charges, specific gravity of 10-pilot cells should be recorded daily to find out the actual charging state.
- Maintain correct level of electrolyte in the cells. High level may split electrolyte.
- Use of impure water is fatal to the battery and may cause total breakdown. So, it is always recommended to use only distilled water to top up.
- The weak cell should be attended without delay.
- Never allow cell voltage to go below 1.70 V.
- During idle periods a freshening charge must be given at normal charge rate.
- Always charge the battery to its full capacity. Never leave the battery in discharged or half discharged condition.

6.16 PRECAUTIONS IN HANDLING ACID

In case acid falls on the ground or if there is any leakage, it should be dried up first with wooden dust and then washed with saturated solution of sodium bicarbonate. After that the floor should be washed with warm water until the stains of acid are removed. Ammonia should never be poured on the acid stain.

In case acid falls on cloth, it may be cleaned with saturated solution of sodium bicarbonate and running water.

If acid falls on the skin, apply sodium bicarbonate and then wash with warm water and if burn is more, apply olive oil also.

Unfortunately, if acid falls in eyes, then wash the eyes with 5% solution of sodium bicarbonate and then with warm water.

6.17 WEAK CELLS AND BATTERY CHARGING

Weak cells do not pick up specific gravity in spite of repeated charging. These must be given special charging. The week cells must be removed from the battery and charged separately at normal charging rate until gassing point is reached and then reduced to half the normal rate. As the cells appear to be fully charged and indicates specific gravity of the acid stationary, the charging should be stopped for an hour. After one hour, charging should be resumed at half normal rate until free gassing again takes place. A further stop of one hour's duration should be made and the charge should again be resumed still at half the normal rate. These stops of one-hour duration alternated by charges should be repeated until gassing starts simultaneously with switching on the charging current. If in spite of this special charging, the cell does not pick up, the cell is required to be opened up and washed with distilled water and again placed in the container.

6.18 CELL TROUBLES AND REMEDIES

Symptoms	Probable causes	Remedies
Excessive gassing	Floating voltage set too high	Reduce floating voltage
Progressive decline into value of specific gravity during floating	Floating voltage set too low	Increase floating voltage 2.16 to 2.2 V per cell
Cell getting of discharged	Leakage of current through grounded circuit of current carrying conductors in battery or charger circuit	Electrolyte or grounding
Rapid fall in cell voltage	Loose connections or corroded terminals	Clean corroded parts with warm water and coat with petroleum jelly
Continued low reading of specific gravity, loss of capacity after full charge low open circuit voltage	(a) Internal short circuit (b) Sulphation	(a) Examine plates for accidental contacts missing or sheered separators (b) Weak cells should be given special charge

6.19 BATTERY CHARGING EQUIPMENT

Battery charging equipment comprise float charger and boost charger.

Float Charger

Float charger has only constant voltage (CV) mode and it can supply trickle-charging current to battery and the station load. Three-phase AC supply is fed to the charger and it gives a DC stabilized output at rated full load current. The variation in DC output voltage is limited to ± 1% for 0 to 100% load variation and simultaneous AC voltage variation of ± 10% and frequency variation of ± 5% from 50 Hz.

A full wave bridge rectifier made up of SCRs rectifies the AC supply to obtain DC supply.

Working

The float charger circuit works on AC phase control principle. The SCR a semiconductor device with three terminals, i.e., anode, cathode and gate is used to convert AC supply to DC supply. The main load current is carried by anode and cathode, while the control current flows through gate and cathode. SCR blocks the forward voltage when gate is not supplied with any current. It goes into conduction, when the gate current reaches at specified level.

So, changing the gate current or pulse supplied to the SCR controls the SCR to put it into conduction state.

Once the SCR in triggered, it remains in conduction till anode current is reduced to zero or reverse voltage is applied to it. Thus, changing the instant of firing/triggering the SCR can control output voltage of rectifier-bridge. Output voltage can be adjusted by a potentiometer.

Ripple content of the output of the float charger is limited by a filter circuit comprising of choke and condenser. Important components of a charger are shown in the Figure 6.2. The

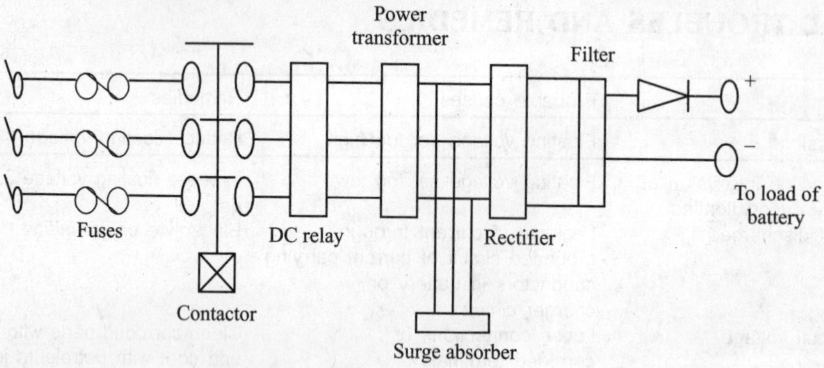

Fig. 6.2. Important components of battery charger.

output voltage of the float charger is maintained constant by means of solid state SCR controller at 2.16 V per cell ± 1% for AC input variation of ± 10% from 415 V, frequency variation of ± 3% from 50 Hz and simultaneous load variation of 0 to 100%. The output voltage of the boost charger can be manually varied by potentiometer mounted on front panel steplessly. Voltage variation is not provided in the manual mode.

Boost Charger

Boost charger has both modes of operation, constant voltage (CV) mode and constant current (CC) mode. In the constant voltage mode it works in the same way as float charger works. And in the constant current mode, it is used to boost charge the battery. The boost charger gives a DC stabilized output current at rated full load current.

In this type of charger, a full wave rectifier having SCRs, diodes and RC suppression networks does rectification. A solid state SCR controller adjusts the triggering angle of SCR and maintains the current constant.

The output current of the charger gets stabilized within ± 2% for 0 to 100% of load variation against AC input voltage variation of ± 10% from 415 volts and frequency variation of ± 5% from 50 Hz.

Working

In the event of power failure, battery should readily meet the various DC loads. When power resumes, boost charger is put 'ON' in CC mode for recharging the battery. Setting of the constant current can be done from 30% to 100%. Once the current is set, the same shall remain constant within ± 2%. Higher current is required to be set during the initial operation of recharging. After the gassing occurs, the required low current is then set.

In CC mode, boost charger is capable of recharging 110 cells at any constant current, selected from minimum current to maximum rated current. Voltage may go up to 2.7 V per cell in a maximum ambient temperature of 50°C. The set constant current shall remain constant within ± 2% for AC input voltage variation of ± 10% from 415 V and frequency variation of ± 3% from 50 Hz.

In CV mode, the boost charger is capable of supplying a DC continuous load (up to rated current capacity of boost charger) at 2.16 V per cell for floating the battery with voltage stability of ± 1% for AC input voltage variation of ± 10% and frequency variation of ± 3% and simultaneous load variation from 0 to 100%.

The output voltage/current of the charger can be manually varied by means of potentiometer mounted on the front panel. The additional feature of the boost charger is to work as a float charger. Any delay in setting right the defective float charger, does not affect the DC system.

6.20 SCR CONTROLLER

The controller consists of two cards housing the following circuits:

1. Power supply circuit
2. Reference and soft start circuit
3. Amplifier for current signal
4. Voltage regulating amplifier
5. Current regulating amplifier
6. Trigger pulse generating circuits
7. Phase sequence and phase failure circuits
8. Auto/manual facility
9. Current limit checking

Operation of the Controller

Controllers are required to control the means of feedback like DC current and DC voltage.

Current Control

The inner feedback loop of the controller is made for current control. The current reference and the actual current signal are compared. The current regulating amplifiers magnifies the difference between the two currents, the error. This change is such that it reduces the difference between the reference and the feedback tends to be equal to the reference. Thus, the inner loop gives a full control on the current. Shifting of the trigger pulses caused by the output of amplified errors gives rise to the change in the rectifier output.

The actual current also gets limited to a fixed level corresponding to the reference limit if there is a limit on the current reference. This loop provides adjustable current limiting feature.

Voltage Control

The voltage controller achieves voltage regulation. Comparison of the voltage signal with the reference signal is done and the voltage-regulating amplifier amplifies the error. The output of this amplifier is given as reference to the inner loop. Depending on error, there is certain DC current which causes a change in the DC voltage at the output terminals. The change is such as to reduce the error and thus brings the voltage feedback equal to the reference.

EXERCISES

1. What are the advantages of secondary batteries over primary batteries?
2. Describe the method of charging of lead acid battery.
3. What do you understand by 'Sulphation'?
4. What is the separator in a battery and of which material is this made?
5. How would you proceed with 'initial charging' of the battery?
6. Describe the meaning of 'Ampere-hour' for a battery.
7. Write a note on battery charging equipment.
8. What are the precautions taken for keeping a battery in good condition?
9. Describe in brief the cell troubles and remedies therefore.
10. What are the common faults in the lead acid battery cell? Describe the remedial measures.

7

Commissioning, Maintenance & Testing of Power Transformers

7.1 INTRODUCTION

Proper maintenance of any equipment ensures certain degree of reliability and safety. Minimum expenditure and least interruption of power supply are achieved if well-planned and timely maintenance are carried out. Excessive maintenance is not advisable because it involves unnecessary cost. Whereas, insufficient maintenance, generally leads to disruption of electrical power supply or accidents. Protective relays are installed in the system to alarm and warn about incipient fault and faulty section is immediately disconnected. Maintenance of the electrical equipment is required to:

- Ensure safety of working personnel
- Ensure highest availability
- Ensure serviceability
- Extend useful life

Maintenance is broadly categorized as:

- Corrective or Predictive maintenance
- Preventive maintenance

Corrective maintenance implies repairs or the measures to restore damaged part or components of the system to their normal functional state. But corrective or predictive maintenance refers at high risk of unexpected failures.

On the other hand, preventive maintenance is based on the assumption that the failures and deterioration processes are dependent on the operating time or time in use. Preventive maintenance includes:

- Periodical change or oil filtration of deteriorated transformer oil.
- Cleaning of OLTC (on load tap changer) at regular interval.
- Tan-delta and insulation resistance measurement of the transformer oil.

Preventive maintenance is considered better upkeepment option because of the following advantages:

- Preventive maintenance strategy consists of planned overhauls.
- It helps in eliminating secondary damages.
- It prevents unexpected breakdowns.
- Preventive maintenance provides extended replacement time for parts of the equipment.
- It needs short repair time because work is carried in a planned manner.
- Since it is planned repair work so personnel safety is good.
- Planned and timely maintenance reduces need of spare parts stock.
- It helps in maintaining reliability of the system.
- System qualifies for better productivity with better product quality.

7.2 TRANSFORMER FAILURE

Transformers are used in every section of the electrical system, i.e., generation, transmission, distribution and utilization. So a reliable, safe and uninterrupted electric supply mostly depends on the trouble free operation of the transformers. Since there is no rotating parts, bearing, etc. and transformers are the most efficient machines with around 95% efficiency. High capacity power transformers are 99% efficient. Transformers are able to work for as long as 25 years or more if operated and maintained properly.

Causes of Failures

The main reasons of transformer breakdown may be one or more of the following:

- Faulty design and/or manufacturing
- Improper installation and commissioning
- Poor attention on operation and maintenance
- Negligence and inadequate supervision
- Fault current and failure of protective devices
- Deterioration due to aging
- Accidents

Causes with failure percentage of in-service transformers are listed hereunder:

Causes	% of failure
Improper design	36
Incorrect manufacturing	28
Material defects	13
Negligence in maintenance	7
Lightning surges	4
Short circuits	2

Factors Affecting Life of Transformer

The main components of the power transformer are:

- Oil
- Magnetic system (Laminated iron core)
- Electrical system (Windings)
- Insulation

Oil is one, which provides cooling and insulation both to the transformer winding. This is the reason that transformer oil is considered the most vital organs of the transformers and determines life of the transformers. Good transformer oil helps the transformer to give efficient service and has following properties:

- High breakdown voltage
- High volume resistivity
- Low dissipation factor
- Chemically stable and non-reactive to any constructional material of the transformer

Moisture, solid suspended particles, fibres, oxygen, voltage surges, acid, frequency variations are the factors, which deteriorates insulating and cooling properties of the transformer oil.

The laminated core made up of CRGO with 3% silicon steel provides magnetic circuit and is another important part of the transformer. Core transfers energy from one circuit to another circuit with the help of windings. Its success depends on the proper size selection according to rating of the transformer. Sometimes core gets damaged partially during transportation. A careful protection is needed for protecting the windings, the vital organ of the transformer. Performance of the windings depends on the short-circuit current rating, temperature rise, impedance of the windings and voltage application to the transformers. Other parts of the transformer, which also need careful attention, are bushings, cable-box, conservator-tank and breather, tap changer and protective devices.

7.3 CONDITION MONITORING OF THE POWER TRANSFORMERS

Condition monitoring of the equipment is a regular feature. This is a process of checking the conditions of the equipment in a regular interval. Components of the equipment experience service abnormalities during their operation. If those incipient abnormalities are not properly taken, they may lead to failure. Periodical condition monitoring with accurate and quantifiable measurements is essential part of maintenance schedule. Few examples of failures, important for condition monitoring are:

- String insulators used in power transmission lines remain in regular tension. After certain time, due to fatigue effect string may break, so periodic condition monitoring of the insulators is essential.
- Wind excites Aeolian vibrations on the transmission line and reduces the flexible strength of the line. Excessive vibrations may cause fatigue failure of the transmission lines.
- Condition based maintenance is performed in the following steps with the help of recorded data.

First step: Detection of the spot where any adverse change in the operating condition is initiated.

Second step: Proper diagnosis and estimation of the upcoming state of the system or its components.

Third step: The final step is the maintenance planning and execution means to repair the faulty sections.

Advantages of the condition monitoring and thereafter performing the maintenance are:

- It avoids unexpected shutdown effectively.
- Unnecessary repairs as compared to periodic maintenance are not required which saves the money and time both.
- In the case of incipient failures, condition based monitoring reduces the stress on the affected components.
- Conditions checking of the components after the repair work prevent new failures.

Condition Based Maintenance (CBM) techniques are:

- Recovery Voltage Measurement (RVM technique)
- Thermo Vision Scanning (TVS technique)
- Dissolved Gas Analysis (DGA technique)
- Partial Discharge Measurement (PDM technique)
- Furfural Analysis (FFA technique)
- Frequency Response Analysis (FRA technique)

Power transformers in operation are examined periodically to estimate its component conditions and maintenance requirement. Following periodical condition monitoring is carried out.

- Insulation resistance and winding resistance measurement
- Winding and transformer oil temperature measurement
- Moisture and acid content in oil measurement
- Dielectric strength of oil measurement
- Dissipation factor (Tan-Delta) measurement of oil
- Measurement of oil resistivity at 90°
- Interfacial tension measurement of oil

The advantages of condition-based maintenance include damage prevention by reducing stress on the components, effective protection from unexpected shutdown, prevention of chances of new failures by proper checking condition after maintenance. Limitations of condition-based maintenance include error in some degree of subjective judgement regarding exact timing of maintenance, difficulty in proving benefits of condition monitoring better than cost of the instruments.

7.4 TRANSFORMER MAINTENANCE

In order to avoid unnecessary failure of the transformer, effective upkeepment and preventive maintenance of each component of it is necessary. This extends the life of the transformer. Routine check up and then preventive maintenance is described hereunder:

7.4.1 Transformer-oil Checkup and Remedial Measures

Being integral part of the transformer, identification of the incipient fault and degradation and thereafter, immediate redressal of the oil tank and its associated components is an essential activity. Following are the requirements for proper upkeepment of the transformer oil:

- Measurement of the oil level should be carried out frequently and investigation for any leakage of the oil from the oil tank or conservator should also be performed regularly. If there is any leakage, it should be repaired without delay and oil should be topped up in the tank up to desired level.
- Mixing up of the used oil is not recommendable. Always top up oil of prescribed quality.
- Sample oil from the transformer in operation should be taken regularly and tested its dielectric strength.
- Purification and dehydration of the transformer oil should be carried out before and after the rainy season every year. After monitoring the condition of the transformer oil, dehydration may be done at any time according to the requirement. After purification and dehydration, following quantities should be measured and ensured that these are within specified limit otherwise oil should be replaced:
  (i) Resistivity
  (ii) Dielectric strength
  (iii) Tan-Delta
  (iv) Moisture content
  (v) Acidity

7.4.2 Tank-body Checkup and Maintenance

Tank of the transformer houses core, windings and cooling and insulating oil. Inner surface of the tank made of iron sheets is painted with oil resistive paint. Outer surface of the tank is treated with primer followed by anticorrosive paint. One side, either top or bottom is provided with gasketted joints to take out core and windings whenever required be inspected or repaired.

During inspection, if there is any rusted portion found, it should be cleaned properly and proper coating with recommended quality of paint should be provided to prevent the corrosion. Spray painting is advisable. It is must for radiators.

Any type of damage or leakage must be attended and redressed immediately. Oil leakage takes place from drain plugs in radiators, oil filled bushings, gas and oil actuated relay's oil seals, expansion vessel and valves in the oil management and cooling system. Common leakage problem initiates from welded or gasketted joints. Gasket and joints must be tight enough and if leakage continues, replace the gasket.

7.4.3 Core and Windings Checkup and Maintenance

Core and windings of the transformers should be inspected properly. Whenever a major inspection is felt, core and windings should be lifted from the tank for visual inspection carefully and inspected as per recommended procedure.

Following parts of the transformer should be disconnected before lifting the core and windings.

- Windings from the bushings and cable boxes inside the tank
- On-load tap changer connection
- Earthing strips between the core and the tank

After inspection and required treatment, insulation test for the core should be performed. The terminal box of the transformer contains three terminals. Terminals are protected by a cover that can be either open are removed during inspection. Out of these terminals, terminal marked 'G' is for ground (body of the tank) connection, terminal marked 'CC' is core clamp connection and the terminal marked 'CL' is for connecting the core laminations. AC insulation testing method should not be applied to measure insulation between individual core laminations otherwise core laminations may get damaged. Before disconnecting the earthing connection from core clamp and tank, link must be removed first. Apply Megger of 2.5 kV capacities between CL and CC & G terminals to measure the core insulation. The insulation requirement between CL and (CC + G) terminals after one minute test is minimum 100 M-Ohms for old transformers and more than 1000 M-Ohms for newly commissioned transformers.

Any break or crack in winding is hazardous for the transformer. It should be redressed on priority. Core laminations if found irregular should immediately be corrected.

7.4.4 Conservator Checkup and Maintenance

It is a tank mounted on the top of the transformer stores transformer oil for cooling and provides insulation to the winding. Oil circulates from conservator to the body tank through radiator. It helps in expansion and contraction of oil due to heating and cooling processes because of change in load. Buchholz relay is placed inside the tank.

Two types of conservators are used; air cell-sealed conservator and diaphragm-sealed conservator. Valve fitted at the bottom for draining the oil should be checked properly for any leakage. During collecting the sample from oil tank for testing, care must be taken to run off contaminants present. By draining oil up to the lower level of the conservator tank through drain valve and applying pressure of 0.5 to 1.0 psig by dehydration plant and maintaining the pressure at the same temperature for twenty-four hours with the vent valve open, no loss of pressure indicates that there is no leakage of oil from the conservator. Change in oil level may be due to the effect of temperature for variation of load on transformer or due to leakage of oil. Oil level indicator is provided in the conservator to view the changes in the oil level. Oil gauge glasses must be kept cleaned and in proper shape to view the accurate oil level.

7.4.5 Buchholz Relay Checkup and Maintenance

Buchholz relay immersed in the tank is a gas detector relay. It is used to detect the incipient fault in the transformer such as arcing, partial discharge, local overheating, etc. This is mounted in the pipe connecting main tank and conservator tank. Two float Buchholz relay sets at an inclination of 3 to 7° to the horizontal. Inclination is necessary to ensure that all the rising gases from the main tank are directed to this relay housing which actuates the alarm contact. Even if there is loss of oil up to below the level of the relay, trip contact closes and unit is tripped. Before tripping action takes place alarm warns the operators. A gas-collecting device is also fitted and as oil displacement takes place due to gas pressure, the gas accumulated in the relay is collected in the device. By opening the drain valve of the Buchholz relay, oil will be completely drained otherwise valve is to be kept closed. Buchholz housing, connecting pipe and the gas chamber should be kept full of oil.

Routine inspection of the relay should be carried out in each year. Oil level should be brought below the level of the relay, first. The float should be checked and it should rise and fall freely otherwise remedial measures are required to be taken.

7.4.6 Silica Gel Checkup and Maintenance

Dehydration breather consists of Silica gel is connected with pipe to the top of the transformer. Silica gel is chemically neutral and has very high capability for absorbing the moisture. It absorbs moisture ten per cent of its own weight. Its colour becomes dark blue if impregnated with Cobalt. Breather should be checked regularly. However, its frequency depends on local climate and operating condition. If transformer is under fluctuating load or operating under a humid atmospheric condition, breather should be inspected very frequently. Colour of a healthy and active Silica gel is blue and as absorbing moisture saturates it; its colour gradually changes to pale pink. Heating up to 110°C to 130°C for about eight to ten hours in a pan can reactivate it. It becomes reusable only if entire mass of the Silica gel achieve bright blue colour. If it is not reactivated, replace it.

During inspection it must be seen that the breathing hole at the bottom seal is not blocked. Oil-seal oil level must always be maintained at the level marked in the cup every time the drying medium is changed.

7.4.7 Pressure Relief Valve Checkup and Maintenance

In the event of a short circuit in the windings, spark and arc vaporize the transformer oil and leads to heavy built up of pressure inside the transformer tank. This way excessive pressure inside the transformer tank gets generated due to internal fault or due to poor quality of the transformer oil. The pressure built up may damage the tank. So, releasing the pressure immediately becomes necessary. Diaphragm relief vent provided with Bakelite at the top is mounted on the transformer to relieve the internal pressure. Bakelite diaphragm breaks down due to pressure exceeding puncturing strength of the diaphragm. This is also known as explosion vent. Top of the relief vent is connected with the conservator through an equalizer pipe to equalize the pressure in both the conservator and explosion vent. Punctured diaphragm permits ingress of moistures and pollutes the transformer oil. So, it is essential to inspect and check the diaphragm, fitted at the relief vent regularly and replace if found damaged.

7.4.8 Bushing Checkup and Maintenance

Bushings are basically insulators used in the transformers. It enables to make electrical connection between transformer and incoming and outgoing transmission and distribution lines. Low voltage and high voltage winding leads are brought out of transformer via bushings. There are porcelain type and condenser type bushings. Plain porcelain bushings are used for the transformer rated up to 12 kV. The other porcelain-based bushings are oil filled type, which may be used up to 33 kV transformers.

Condenser bushings are capacitance-graded bushings. Grading is done to make uniform distribution of axial and radial electric field across the bushing. Uniform electric field causes excess electric stress on some portion of the bushing that may create cracks. SRBP (Synthetic Resin Bonded Paper) insulated bushings are useful up to 245 kV. In these bushings internal insulation is a core wound from resin coated paper. Each layer is bonded with previous layer by resin coating. Curing of the resin is done to make bonding perfect. Transformer oil is filled in the annular space between porcelain and capacitor core.

Frequent cleaning of the bushings is required, if placed in humid and polluted localities. Regular measurement of the Tan-Delta and capacitance should be done to ensure that oil is not deteriorated below the acceptable level. Oil level should also be checked frequently.

Insulation resistance of the bushing should not be less than 10^{10} ohms at any point of time, however, required value is 5×10^{10} ohms.

7.4.9 Tap Changer Checkup and Maintenance

Output voltage of the transformer at which electrical power is to be fed to the consumers is kept within prescribed limit with the help of Tap changer. System voltage varies, if turns ratio changes. Basically sliding the tips of the Tap changer changes number of turns of the transformer winding. Off-circuit Tap Changer and On-load Tap Changer (OLTC) are two types of Tap changers in use.

Shutting off and deenergizing the transformer is necessary before changing the tap by Off-circuit Tap changer. A mechanical lock is provided on it, which operates only by known and authorized person. Handle of the Tap changer should not be kept unlocked or kept halfway otherwise there is possibility of unexpected and unwanted operation.

OLTC (On-load Tap Changer) is used in large transformers and for its operation; deenergization of the transformer is not needed. It may have place either on HV winding or on LV winding depending upon the requirement. Tap selector, Diverter switch, Operating mechanism and Motor drives are the main components of an OLTC. Tap selector consists of a number of contacts arranged in a curve. Diverter switch consists of impedance or transition resistance. In some cases tap selector and diverter switch are combined in the selector switch. Selector switch is a device, which makes, carry and break the current. It combines the duty of a tap selector (selects proper tapping connection) and a diverter switch (diverts the current from the tap in use to the tap to be used). Driving mechanism is the system by which the Tap changer is actuated.

A frequent check up of Tap changer is necessary because it is a current interrupter. Before application of electrical operation hand operation check should be performed. Tap changer should move freely up and down in full range. Interlock should also be checked to ensure that it is functioning properly. Operation checking of Tap changer is required to be carried out annually and preventive maintenance after every three years. In the very high rating transformer, Tap changer is driven by motor drive mechanism system. This consists of an electric motor and speed reduction gearbox with control and indication devices. Checking of the tap positions should agree and reach simultaneously at motor drive unit head.

7.4.10 Checkup and Maintenance of Temperature Indicator

Oil temperature and winding temperatures are required to be measured and recorded properly time to time. Temperature indicators work on the principle of liquid expansion of the liquid being sealed in the bellows. Temperature rise shown by the indicators may be:

- Due to faulty thermometer
- Deteriorated cooling ability of the cooling system

Faulty thermometers must be calibrated with standard thermometer. Current transformer leads must be short-circuited in control cabinet before disconnecting the winding temperature indicator for testing. The fastener of the capillary tube should be checked and tightened if found loose. Dial glasses should be kept clean. Broken glasses should be replaced immediately otherwise it may lead to the damage of the instrument.

Oil is filled up in a pocket in which thermometer is kept. At least once in a year, level of the oil should be checked. If oil level is found below the prescribed level, top up required amount of it. Winding temperature should not increase beyond 150°C.

7.4.11 Checkup and Maintenance of Cooling System

In order to control temperature of the transformer and keep it within permissible limit, cooling system is provided. Types of cooling system depend on the rating of the transformer. Oil Natural and Air Forced (ONAF), Oil Forced and Air Forced (OFAF) and Oil Forced and Water Forced (OFWF) cooling systems dissipate heat efficiently by using fans and pumps. These systems are employed to cool hot spots in the windings of the temperature.

Exhaust fans and other cooling systems employed at the surface of the transformer should be washed and cleaned regularly because there are chances to accumulate dust, insects, leaves, etc. on these.

Without shutting down the transformer, following ON-LINE inspection is required to be done:

- Fans are running at designed speed means there is no blockage of airways.
- No moisture ingress in the fans motor.
- No contamination of motor and fan blades.
- Bearing lubrication is enough.
- Ensure that fan rotation is in correct direction.
- Check and take remedial measures if there is corrosion in any part.
- See if there is unusual noise. Check the bearings and tightness of the blades and correct these.

Greasing of the bearing after every 1000 running hours is recommendable.

Oil pump should also be checked regularly for:

- Correct rotation
- Abnormal vibrations
- Unusual noise
- Electrical problem
- Corrosion on any part of the pump

7.4.12 Checkup and Maintenance of Kiosk and Marshalling Cubicle

Kiosk and Marshalling cubicle of transformers covers the main working components of the transformer. Under checks of these items, following should be carried out:

- Checking of the condition and redressal of primers and paint
- Condition of the hinges, door handles and opening of doors and working of door switches must be inspected and corrected if there is any fault. Door seals should also be checked
- Checking of the condition and redressal of thermostat and heater working condition
- Checking of the condition and redressal of lighting system
- Checking of the condition and redressal of equipment mountings
- Checking of the condition and redressal of isolation trip signal
- Checking of the condition and redressal of HRC fuses
- Checking of the condition and redressal of the local alarm system
- Checking of the condition and redressal of the cable terminations

7.4.13 Checkup and Maintenance of Valves

Each valve is expected to operate freely. If valves are properly greased and suitably placed, they continue free operation. Greasing is done on the threaded stem of the valve. Greasing provides lubrication and protects from the adverse atmospheric conditions to avoid rusting. Especially during the major overhaul, all valves should be checked and greased.

Before checking the valve conditions, transformers and associated equipment must be switched off and kept out of service.

7.4.14 Checkup and Maintenance of Earthing

After every six months, condition of the earthing provided to the transformer should be checked and then earth resistance measurement should be carried out. Corrective measures should be taken if earth resistance is found beyond permissible limit. Generally, two sets of earthing on the neutral side of transformer winding is provided. All the metal parts like transformer tank and accessories such as Marshalling Kiosk, oil-pump motor, fan motor, etc. are earthed properly and a good low earth resistance is necessary for adequately protecting the transformer.

7.4.15 Checkup and Maintenance of Cable Sealing Boxes

Cable sealing box is used for the termination of the leads of low voltage and contains insulating medium with bushing. Terminals of the lead are connected with the sealing box. Cable sealing boxes receive and protect the end of the metal-sheathed cables. M-seal or Rachem cable jointing kits are used. Bituminous compound of the seals sometimes gets cracked and then it should be re-compounded otherwise cracks accumulates water around the plug, which may lead to short circuit. Some joints are gasketted and covered by taps. These joints should also be checked and tightened according to the requirement.

Note: One set of different types of bushings, one set of cooling fan, sufficient stock of transformer oil; one set of limb winding, etc. should be kept ready stock as spare.

7.5 MAINTENANCE AND CHECKUP

(i) Preventive Maintenance

Symptoms	Parts to be checked	Remedial action
1. Oil is getting hot	Thermometer, Short circuit in the windings, Overloading	Reduce the load if overloading is there otherwise shutdown the transformer
2. Windings are getting hot	Short circuit in the windings, Overloading	Reduce the load if overloading is there otherwise see the winding inter-turn and repair for any short circuit.
3. Abnormal noise	Frequency of the supply voltage	Shutdown the transformer if supply frequency is beyond-permissible limit
4. Load-amperes and/ or voltage high	Check against rated figures	Reduce the load if overloading is thereby performing load-shedding otherwise shutdown the transformer
5. Low oil level in the tank	Check oil temperature and oil leakage	Top up the oil. Check the oil system for any leakage
6. Low oil level in the bushings	Check oil level from gauge glass	Top up the oil. Check the oil system for any leakage
7. Low oil level in the conservator	Check oil level from gauge glass	Top up the oil. Check the oil system for any leakage
8. Leakage of oil in the cooler	Check oil level from gauge glass	Top up the oil. Check the oil system for any leakage

Routine health checkup of the high-voltage transformers are required to monitor the conditions of its different parts. Few parts are required to be inspected on hourly basis and take up proper measurement of the related quantities. Similarly, inspections are carried on hourly, daily, monthly, quarterly and half yearly basis. If there any abnormality found, corrective measures are required to be taken immediately. Preventive checkup and preventive maintenance schedule of the transformers and its auxiliaries are essential in order to ensure an uninterrupted power supply to its consumers. Service conditions, location, capacity of the equipment govern the frequency of checkup and related redressal. These are described hereunder.

(ii) Daily Checkup

Symptoms	Parts to be checked	Remedial action
1. Loose relief vent-diaphragm	Relief vent-diaphragm	Tighten and ensure that relief vent-diaphragm is intact
2. Loose dehydrating breather oil seal	Check oil level in oil cup	Top up the oil. Check the oil system for any leakage

(iii) Monthly Checkup

Symptoms	Parts to be checked	Remedial action
1. Silica gel colour gets pink	Silica gel	Change the silica gel first by spare stock and then the old silica gel may be reactivated for future use
2. Cracks and/or dirt deposits on bushings	Bushings	Clean the bushing and if needed replace
3. Noise from the bearings of cooler fans, motors, on-load tap changer	Control and interlocks of the gearbox, Gear-box oil level and contacts, Bearings	Replace the worn out and/ or burnt contacts and the other defective parts
4. Voltage output not as per prescription	automatic control of the on-load tap changer	All the circuits should be checked independently and faulty sections should be corrected

(iv) Half-yearly Checkup

Symptoms	Parts to be checked	Remedial action
1. Problem with on-load tap changer driving system	All moving parts, motor, contacts, break shoes, etc.	Adjust, clean or replace the defective part
2. Earth resistance related troubles	Earth system	Measure the earth resistance and take corrective measure
3. Blockage of ventilation	Ventilators	Ensure that air passages are free
4. Water content in transformer oil	Transformer oil	Dehydrate the transformer oil

(v) Yearly Checkup

Symptoms	Parts to be checked	Remedial action
1. Low breakdown voltage of Transformer oil	Dielectric strength	Filter and dehydrate the oil or replace it
2. Trouble in oil filled bushings	Oil	Filter the oil or replace it
3. Loose gasket joints	Gaskets	Tighten the bolts and ensure there is not uneven pressure

4. Trouble in cable boxes	Sealing	Replace leaking gaskets, examine compound for cracks
5. Cracks, dirt on surge diverters	Surge diverters	Clean or replace
6. Mal-operation of relays, alarms	Relay accuracy	Take corrective measures

(vi) Two-yearly Checkup

Symptoms	Parts to be checked	Remedial action
1. Rusting of oil conservator	Internal inspection of the conservator	Thorough cleaning, welding (if required) and painting
2. Mal-operation of Buchholz relay	Mechanical inspection of the Buchholz relay	Adjustment of the floats, switches, etc.

7.6 INSTALLATION OF THE TRANSFORMER

Before installing a transformer, foundation preparation at site is required. After preparing the foundation of suitable dimension according to the rating and size of the transformer thorough inspection and recommended testing are done. These are described hereunder:

7.6.1 Foundation at Site

An energized transformer exerts pressure on its foundation. Concrete made foundation with suitable size and plinth level must be strong enough to bear the complete weight of the energized transformer. High rating transformers are moved with the help of wheels. A rail track is Safety, Commissioning, provided on which rollers move. Wheels are locked after placing the transformer at its final position. Locking of the rollers is essential otherwise transformer may move from its final position.

7.6.2 Facility for Storage and Drainage of Oil

Facility for storage of the oil and drainage from the transformer tank must be provided for both indoor and outdoor installations. Soak pit is made for the transformers of oil capacity 9000 litres or more for indoor installation. Soak pit is filled with sand and gravel. Sufficient slope is required for proper flow of oil to the soak pit. Drainage facility is required for outdoor installation for transformers of oil capacity 9000 litres or more.

7.6.3 Cabling

Two types of cables, namely, power cables (high current rating) and control cables (low current rating) are used. Power cables connect the secondary terminals. Control cables connect the relays, measuring and communication equipment. Power cables and control cables should never place in the same trench. If there is cable box fitted with the transformer, it should be filled with oil and sheathed cable should be used. For low and medium voltages, bare conductor (flexible type ACSR) is used for outdoor substation and cable connection is generally used for indoor substation. A wooden support with angle iron for low-tension cable is required to avoid bending of cables coming out of the transformer. Wooden support is necessary otherwise bushings may get cracked.

7.6.4 Switchgear and Protection

Switchgear and protective devices are placed in a different room located away from the power transformer. These are installed in control room as a part of control panel. Control cables used for connecting the switchgear and protective devices and power cables should run through separate trenches and conduits.

7.6.5 Fire Protection

Multi-fire protection system is used for protecting the large transformers from the fire hazards. In this system, a pipe is fixed surrounding the transformer, which is connected with storage point of the fire-extinguisher gas. As fire takes place temperature rises and a valve opens the gas opening to flow the extinguisher through pipes. Many nozzles are fixed with the pipe and gases come out from these nozzles directed towards transformer, which extinguishes the arc. For light level fire, Carbon Tetrachloride (CCl_4) and foam type fire extinguisher and sand filled bucket are also kept ready for the application.

7.6.6 Commissioning of the Transformers

After preparing the site and other associated facilities mentioned above, following steps should be taken before final commissioning of the power transformers:

- Fitness of the accessories like conservator, breather, explosion vents, Buchholz relay, temperature indicators, bushings, tap changer, cooling equipment
- Filling of the transformer oil
- Drying out and degassing of the oil
- Breather charging and filling of the silica gel
- Cleaning of the bushings
- Earthing of the tank and covers
- CT and PT connections with alarms, measuring instruments and protective devices
- Setting of the relays
- Cables and conductors connections with making of the leak proof joints

After ensuring that all the parts and accessories of the transformer are ready to provide their services efficiently, leave the oil for twenty-four hours to set for charging the transformer. During this process air should be released after an interval of six hours and voltage buildup should be allowed to its full level in four to six hours.

During commissioning process of the transformers, following precautions are recommended which must be followed:

- Disconnect the primary and secondary terminals of the transformer from supply
- Ground the line terminals
- Disconnecting switches must be locked in open and visible position
- Ensure that fuses and cutouts are put in secure place so that they are not reused accidentally
- Internal pressure of the tank should be relieved by opening the transformer tank top
- While working on the transformer there should always be present two or more technically sound persons

7.7 BDV (BREAKDOWN VOLTAGE) TEST OF TRANSFORMER OIL

Before putting the oil in the transformer tank and conservator, breakdown voltage test determines the electric strength and extent of quality of the oil. BDV test is required to be carried out before and after filtering, degassing and dehydrating the oil. During taking the oil sample from the tank for BDV test, first dust and contaminants of the valves must be removed then allow some amount of the oil to flow into a separate container. Bottom valve is used for collecting the oil sample.

For this test, sample is placed in the container of the test set in which two identical brass or copper-made sphere gaps are separated by few centimetre (2.5 mm). The gap between the spheres is set to an accuracy of $\pm$ 1 mm. An electric field with continuously increasing voltage till the breakdown takes place is applied across the terminals of the sphere gaps.

Dividing the gap between spheres from applied voltage at breakdown does breakdown voltage calculation. The temperature of the test cell should be equal to the ambient and be recorded. Test is repeated and carried out at least six times on the same sample. The arithmetic mean of the six test results may be considered as accepted result for electric strength. Following acceptance criteria according to the voltage class of transformer may be considered:

Breakdown voltage	System voltage
40 kV	Less than 72.5 kV
50 kV	Less than 145 kV
60 kV	Less than 175 kV

Test cell also known as test vessel is made up of transparent material like plastic or glass. It should also be non-absorbent. It is a closed cell of having oil volume 300 ml to 500 ml.

After each breakdown, the oil should be stirred between the electrodes with the help of a clean, dry glass rod. Stir the oil in cell gently so that formation of air bubbles is avoided.

7.8 PD (PARTIAL DISCHARGE) DETECTION

A localized electrical discharge, which partially bridges the insulation between conductors, is known as partial discharge. It is initiated because of ionization process in cavities inside solid insulation, gas bubbles in liquid insulation or across dielectric surfaces. It may happen in electrical insulation due to:

- Improper manufacturing of the insulation
- Wet condition of the insulation
- Presence of abnormal over voltages
- Presence of sharp edges on the conducting part, which generates high stress areas

Partial discharges involve small amount of energy but its existence for longer duration may lead to progressive deterioration of the dielectric properties of the insulation. It is required to detect and test the partial discharges in transformer bushings and its oil.

The inception voltage, voltage at which partial discharge initiates plays an important role in partial discharge detection and it is always higher than the normal rated voltage.

The partial discharge detector carries detection of the partial discharges. Recognition of the partial discharge magnitude and inception voltage is done by comparing the blueprints either obtained from the manufacture or by previous tests. Following precautions should be taken during partial discharge detection:

- High voltage connections must be kept discharge free
- Sharp points on the conductors must be properly shielded by suitable shield or corona ring
- External disturbances like supply mains or earthing system interferences must be avoided
- Electromagnetic radiations should not be present there
- Noises caused by the bad contacts and discharge in the test circuit should not be present
- Unnecessary metallic bodies should not be near testing site otherwise they behave like floating potential points
- Electromagnetic fields of rectifiers, radio transmitters and rotary machines in the surroundings must be properly shielded

7.9 DRYING OUT OF THE TRANSFORMER

Transformer oil and insulation absorbs moisture if kept unused for longer duration because these are hygroscopic. So before energizing the transformer, drying out of it and removing the moisture from oil, winding insulation and other internal parts becomes essential otherwise insulation may fail. Moisture with oxygen causes oxidation and deteriorates physical, chemical and electrical properties of the oil like dielectric strength. Methods of drying out a transformer are:

- Dehydrating the oil by filtering and dehydration plant
- Filtering and degassing the oil with an filtering plant using a high vacuum filter pump
- Drying core and windings by using oven or heating chamber and heaters
- Drying by short circuit method

Which method is to be applied depends on the rating of the transformer, availability of the equipment and site. Transformers above one MVA require to be dried out by using process accelerated by applying vacuum in the portion between the oil and the tank cover. Transformers up to one MVA are dried out without using vacuum.

Generally, first two methods are very common for drying out the transformers. While using oven or heating chamber and heaters, transformer tank should be covered with fire resistant material like asbestos sheet, glass sheet, etc. Thermostat must be correct to show correct temperature. The temperature of the transformer oil should be maintained at 800°C and winding temperature at 900°C.

For accelerated drying process, external electric heaters are provided surrounding the tank. There should not be any interruption during the drying out process.

Estimated duration of drying out of the transformers is:

Transformer rating	No. of days
11 kV	1 to 6
220 kV	10 to 30
400 kV	15 to 40

Following precautions should be observed essentially during the drying out process:

- Proper ventilation arrangement should be done to remove the moisture produced by transformer oil
- Transformer should not be left unattended during drying process

7.10 FILTRATION AND DEHYDRATION OF TRANSFORMER OIL

When water content in the transformer oil increases beyond permissible limit and percentage of gas content exceeds the level of normal operation dehydration of the oil becomes necessary. A transformer oil filtration plant is used for the purpose. Filtration or purification is carried out to remove dirt, sludge and other types of foreign particles and impurities from the oil. Before starting the filtration and dehydration plant, ensure that plant is properly earthed. Check that all the switches of the control panel are in off condition. Do not forget connecting the cooler water supply. After this connect the inlet and outlet hoses to the respective inlet and outlet valves of the transformer and the plant.

Evacuation of the plant is required before starting the oil processing. For proper and effective oil filtration, first the inlet to the machine is taken from the bottom of the transformer tank and outlet hose connection is given to the top of the tank. After every effective filtration of twelve hours this connection is changed. For proper removal of water, hose connections should be airtight and cool water should be circulated through the condenser.

In the transformer oil filtration plant, oil first enters in the preliminary filter. Preliminary filter is made up of perforated coarse filter and provided with a magnetic strainer. After, preliminary filtering oil moves to flow through 200 kW heaters to heat up the oil. A thermostat is set to ensure that temperature of the oil at outlet should not increase beyond 600°C. Now oil passes to the degassing chamber in which oil is subjected to a very high vacuum that separates water and gas. For efficient moisture condensation, a coil type vapour condenser is provided in the vacuum pipeline where cool water is circulated for better condensation of the moisture.

BDV test is performed after every twelve hours of the oil processing. Moisture content is also checked. After receiving desired values, processing may be stopped.

EXERCISE

1. Describe the preventive maintenance schedule of a power transformer.
2. Explain filtering of transformer oil.
3. Discuss need and importance of drying out of the transformers.
4. What do you understand by Partial Discharge? Discuss PD detection in a transformer.
5. Why BDV test is required?
6. Describe the statement 'Preventive maintenance is considered better upkeepment option'.
7. What are the main causes of transformer failures? Discuss the factors affecting life of the transformers.
8. Write a note on condition monitoring of the transformers.
9. Write a detailed note on checkup and maintenance of different parts of a power transformer.
10. Describe installation and commissioning of the power transformers.

8

Commissioning, Maintenance & Testing of Electric Motors

8.1 INTRODUCTION

About seventy per cent consumption of electricity in the industrial sector account for electric motors. Motors drive fans, compressors, machines, pumps and many other types of equipment. Agricultural sector also uses electric motors for pumping irrigation water. Electric motors are broadly categorized as direct current (DC) motors and alternating current (AC) motors. The third category of motors is universal motors, which are capable to run by feeding either AC or DC current. Further, DC motors are classified according to their construction as series, shunt and compound motors. Permanent magnet motors are also DC motors.

AC motors are of two types based on supply, i.e., single-phase and polyphase motors. Three-phase squirrel cage motors are mostly used in industries and in agricultural sector.

The main objective of preventive maintenance of the electric motor is to keep the motors up-to-date and in fit to use condition in order to achieve best efficiency. Efficiency of the motors is the ratio of the mechanical energy delivered at the rotating shaft to the electric energy input at its terminals. The overall efficiency and the performance of the motors are monitored and governed by the following facts and factors:

- Ambient operating conditions
- Losses
- Routine check up and preventive maintenance
- Quality of the power supply
- Rewinding of the burned-out motors
- Aging
- Types and quantity of the load
- Energy efficient measures like balanced phase voltages, proper starting control strategy, suitable cooling and ventilation system, good power factor
- Proper selection of pulleys, belt, gears and bearings
- Types of enclosures whether these are capable to prevent dust, contaminants and pollutants

8.2 FAULTS IN ELECTRIC MOTORS

Faults in electric motors may be categorized as faults in electrical system as well as faults in mechanical systems of the motors. Probable faults are listed below:

Fault	Reason
	Electrical faults
Motor stalls	Poor electrical contact connections, Open circuit, Low terminal voltage, Overloading, Mechanical locking at air gap or in bearing, Improper insertion of the control resistance
Motor start-up failure	Low voltage, one phase open, No power supply, Overloading, Defective control system, Defect in rotor, Mechanical locking at air gap or in bearing, Wrong stator coil connection
Overheating of the motor during under load running	Rotor rubs in the stator bore, Improper voltage, Shorted stator coils, unbalanced terminal voltage, Grounded coil, Single phasing, Overloading, Clogged blower or air shield with dirt which prevent proper ventilation for motor
Motor runs but dies down	Overloading, Power failure, Insufficient pull up torque, Motor takes much time to accelerate, Low applied voltage, Excess loading, Poor circuit connection, Defect in squirrel cage rotor, High inertia of the load
Motor failure to come up to the rated speed	Primary circuit open, Low voltage, Improper secondary resistance in wound rotor, Improper bedding of brushes on rings, Defective or broken rotor bars
Wrong rotation	Reverse phase sequence
Speed reduces after increase in motor loading	Wrong connection of the motor starter, Defective contacts, Low voltage, Overloading, Phase lead open
Line current unbalancing in 3-phase motors during normal operation	Unequal phase resistances, Winding short-circuiting, Unequal terminal voltages, Short circuit in stator winding turns, Soldering failure, Single phasing, Heating of brushes, Poor rotor contacts of resistance wound rotor
Low insulation resistance	Loose terminal bolts, Drenched condition, Motor submerged in flood water
Motor oscillation	Rotor rubbing in stator bore, Fluctuation in supply voltage as well as in load, Loose cable contacts
Rotor winding getting loose	Low wire tension and defects in soldering
Breaker tripping on starting	Phase-lead open circuited, Rotor rubs stator bore, Overloading, Locked rotor, Winding short circuited, Improper tripping setting, Low voltage, Unmatched motor starter characteristics
Sparking at slip rings	Overloading, Poor maintenance, Slip rings out of teeth, Rough dirty or oily slip rings, Ridged slip rings, Improper brush pressure, Improper selection of brushes
Noise	Non-uniform air gap, Rotor unbalancing, Noise by driving system, Loose bearing, Defect in bearing

Fault	Reason
	Mechanical faults
Motor vibration	Misalignment of the motor shaft and improper coupling, Weak foundation, Unbalancing in the driven machines, Too hot bearing, Defects in the bearings, Unbalancing in the rotor
Scraping noise	Fan strikes with insulation, Loose motor on bed plate, Fan rubs the end shield
Hot bearings	Misalignment, Pulley too far away, Excessive belt pulling, Bent shaft, Low pulley diameter
Oil leakage in the plugs	Loose plug cover, Overflow plug, Cracked overflow plug
Knocking in bearings	Much bearing play, Rough races on the bearings, Foreign matters in the bearing
Shaft does not rotate freely with load	Misalignment, Excessive belt pull, Contaminated lubricants, Improper grease in the bearings, Rotor rubs in the stator bore
Hot bearing balls	Rough races on the balls, Overloading on the balls, Heat from the external source, Foreign matters in the bearing, Excessive lubricants, Contaminated lubricants, Insufficient grease

8.3 MAINTENANCE TECHNIQUES

Plant engineers adapt following techniques for maintaining the performance of electric motors to ensure lowest downtime.

(i) Reactive Maintenance

Maintenance or repair performed after occurrence of breakdown is known as reactive maintenance. The main features of this type of maintenance are:

- Maintenance components must be kept ready
- There are many risks
- In most of the cases, reactive maintenance are more repetitively needed in comparison to planned maintenance

(ii) Preventive Maintenance

Preplanned cyclic maintenance of the plants and machineries comes under preventive maintenance categories. This is required to be done to prevent the fault occurrence and successive breakdown. The main features of this type of maintenance are:

- Action follows tight schedule
- Functionaries know their job
- Prescheduled and planned outages are operational
- Low power demand time is selected for the maintenance work

Sequence of the motors for maintenance can be chosen according to reliability requirement.

(iii) Predictive Maintenance

Predictive maintenance is performed on the basis of information gathered from the condition-based monitoring. Predictive maintenance procedures can be modified and optimized when maintenance engineers plan predictive maintenance of the plants or systems.

In order to keep overall efficiency of the plant high, plant components are monitored on ON-LINE. Advantage of condition-based ON-LINE monitoring system and predictive maintenance is to greatly achieve extended life and overall plant performance.

8.4 SUGGESTIONS TO ACHIEVE BEST PERFORMANCE OF THE MOTORS

1. In practice, most of the time motors are compelled to run the machineries till breakdown occurred and then only get repaired. This must be avoided and preventive maintenance techniques should be adopted.
2. During the planning and procurement stage, correct specification, as requirement should be spelled out.
3. In the case if motors of the same plants are procured from the different manufacturers, the spare parts should be stocked for similar make motors.
4. Working personnel should be trained by the manufacturer to handle the spare parts and upkeepment of the motors.
5. All the metallic parts are ensured to be perfectly connected with the security terminals.
6. The coupling and belt drives must be properly shielded and protected against touching.
7. Motor must not operate without guards foreseen by the motor construction.
8. Supply conductors must be properly protected from any type of damaging.
9. Fluctuation in the ambient temperature generates vapour condensation in the body of the motor. Water should be periodically removed.

8.5 PERIODICAL CHECKS

Malfunctions arise in the electric motors must be rectified immediately. Periodical checks help to identify and locate the incipient faults as well to take up preventive measures. Its frequency depends on the environment and working condition of the motors, for example, if a motor works in the dusty room, after every three months (sometimes more frequently), minor checkup requires. Minor check comprises:

- Inspection and motor cleaning externally
- Inspection of supply mains, cables and conductors
- Checking of contacts, screw and proper tightening
- Stator winding insulation resistance measurement

In normal working conditions, minor check and inspection must be carried out in every six months.

A detailed general check and inspection must be carried out every year. In this dismantling of the motor is allowed to clean and correct internal parts. A monthly inspection is carried to inspect, clean and correct the defects without dismantling the motor.

8.5.1 Routine Inspection of the Motor Parts

Stator

- Measurement and recording of the voltage across stator terminals and current in the stator winding.
- Measurement and recording of the vibration and temperature of the stator frame.
- Presence of explosive or corrosive gas.
- In addition to above ambient and outlet air temperatures should also be recorded.

Bearing

- Noise measurement
- Vibration measurement
- Temperature measurement

For sleeve type ball bearings, coolant water and oil must be inspected for any leakage.

Slip Ring and Brush Holder

- Presence of any spark from the brush
- Wearing of the brush

8.5.2 Monthly Inspection of the Motor Parts

Stator

- Measurement and recording of the insulation resistance.
- Draining of water if any

Bearing

Retightening of the bolts and nuts.

Roller Type Ball

Greasing up.

Sleeve Type Ball

Inspection of the extent of turbidity in lubricating oil and abrasion.

Slip Ring and Brush Holder

- Inspection of the degree of wearing of the slip rings and the brushes.
- Inspection of the short circuit devices.

8.5.3 Yearly Inspection of the Motor Parts

Stator

- Dust removal.
- Stator slots and core looseness check.

Rotor

Cage type

- Dust removal.
- Rotor bar deformation.

Wound type

- Measure insulation resistance before and after disassembly.
- Check looseness of brazing at the coil connection.

Bearing

Check damage on the bearing.

Roller Type Ball

Wash and inspect carefully.

Sleeve Type Ball

Inspection of the extent of abrasion and clean oil box.

8.6 MAINTENANCE TEST EQUIPMENT

Test equipment	Use
Thermometer	Temperature measurement of the motor surface, windings, etc.
Portable voltmeter	Voltage measurement and recording
Portable ammeter	Current measurement and recording
Megger	Insulation resistance measurement
Kelvin bridge	Low resistance measurement of the winding
Wheatstone bridge	High resistance measurement of the winding
Vibration meter with frequency analyzer	Vibration measurement and frequency analysis
Tachometer, Stroboscope	Speed measurement
Tan-Delta measuring set	Dielectric loss angle measurement
High voltage test set	High voltage (upto 1.5 times rated voltage) withstand test when overhauling is taken up
Noise meter	Noise measurement

8.7 ELECTRIC MOTOR MAINTENANCE

Unusual service conditions, poor maintenance schedule, inadequate protection and wrong application may lead maloperation of the electric motors. Above all maintenance aspect is most important and requires major attention.

The present-day trend is to build much smaller and lighter motors. Heavy motors are being avoided because of their demerits. Smaller and lighter motors are designed very precisely to the exact ratings and any lapse in the maintenance affects motors performance. A well-scheduled maintenance scheme senses and locates the problems before they develop into breakdown or cause damage to the machine and its system.

Types of failures generally occur in the induction motors are given in the table below.

Type of failure	Percentage
Insulation failure	60%
Rotor bar failure	15%
Failure due to mechanical reasons, i.e., bearings, rubbing of rotor with stator, etc.	10%
Other reasons	15%

8.7.1 Nature of Failure in Motors

Types of faults in induction motors are discussed hereunder.

- Insulation failures occur due to imperfection in the insulation material during its development. The other cause of failure is contamination and pollutants available surrounding the insulation. Oil, grease, fuel ash, residues from flue gases, dust, etc. get deposited on the end part of the winding. In the extreme polluted conditions, sometimes contaminants get deposited across the entire enclosure of the motors. These contaminated layers on the metallic parts of the motor when get wet due to moisture becomes good conductor of electricity. Conducting path due to contamination may cause failure to ground at the time of motor starting. This may also cause short circuit in the windings of the motor. This generally happens, when motors are started after putting it out of operation for few days. In the slip ring motors, contamination deposits creates much problems as compared to wound rotor motors. Regular contaminants deposition reduces age of the insulation.

- Insulation life gets reduced due to overheating. Too many frequent starts and overloading of the motor overheats the motor winding.
- Due to continuous rotation and vibration in high-speed motors, rotor bars sometimes fail.
- Air gaps of the induction motors are kept small and a little imbalance creates radial movement and rotor starts rubbing the stator. Continuous rubbing damages stator insulation and cores. Coupling misalignment, stator frame or bed plate misalignment also cause rubbing.
- Action of the fast interrupting circuit breakers of the high-tension motors causes switching surges. Switching surges are short duration over voltages. These over voltages generate stress on the inter-turn insulation, which may lead to puncture or breakdown of the insulation.
- Some of the other factors which may contribute to the motor failures may be deteriorating environmental conditions, improper soldered frame end winding joints, terminal box failure, blocking of ventilating system, magnetic wedge failures, etc.

8.7.2 Preventive Maintenance

Routine health checkup of the high-tension motors are required to monitor the conditions of its different parts. Few parts are required to be inspected on hourly basis and take up proper measurement of the related quantities. Similarly, inspections are carried on daily, weekly, monthly, quarterly and half-yearly basis. If there any abnormality found, corrective measures are immediately taken.

(i) Hourly Checkup

Symptoms	Parts to be checked	Remedial action
1. Motor is getting hot	Voltage and current measurement	Reduce the load if overloading is there.
2. Variation in winding temperature	Winding temperature measurement	See the winding inter-turn and repair for any short circuit.
3. Abnormal noise	Bearing temperature	Stop the motor if temperature continuously rises and correct the bearings.
4. Rubbing sound	Rotor shaft alignment	Stop the motor and rectify the fault.
5. Improper oil level and flow of oil to the bearing	Oil level and flow in the bearing	Top up the oil. Check the oil system for any leakage.

(ii) Daily Checkup

Symptoms	Parts to be checked	Remedial action
1. Abnormal rise in body temperature	Cooling system	Correct the cooling system.
2. Much temperature difference in water inlet and outlet of motor cooler	Water flow of the motor cooler	Correct the water pressure and cooler chocking.
3. Colour change of lube oil	Lube oil colour	Change the oil if its colour has become black.
4. Oil leakage	Bearings, connecting pipes, etc.	Rectify the loose connections and cracks if any.
5. Vibrations	Bearings and shaft	Remedial actions should be taken immediately if vibration is beyond permissible limit.

(iii) Weekly Checkup

Symptoms	Parts to be checked	Remedial action
1. Too much vibrations	Visual checks for proper tightness of foundation bolts and coupling bolts	Stop the motor if any bolt is found loose or vibration is beyond permissible limit.
2. Lube oil colour change	Lube oil check for colour and quantity	Top up the lube oil up to the proper level or if colour of the oil is found black change the oil completely.
3. Heating (High leakage current flow)	Insulation resistance measurement	Take corrective measure.

(iv) Monthly Checkup

Symptoms	Parts to be checked	Remedial action
1. Sound from bearings	(a) Ball & roller type bearing (b) Sleeve type bearing	Check alignment and fill up the grease. For excessive wear or sludge scrap it and then refit.
2. Dampness and dirty surroundings	Check insulation resistance, Winding insulation, Measure vibrations, bearings and body temperature	Improve working conditions and rectify the shortcomings.

(v) Quarterly Checkup

Symptoms	Parts to be checked	Remedial action
1. Loose belts, improper ventilation	Motor cover and its fan	Tighten the cover bolts and ensure proper ventilation to the motor internal parts.
2. Crack on foundation and loose foundation bolts	Motor frame and foundation	Repair the foundation cracks and tighten the foundation bolts.

(vi) Half-yearly Checkup

Symptoms	Parts to be checked	Remedial action
1. After removing the end covers overheating, cracked and roasted insulation and varnish are observed	Stator winding	Take necessary action for the overheating. Deposits should be removed with the help of dry compressed air.
2. Improper functioning of protective devices and alarm	Protective devices and alarm	Repair or replace.
3. Non-uniform air gap which may be due to worn out bearing	Measurement of the air gap	Take corrective measures if air gap is found non-uniform.

8.8 INSTALLATION OF ELECTRIC MOTOR

8.8.1 Preparation

Before setting up the motor on its foundation at site, following preparations are required:

- Generally, protection is provided for the bearings. Remove the protection covers.
- Rotate and set the terminal box of the motor to the best suitable position so as supply cable could be installed easily. The position of the terminal box should be chosen free from humidity penetrating inside the connection box.
- Check the possible damages caused during transportation and storage. Repair the damage if any.
- Measure insulation resistance of the metallic frame and covers. Ensure that the insulation resistance is within acceptable limit (it should always be greater than 10 Mega-ohms).
- Humidity from the winding must be removed by drying up process. Temperature of any part of the motor should not exceed 120°C at any point of time.

8.8.2 Erection of the Motor

During erection process, shaft end pivot is fitted with a well-balanced coupling, pulley or a gear according to the requirement. Steps for erecting an electric motor are:

- Protective paint from the shaft is to be removed first.
- Grease or recommended oil coating is applied on the clean shaft.
- A heated up (say up to 85°C) coupling or pulley or gear is pressed and fitted on the pivot.
- Care should be taken to avoid blow during mounting the motor, which may damage the bearings.
- A clearance of at least 0.1 mm must be kept between the coupling halves.
- Excessive tensioning of the belt in the belt drives must be avoided.

Proper material for the belt should be chosen which could sometimes be used in explosion risk area.

Note: Before motor connection to the mains, mains voltage and motor rating should match. Terminals and winding connections are designated in the plate attached with motor's body. Connection diagram is also put available on the plate. In the terminal box, two neutral grounding terminals and one grounding terminal are provided on the frame foot or flange disk. Grounding as per existing regulations must be done.

8.8.3 Testing of the Electric Motors

As per IS-325, following two types of tests are required to be carried out:

- Acceptance and Routine tests
- Type tests

Following two types of the tests are also carried out in addition to the above tests.

- Additional tests
- Maintenance tests

Acceptance and Routine Tests

Acceptance and Routine tests are required to be conducted on the all the motors. Following sequences of the tests should be followed:

- Measurement of the insulation resistance
- Measurement of the winding resistance of the stator and the wound rotor
- No load test at rated voltage to determine input current, power and speed
- Voltage, current and input power measurement in the locked rotor conditions. Voltage to be kept low
- Reduced voltage running tests for squirrel cage motors
- Open circuit voltage ratio of stator and rotor windings for slip ring motors
- High voltage tests

Type Tests

Following sequences of the tests should be carried out:

- Similar tests at the serial number (i), (ii), (iii) and (vii) of the Acceptance and Routine tests
- Open circuit voltage running test at no load to check the ability of the motor to run up to full speed on no load. In both directions of rotation with rated line voltage applied to the motor
- Similar test as mentioned at serial number (iv) of the Acceptance and Routine tests
- Full load test to determine efficiency, power factor and slip
- Temperature rise test
- Torque and momentary overload test

Additional Test

Following additional tests are also requiring ensuring the perfect operation of the motors.

- Tan-Delta test
- Impulse test
- Noise level test on no-load
- Vibration test on no-load
- Degree of protection test

Maintenance Test

- Visual inspection of the machine
- Drying out test
- Measurement of the winding resistance
- Polarity test
- Measurement of the insulation resistance
- Polarization index test for the high voltage winding

8.9 DRYING THE INDUCTION MOTOR

Drying of any electrical machine at the time of commissioning is essential because moisture present inside the machine may reduce the insulation resistance, which may cause short circuit and damage to the machine. Drying the induction motors is done with the help of magnetization current. Following preparations are recommended before applying magnetization current for heating the motor for drying purpose.

- Before going for drying, machine should be examined and cleaned completely and ensure that there is no dust or any contaminant at any part of the machine.
- Record the ambient temperature.
- Conduct insulation resistance and winding resistance test.
- Machine should be partially covered with the cloth or insulation sheet (may be asbestos) to avoid heat dissipation.

Initially, low magnetizing current should be allowed to pass through the machine. Magnetizing current should be increased slowly in steps. Current and voltage must be monitored throughout the drying process with the help of the ammeter and voltmeter. Temperature measurement should be done in regular interval and it should not increase 75°C at any point of time. During measurement of the temperature and insulation resistance, drying may be stopped and after measurement completes it should be restarted. Once steady insulation resistance is achieved, drying process should be stopped.

8.10 VISUAL INSPECTION OF THE MACHINE

While go for either commissioning or maintenance, following procedure should be adapted.

- At first machine should be identified whether machine is new, failed or required to be overhauled.
- Next step should be to inspect the mechanical joints and connections to ensure that those are not loose.
- Check the rotating parts free to rotate by rotating the revolving parts slowly with hand.
- Cable connections with motor terminals should be inspected by opening the terminal box.
- Check thoroughly the starters and relays and their components.
- Remove the end covers of the motors and access the winding condition by visual inspection. Check the winding for:
  a) Presence of foreign or unwanted materials
  b) Surface contamination due to moisture, dust, carbon, oil, and grease
  c) Cracks, overheating or any other type of mechanical damages
  d) Cooling ducts or passages for blockage
  e) Check the conditions of the insulation tapes, and stator winding

8.11 INSULATION RESISTANCE MEASUREMENT OF THE WINDING

For measuring the insulation resistance of the motor winding, a Megger, a multimeter, short cable and discharge rods are required. Before going for measuring the insulation resistance, following preparation is recommended.

- Switch off the power supply and remove cover of the terminal box of the motor
- Stored charges due to capacitive effect should be discharged
- Cable or wire connection with motor terminal studs should be removed
- Dust or contaminant deposited on the terminal stud should be cleaned by proper solvent, i.e., white spirit, naphtha, carbon tetrachloride (CTC) or cloth
- Metallic body of the motor should be connected to the earth
- Temperature of the motor should also be recorded before taking the reading

For taking the reading of each winding, remove the star or delta connection of the induction motor. Ensure that each winding is continuing with the help of continuity testing. Perform insulation resistance measurement at least twice separately for each phase winding. Insulation resistance measurement is required to be conducted for winding phase to earth as well as between the phases. Same procedure is used for measuring the insulation resistance of the direct current motors. In the beginning, remove the interlinking connections between the windings and take the steps given as above for measuring the insulation resistance of the induction motors.

EXERCISES

1. What are the main objectives of the preventive maintenance of the electrical machines?
2. Write the facts and factors by which overall efficiency and the performance of the motors are monitored and governed.

3. Describe different types of electrical and mechanical faults with reason, which may occur in electric motors.
4. What are the maintenance techniques, generally adapted by electrical maintenance engineers?
5. What should be the periodical checks suggestions to achieve best performance of the motors?
6. Describe advantages and need of the periodic maintenance checks of the electrical machines.
7. Give the preventive maintenance schedule of electric motors.
8. State and describe various steps in installation and commissioning of induction motors.
9. What is significance of preventive maintenance and servicing of the electric motors? State the activities in preventive maintenance of motors.
10. What is insulation resistance? Describe method of measuring insulation resistance of an electric motor.
11. Write detailed note on the following:
    a) Installation of the electric motor
    b) Testing of the electric motor
    c) Drying of the electric motor
    d) Visual inspection of the electric motor

9

Commissioning, Maintenance & Testing of Alternators

9.1 INTRODUCTION

Before commissioning alternator to its location, the first step is to bring the machine near foundation already prepared. The cases of the packing should be checked and matched with the dispatch details. The warning and 'How to handle' instructions should be followed strictly. After unpacking, any damage or loss of packages during the transportation or transit must be intimated to the supplier, insurance company or the manufacturer. Alternators or any accessory not found as per required specification must be returned to the supplier.

9.2 INSPECTION

If packing of the delivered machine is received as per dispatched documents, the next step is to unpack the machine and its accessories. After unpacking and detail checking, alternator and its accessories should be cleaned perfectly with the help of the brushes. All the nuts and bolts should be checked carefully, to ensure that nothing has become loose during transportation, loading and unloading. Ensure that no parts of packing paper, small pieces of thermo coal, wood wool, dust, etc. are in any section of the alternator or its accessories.

In some cases, alternators are placed idle in non-working condition for longer duration. It may also happen that machine might have stored in a damp location. In such cases, insulation gets weakened and putting the machine in operation may cause short-circuiting. Following steps should be followed during inspection of an alternator:

- In order to check the condition of the alternator and accessories, external inspection is carried out.
- By opening the terminal lids, terminals are checked whether these are tight or not.
- To remove dust, blower should be used.
- Turn the rotor by hand to check whether it is moving freely or not.

- Redress the fault observed during the inspection.
- Proper greasing should be done for the bearings.
- Measurement of the insulation resistance between windings and frame should be tested and matched with the data provided by the manufacturers. If insulation resistance is found lower, drying of alternator is essential.

Note: Alternator and its accessories are recommended to be coated with anticorrosive grease and wrapped in moisture resistant material like wax paper. Storeroom's temperature should be maintained above 15°C throughout its storage. Drying out of the alternator is described in the later part of this chapter.

9.3 INSTALLATION

The activities involved in the alternator installation process are:

- Finalization of location and layout diagram
- Identification of position for alternator
- Foundation making
- Levelling and alignment
- Grouting
- Fitting of the accessories
- Final levelling

The location should be selected such as to have sufficient margins all around the foundation for facilitating regular inspection, repairing, etc. Alternator after installation should not obstruct the maintenance work or routine inspection of itself and its accessories like control panel, switching panel, busbars, switchgear and protection equipment, etc.

Laying should be done according to the layout diagram. Laying means marking of the foundation plan on which excavation of the soil is done as per specification. Concrete structure (RCC) of the foundation is made to protect the alternator from the vibrations generated by running of the alternator as well as turbines. For placing of the cable ducts and pipes to carry cables and wires, provisions are made along with the foundation. Foundation bolts are fixed on the bedplates provided in the foundation. Sawteeth type bolts or an eye shape type bolts are used at the lower portion of the concrete foundation just to provide proper grip in the foundation.

Once the foundation is ready, positioning of the alternator is done with the help of pulley blocks or overhead cranes or by chain hoist because alternators are very heavy. The expert people should do positioning of the alternator. 0.2 mm to 2 mm thick strips of sheet steel are known as 'shims', which are sometimes required to be inserted under the foot of the alternator. Inserting the shims helps doing proper alignment of the alternator shaft with turbine's shaft.

Grouting means connecting the machine with the foundation by a concrete mixture and cement mortar. Quick setting cement is preferred for grouting. Bedplate is inserted in the hole dug in the foundation and then quick setting cement is continuously poured. Air gap should not be left during the grouting process and machine should not get disturbed for few days after grouting so that cement will find time to get set.

Three-phase and neutral outgoing cables are the power cables. Requirement of the control cables include for winding temperature, bearing temperature, measuring and protective instruments connections. All the external power cables including control cables are laid and properly connected to the terminal box of the alternator. Power and control cables must be laid in separate trenches or racks to avoid electromagnetic interference.

There are two types of panels one power panel and other control panel. Power cables terminate at power panel where circuit breaker and automatic voltage controller are housed. Control cables terminate at control panel where wattmeter, power factor meter, frequency meter, voltmeter, ammeter and protective relays, i.e., over voltage relay, over current relay, earth fault relay, etc. are housed.

Now, alternator is ready to be commissioned.

9.4 COMMISSIONING

Before final commissioning of the alternator and declare to fit for providing services, following checks are required to be performed.

- At the outset check all the connections (cable connections and control wiring) as per layout diagram and drawing. Measure insulation resistance of the cables.
- Ensure that earthing is proper and measure earth resistance.
- By removing potential transformer fuses and shorting current transformers measure insulation resistance of the alternator.
- Test and ensure that relays are fit to operate at its lowest set values. Check and ensure that circuit breakers are operating mechanically correct and trip contacts of the relays are also mechanically all right.
- Check and ensure that polarities of the CT and PT are correct.
- Test and ensure that polarization index (PI) of the alternator is within prescribed limit.
- Perform primary and secondary injection tests. Injecting rated current into busbars carries out primary injection test. Secondary injection test is carried out by injecting current into secondary of the current transformers.
- Perform vibration test, bearing temperature test, cooling and air-circulation test, and no-load temperature rise test. Perform these tests by running the alternator on no-load at rated speed without excitation for few hours.
- Now after successful no-load unexcited run, switch 'ON' the excitation. Note down voltages of all the phases, excitation voltage and current, vibration, bearing temperature and no-load temperature. This should be done for few hours to ensure that values recorded are in order.
- At the last, load test is performed for 72 hours. This test is also known as 'Performance Guarantee Test (PGT)'. In this test alternator is loaded externally by closing circuit breaker. At first 10 to 20% load should be provided and all the critical parameters used for evaluating the alternator performance should be noted. The procedure repetition should take place with increasing load in steps till full load is applied. After performing successful PGT test, alternator is handed over for routine operation.

9.5 DRYING OF ALTERNATOR

Economic reasons and to cater large power demand necessitates to go for compact and extensively enclosed design of modern generators. The windings of the alternators may absorb little moisture during standstill. However, quantity of the moisture absorption depends on the atmospheric conditions at the site. Moisture reduces the insulation resistance of the winding (IR).

So, before first commissioning and after a long shut down of the machine, alternator windings must be dried out. If alternator is stored in a hot and dry climate, drying may not require because windings do not absorb much moisture. So, before deciding the dry out, PI (Polarization Index) test should be performed.

Polarization Index is the ratio of insulation resistance measured at 10 minutes to that of measured at one minute. The minimum value of PI for class-A insulation is 1.5 and for class-B and E insulation is 2. If the value of PI is less than 1.5, drying is must otherwise it may get short-circuited. For the value of PI between 1.5 and 2, it is recommended to dry out the windings on priority.

Insulation resistance of the stator and rotor windings should not be less than one mega-ohm at 75°C. Drying out is the process of heating and cooling of the windings, which may take three days. In the few modern generators, short-circuiting is not required for drying. In these alternators, the windings are heated to get dried because of friction and windage loss. For this purpose, alternator is rotated at rated speed without applying excitation. The methods of drying of the alternators are:

- By hot air circulation
- By space heaters
- By feeding DC current to the stator
- By short circuit operation

Drying by Hot Air Circulation

This method is used to dry small alternators at standstill. Moisture of the windings is removed by circulating clean, dust-free hot air into the windings. Hot air can be sent into the windings by heat convectors guided diagonally through the generator by hosepipes through openings in the end shields. To avoid water precipitation, temperature of the discharged air from the generator should be 10°C above the ambient temperature. Inlet air temperature should not increase beyond 80°C.

Drying by Space Heaters

Large size alternators have inbuilt space heaters embedded in the stator winding. When alternator main circuit breaker gets 'OFF', space heaters get 'ON'. Space heaters do not allow moisture to condense and settle on the windings by keeping the interior of the alternator hot.

Drying by Feeding DC Current to the Stator

In this method, all the three-phase windings are connected in series. DC current of 50% of stator rated current is allowed to pass through it. Welding set may be used to provide DC current. Passing the DC current, temperature rises gradually, which should not be allowed to rise beyond 90°C. This method is also used to dry small alternators.

Drying in Short Circuit Operation

Shorting stator windings and applying low excitation method is most common for drying alternators. In this method, all the three phases of the stator are short-circuited at the alternator terminals. Guide vanes of the turbine are kept open and alternator is run at full speed. Excitation is so adjusted that full load current circulates in the short-circuited stator windings. Forty per cent of the rated current should be allowed for first two hours and then the full load current. Temperature of the windings must be monitored by the thermometers and current should be so adjusted so as windings temperature should rise gradually and does not rise beyond 90°C.

In the beginning of the drying process insulation resistance decreases quickly and after some time, it starts increasing. Drying process should be stopped if winding resistance measured periodically is above minimum value and PI value is obtained above 1.5. However, it is recommended to stop drying process, if three consecutive hourly readings of insulation resistance remain same.

After cooling the alternator up to the ambient temperature, it is ready for switching 'ON'.

9.6 MAINTENANCE OF ALTERNATOR

A well-planned maintenance of the alternator is necessary to be carried out at least once in a year and complete overhauling once in three years. Overhaul of the machine consists of replacement of the main components, units and assemblies. It may involve rewinding of the stator and rotor, partial or complete dismantling of the alternators.

The main reasons of frequent maintenance of the machines may be summarized as:

- Defective design
- Use of improper or poor quality material
- Non-recommendable operating conditions
- Poor maintenance
- Underrating, overrating or misuse of the machine

A scheduled checkup and redressal of the faults ensures longer life for the alternator. This also ensures uninterrupted power supply to the consumers. Alternator is a rotating machine and owes to wear and tear in its components. Wear and tear in the components of the alternator causes degradation of the material and speeds up ageing process. Immediate attention should be paid as and when signs of damage are shown. Damage must be timely repaired. Effective system of maintenance is required and it includes:

- Timely inspection and servicing of the components

- Scheduled periodic repair of different components of the alternator
- According to the operating conditions, condition monitoring and recommended testings in specific time interval

Advantages of Preventive Maintenance

Preventive or scheduled maintenance has definite advantages described below:

- Standby equipment requirements reduces
- Preventive maintenance makes safer movement of the workmen near the turbine and alternator
- It reduces chances of breakdown
- It helps to determine the need for minor and major repairs
- Correct information regarding conditions of the equipment provides flexibility to plan proper operation of the electric power generation
- Less wear and tear in the rotating parts
- Life of the alternator gets increased
- Premature failures are avoided by performing scheduled maintenance of the alternators

9.7 PREVENTIVE MAINTENANCE SCHEDULE FOR THE ALTERNATOR

9.7.1 Seasonal Checkups

- Dismantle end-covers of the alternator. Clean and inspect end-windings, rotor-fans, and slip rings. Provide varnishing to the windings and grease the fans.
- Dismantle bearing covers and checkup journals, shells, etc. Grease the bearings.
- Clean the exciter, check the brushes, and tighten the nut and bolts and fasteners.
- Clean the filters, coolers and other accessories.
- Check the power cable and control cable terminations and tighten them.

9.7.2 Yearly Checkups

- Repeat the checkups mentioned in the seasonal checkups.
- Check the oil and grease in the bearings. If there is leakage correct it.
- Clean the exciter, rotor, stator, armature-commutator and bearings with the help of compressed air.
- Check AVR (Automatic Voltage Regulator) along with field discharge resistance and field breaker.
- Check all the rotor windings.

9.7.3 Three-Yearly Overhaul

- Check the bearing clearances. If found abnormal correct or change the bearings. Fill up the bearing oil and or grease it.

- Check the various exciter and generator air gaps.
- Clean the exciter, rotor, stator, armature-commutator and bearings with the help of compressed air followed by special cleaning fluids.
- Measure the insulation resistance of the exciter and stator and ensure that these are within limits otherwise dry-out the alternator.
- Measure the dissipation constant (tan-delta) of the stator winding.
- Individual phase resistances should be as close as possible and should not vary beyond 2% from the commissioning test values.
- Check all the control circuits and ensure that all the protective devices are working well.
- Check AVR (Automatic Voltage Regulator) along with field discharge resistance and field breaker.
- Check all the rotor windings. If found necessary, rewind these.

EXERCISE

1. Describe importance and need of routine inspection of the alternators.
2. Explain the procedure of installation and commissioning of alternators.
3. Describe the preventive maintenance schedule of the alternators.
4. Describe various tests performed on synchronous machines.
5. Write a note on the 'Maintenance of Alternators'.
6. Write advantages of the preventive maintenance of the alternators.
7. Describe various methods of drying the alternators.

10
Maintenance of DG Sets

10.1 INTRODUCTION

Whenever we talk about malfunctioning of a diesel-generating set, we should understand that not all the problems are of engineering nature, several arise from communication gaps or sheer misunderstanding.

The engine operator is responsible for taking care of engine while engine is being operated. The set rules, which operator/supervisor must observe and follow to get best service from the DG sets. Ten maintenance steps are given below.

- Keep dirt out of the engine.
- Maintain a lubricating film on all bearing surfaces
- Regulate the engine's fuel
- Control operating temperatures
- Guard against corrosion
- Let the engine breathe
- Prevent overspeeding
- Know your engine condition
- Correct troubles while they are simple
- Schedule and control your routine maintenance

10.2 MAINTENANCE SCHEDULE

A diesel engine requires regular scheduled maintenance to keep it fit for providing its services efficiently. Preventive maintenance is the easiest and least expensive type of maintenance. Preplanned maintenance schedule permits the department to do the work at a convenient time. Prevention is better than cure.

In the contaminated or pollution conditions, some checks are required to be performed more frequently because; actual operating environment of the engine governs the maintenance schedule. Generally, maintenance schedule check-sheet is used as a guide until adequate experience is obtained to establish a schedule to meet a specific operation. Some checks have to be performed more often under heavy dust, humid or other special conditions. Regular routine maintenance keeps the operating costs lower.

10.3 ENGINE STORAGE

If an engine remains out of the service and its use is not immediately forthcoming, some precautions should be taken to prevent rust. The rate of corrosion varies with the climate conditions. The procedure covers the basic method of preparing the engine for temporary/long storage.

- Engine starting must be done by using diesel fuel, and the speed should gradually increase to a fast idle (say up to 1200 rpm), however it depends upon make of the engine. It should be operated at no-load till the engine gets thoroughly warm.
- As the engine stops, drain all lubricating oil from oil sump and refill the engine oil sump with the recommended lubricant.
- Drain coolant from cooling system (engine, radiator and cooling water pump) and thoroughly flush with clean water and high-pressure radiator cleaner. System is then to be refilled with 10:1 mixture of water and high-pressure radiator protector.
- Start the engine with the fuel line to the filter and engine pulling fuel from the can with regular fuel. The injector-drain line flows into the container with regular fuel. As the engine starts and run smoothly at idle, switch the fuel to the container with the preservative oil. Operate engine five to ten minutes on no-load till the preservative oil coming out from injector returns. And then stop the engine.
- Drain the oil sump, rule filters and reinstall the drain plugs. New oil may be added up or sump may remain empty until engine is ready for use.
- Turn fuel pump manual shutoff-valve to 'OFF', so engine will not start. Disconnect inlet and exhaust manifolds of engine (engine must be cool) and spray preservative oil of good quality into air intakes, exhaust and exhaust outlets, engine being turned during the spraying operation.
- Cover all intake manifold openings with tape to prevent entrance of dirt and moisture.
- Cover all engine openings including coolant inlets cylinder block, oil breather and crankcase.
- Do not rotate crankshaft after the above-mentioned operations. The battery, if any, is kept in dry condition.
- All vents (e.g., engine exhaust pipe, breathers, etc.), dynamo, starter, motor, magneto if any, and air cleaners should be carefully sealed with waterproof paper and waterproof adhesive tape.
- Loosen V-belt tension.
- Remove rocker lever covers and spray preservative oil. Over rocker levers, valve-springs and stems, guides, crosshead and push-tubes. Replace the cover.
- Tag engine to indicate that it has been treated with preservatives and should not be turned over until ready to run due to possible reduction of the protective film. The tag should show the date of treatment.

- Store the engine in a place protected from weather and where air is dry and temperature uniform, if possible.
- Periodically inspect engines for rust or corrosion and take corrective action, if necessary.

Note: If the engines are to be stored for longer duration (more than six months), repeat the engine preservative treatment following the above steps after every six months.

10.4 PREPARING A STORED ENGINE FOR SERVICE

When an engine is removed from storage and put into service, the following options listed should be performed.

Cleaning of Engine

- Clean off all the accumulated dirt and contaminants from exterior of the engine.
- Remove all paper over, tape and wrappings.
- Use suitable solvent, cleaner or degreaser to remove rust preventive compound from unpainted surfaces of engine.
- Refill crankcase with clean lubricating oil.
- Flush cooling system.

Inspection

- When an engine has been stored for less than six months, it is necessary to make a routine initial inspection. This inspection includes adjustment of injectors, valves, belts and checking of cylinder head-cap screws, oil filter and connections, air filter, screens and straps.
- Pressurize the lubricating system about 1 kg/square cm including the turbocharger or supercharger prior to starting the engine.
- Flush entire fuel system with clean fuel until all preservative oil is removed. Running the engine on 'No Load Idle' for five minutes with clean high-speed diesel oil can do this.
- Before re-commissioning a stored engine, any foreign matter, which may be collected on screens and strainers during initial operation, must be removed. It is essential for considering the engine properly prepared for future service. Engine is now ready to start for regular operation.

Note: Engines subject to ambient temperatures and below must also be equipped with a lubricating oil heater. While using a lubricating oil heater immersed in oil, the maximum temperature of heater surface in contact should be less than 149°C to minimize formation of hard carbon on the heating element.

Standby units should be operated once a week under a minimum of 25% of rated kW load for at least thirty minutes. During this test, the engine must reach normal operating temperature.

10.5 DAILY MAINTENANCE CHECKS

The engine must be maintained in top mechanical conditions in order to get optimum satisfaction from its use. Comparison and intelligent interpretation of the daily report along with a practical follow up action eliminate practically all running failures and emergency repairs. The report must consist of the following.

- Keep oil level as near 'H' (high) mark as possible and check the oil level with dipstick oil gauge located on the engine. Always operate the engine with oil level between the marks 'H' and 'L'. Never operate the engine with oil level below the "L" (low) mark or above the "H" (high) mark.
- Visually check belts for looseness. If there is evidence of belt slippage adjust it.
- Check fuel system and fuel pump to ensure that there is not any maladjustment or tempering. Check all connections for leaks or damage. Check engine for damage; correct according to the necessity.
- Keep the cooling system filled up to the operating level. Check the coolant level daily at each fuel fill point. Investigate for causes of coolant loss. Check the coolant level only when the system is cool.
- Loosen the fuel tank drain-cock or plug. Drain approximately one cup of fuel to remove water and sediment. Close the drain-cock or plug. A water separator helps to minimize the moisture content in the fuel tank.

10.6 WEEKLY MAINTENANCE CHECKS

Following checks should be performed in each week time. Remedial measures should also be taken as and when required.

Air Cleaner Check

Depending upon the operating conditions, if necessary clean pre-cleaner jar and dry type air cleaner dustpans preferably daily. Under extremely dirty conditions, an air pre-cleaner may be used.

Inlet Air Restriction Check

A mechanical restriction through a dry type air cleaner is also known as vacuum indicator. This unit is generally mounted in air cleaner outlet. The red flag in window gradually rises as cartridge loads with dirt. After changing or replacing cartridge, reset indicator by pushing reset button.

Filter Element Cleaning

The paper of filter element gets weakened as a result of cleaning and can lead to rupture or microscopic damages. Also, inspection of the filter element after cleaning is difficult. This is the reason that many air filter manufacturers discourage the practice of cleaning air cleaner elements. So, it is suggested to replace the filter element for longer engine life before first overhaul. However, if it is decided to clean the filter element, following suggestions should be followed.

- Clean only outer element. Never remove inner element for cleaning. Inner element should be removed only for the replacement.
- Outer element should be removed for the cleaning only when red band appears on the vacuum indicator. It is observed that elements are cleaned frequently to keep the system clean. But this practice leads to damage to the paper element as well as problems associated with handling and too frequently opening and closing the air intake system.

Note: Holes, loose end seals, dented sealing surfaces and other forms of damage require immediate element replacement.

Weekly Cleaning and Inspection of Heavy Duty Pre-cleaner

Weekly cleaning and inspection of heavy duty pre-cleaner is very much essential.

- Clean preheater openings of all the soot, oil film and any other objects those may have become lodged in openings. Remove dust or dirt in lower portion of pre-cleaner and aspirator tube. Inspect inside of air cleaner housing for foreign material.
- Inspect dirty pre-cleaner for soot or oil. If there is soot inside cyclone tubes, checks for leaks in engine exhaust system.
- Exhaust 'blow-back' into air intake and exhaust from other equipment. If pre-cleaner appears 'oily', check for fumes escaping from crankcase breather. Excessive oil mist shortens life of any dry type pre-cleaner.
- Inspect clamps and flexible hose or tube and be sure all fittings are airtight on cleaners with exhaust aspirators.

Draining the Tanks

Condensed moisture in air tanks and lines may freeze and make controls useless during the cold weather. So, it is recommended to drain air tanks to keep all water out of the compressed air system.

10.7 'B' MAINTENANCE CHECKS

During engine operation as time passes, the chemicals from corrosion inhibitors are depleted. So, it is necessary to do the service treatment to replenish the depleted chemicals. At each 'B' Maintenance Checks, perform *all the checks described earlier is called 'A' check.*

Interval of maintaining a proper 'B' maintenance check is very important factor in preserving the integrity of an engine. Lubricating oils contaminate the direct result of engine operation and load factor involved. The amount of contamination generated depends on the amount of fuel, the engine consumes. At each 'B' check interval, it is recommended to change the full-flow filter and the bypass filter. An oil change interval will be generally different for different types of engines and of different ratings.

Change Engine Oil and Filter Element

Lubricating oil must be used as per strict specifications. After proper inspection and testing, if lubricating oil is found deteriorated or degraded, it has to be changed. For changing the oil and filter elements, the following steps are to be followed.

- Loosen centre bolt securing lube oil filter to lubricating oil pump.
- Remove filter element, cut it open and check for source. Discard 'O' ring and element. Insert new filter element into the can.
- Install new rectangular seal on the pilot located on the lube pump.
- Install can and element assembly with its mounting bolt and washers.
- Remove NPTF plug on can, clean oil and replace the plug.
- Torque the can-retaining bolt to 30 to 35 lbs. (41 to 47 Nm).
- Run the engine, check for leaks, recheck engines-oil level, and add oil as necessary to bring the oil level to 'H' mark on the dipstick.

Note: Always allow oil to drain back to the oil pan before checking the level. This may require 15 minutes.

Change Super-lubricating Oil Bypass Filter Element

- Loosen (four) cap screws from head and remove head super L.O. bypass filter.
- Take out element and remove ring sealing between head and shell.
- Replace ring sealing and element. Fill filter with some oil and reassemble.
- Run the engine, check for leaks, shutdown the engine. Add oil as necessary to bring the oil level to the 'H' mark on the dipstick.

Engine Coolant Check

Periodic test of engine coolant is necessary to ensure the engine coolant specifications, i.e., 8.5 to 10.0 pH and 3500-5000 ppm chromate concentration to control the cavitations, corrosion/ erosion to avoid liner pitting. The following instructions may be followed for checking pH of engine coolant.

- Take a coolant sample from engine directly (undiluted) in the beaker up to 25 ml mark.
- Dip one strip of pH paper in the coolant sample and compare the colour with hart on the pH paper pad immediately on removing from the coolant sample.
- The pH should be between 8.5 and 10.

Note:

- Destroy pH controller container after use and wash the hands thoroughly with soap.

- If the system is topped up by water, it leads to dilution of the coolant. In order to maintain the coolant concentration, it is must to top up the system by coolant, not by water.
- Coolant used in cooling system is a mixture of distilled or de-ionized water and corrosion inhibitor. Water quality is important for cooling system performance. Therefore, it is important that high quality water is used which is free from suspended particle or mud. Distilled or de-ionized water is essentially used in cooling system.

10.8 FIRST 1500-HOUR CHECK

At first, 1500 hrs check, perform all the checks described earlier. Then following checks are to be performed.

Adjust Injectors and Valves

- It is essential the injectors and valves be in correct adjustment at all times for the engines to operate properly, one controls engine breathing and other controls fuel delivery to the cylinders.
- Be sure the injector and valve set markings, wherever located, are in proper alignment with the indicator mark.

Engine Temperature

- Conditions provide the necessary stabilization of engine components to allow for an accurate valve and injector adjustment. For example, the engine must be at any stabilized temperature of 60°C or below.
- A resetting after the engine is warm and not recommended. Determine the cylinder in position for valve set. The crossheads and valves will be adjusted on the cylinders to check for valve position.
- Wiggle the valve rocker levers on the two cylinders in question. The crossheads and valves on the cylinder where both levers feel loose are ready to adjust. Use the chart to determine the injector that is ready to adjust.

Note: An over tightened setting on the injector adjusting screw produces increased stress on the injector train and the crank injector lobe which can result in engine damage.

10.9 'C' MAINTENANCE CHECKS

- Change aneroid oil.
- Replace aneroid breather.
- Change hydraulic governor oil.
- Check fan hub, idler and water pump.

- Clean radiator externally.
- Inspect units like alternator, generator, starter, replace as required.
- Check evacuator valve on air cleaner. Change if required.

10.10 'D' MAINTENANCE CHECKS

- Clean and calibrate injectors.
- Clean and calibrate fuel pump.
- Clean and calibrate aneroid.
- Clean cooling system.
- Inspect/check and correct the turbocharger.
- Inspect and correct vibration damper.

Note: If, at any time the engine experiences following types of problems, check vibration damper. Replace if necessary.

(a) Gear train failure.
(b) Accessory drive shaft failure
(c) Crankshaft failure
(d) Damper mounting cap screw failure
(e) Flywheel mounting cap screw failure

  - Visual check of the viscous vibration dampers should be done. For any leakage or physical damage it should be replaced.
  - All air compressors have a small amount of oil carry over, which lubricates the piston rings and moving parts. When this oil is exposed to normal air compressor operating temperatures over a long period of time, varnish or carbon deposits get developed. Generally, after 6000 hrs (run of about two years), the air compressor piston rings should be inspected. If the proper inspections are overlooked, the air compressor piston rings will be affected by high operating temperatures, and will not seal properly. First remove the crankcase breathers from the right bank and left bank. Clean it in an approved cleaning solvent, and dry with compressed air and reinstall.

10.11 SEASONAL MAINTENANCE CHECKS

Some more maintenance checks are also required to perform once or twice each year.

Hose Replacement

Inspection of the oil filter, cooling system hose and hose connections is necessary to ensure that there are no leaks and/or deterioration. Get it cleaned otherwise replace if feel necessary.

Preheater Cold Starting Aid Check

Remove the pipe plug from the manifold, near the glow plug and check the operation of the preheater.

Thermostats and Seals Check

Remove the thermostats from the thermostats housings and check for proper openings and losing temperature.

Steam Cleaning of the Engine

Steam is the most satisfactory method of cleaning a dirty engine or piece of equipment. If steam is not available, use an approved solvent to wash the engine. All electrical components and wiring should be protected from the full force of the cleaner spray nozzle.

Mountings Checking

- An engine mounting bolt get occasionally loose and causes the engine supports and brackets to wear rapidly. Tighten all mounting bolts or nuts and replace any broken or lost bolts or cap screws.
- Tighten all turbocharger mounting cap screws and nuts in order to ensure that they are holding securely. Mounting bolts and supports should be tight so that vibration will be at minimum.
- Fan should be sure it is securely mounted; tighten the cap screws as necessary. Check the fan for wobble or bent blades.
- Check the fan hub and the crankshaft drive pulley to be sure that they are securely mounted. Check the fan hub pulley for loosen nuts or wobble; if necessary, remove the fan pilot hub and tighten the shaft nut. Tighten the bracket cap screws also.
- End clearance is necessary for the crankshaft of a new or newly rebuilt engine. A worn engine must not be operated with more than the worn limit end clearance. If the engine is disassembled for repair, install new thrust rings.
- Check the zinc plugs in the heat exchanger and change the eroded plugs. Chemical reaction of raw water circulated through the heat exchanger decides the level of erosion of the plugs and the frequency of changing the plugs.

EXERCISES

1. Write ten maintenance steps of the diesel generator set.
2. Describe necessary preparations/treatments required to be done before putting the DG set for storage.
3. Write note on ‘preparations for stored engine of DG set to be put into service’.

4. Explain importance and necessity of daily maintenance checks of DG sets.
5. Write short note on:
   (a) Weekly maintenance checks of DG set.
   (b) Seasonal maintenance checks of DG set.
   (c) Daily maintenance checks of DG set.
6. What are the steps and checks, which come under 'B-maintenance checks'? Describe each.
7. What are the steps and checks, which come under 'C-maintenance checks'? Describe each.
8. Describe systematically 'D-maintenance checks' of DG set.

11

Maintenance of Substation Equipment & Transmission Lines

11.1 INTRODUCTION

All the equipment installed in the electric substations play important role for providing reliable power supply. So, preventive maintenance of the system is very much essential not only to maintain a healthy power system network but also to optimize the life and performance of the equipment.

Reliability of service is impossible to define in a quantitative manner. It can be described by factors such as frequency and duration of service interruption. These interruptions may have serious impact on functioning of hospitals, public buildings, industrial processes, etc. Hence, to ensure safe and reliable power supply, preventive maintenance of the equipment of substations, transmission and distribution lines are essential. With systematic maintenance, the life and performance of the equipment can be enhanced and frequency of interruption can be minimized.

Following equipment and systems are required to be maintained properly:

- Transformers
- Reactors
- Circuit breakers
- Current transformers
- Potential transformer/Capacitance voltage transformers
- Isolators and earth switches
- Lightning arrestors
- Busbars
- Insulators
- Jumpers
- Connectors & clamps
- Structures
- Filter capacitors

- Wave traps for PLCC equipment
- Protection systems
- Telephone systems
- Air-conditioning plants
- Battery Bank/Charger
- AC/DC Distribution system
- Fire protection systems
- Diesel generating sets
- HVDC system
- Transmission line
- Earthing system
- Waterlogging
- Heavy equipment cooling system
- Timely pruning of the trees and cutting grasses.

11.2 PREVENTIVE MAINTENANCE SCHEDULE FOR TRANSFORMERS AND REACTORS

Transformers are considered the heart of the substation and with proper care and maintenance its life and performance can be optimized. Proper upkeepment of the transmission substation transformers responsible for supplying the electrical power in huge areas whether the consumers are industrial or household must be done with utmost care and in strict rules and maintenance schedule. Following maintenance schedule is recommended for it.

S/N	Detail of maintenance work to be carried out	Frequency	Remarks
1.	Observation of oil level of conservator and leakage if any.	Daily	
2.	Checking of bushing oil level	Monthly	
3.	Checking of auto starting of cooler oil pumps and fans	Monthly	
4.	Monthly checking colour of silica gel in breather and regenerating if required.	Monthly	
5.	Checking of oil colour, BDV (Breakdown Voltage) test, moisture content and resistivity of main tank oil	Half yearly	
6.	Vibration measurements for reactors only	Half yearly	
7.	Measurements of Acidity and IFT for main tank oil	Yearly	
8.	Tan-Delta measurement for bushings	Yearly	
9.	External cleaning of radiators and all windings	Yearly	
10.	Insulation resistance and tan (delta) measurement of windings	Yearly	
11.	Measurement of winding resistance at all tap positions	Yearly	
12.	Checking of all remote indicators, WTS and Tap position indicators	Yearly	
13.	Testing of pressure relief devices, Buchholz relay, OLTC surge relay and replacement of damaged gaskets of the terminal box	Yearly	
14.	Testing of Buchholz relay by injecting air or nitrogen	Yearly	
15.	Marshalling boxes cleaning, tightening of terminals, checking of contactors, space heaters, illumination, etc.	Yearly	
16.	Checking of oil level and calibration of oil temperature indicator and winding temperature indicators.	Three-yearly	
17.	Checking and topping up of oil level of oil seal of breather	S.O.S.	
18.	Filtration/degassing of main tank and radiation oil.	S.O.S.	

11.3 PREVENTIVE MAINTENANCE SCHEDULE FOR CVT/PT

S/N	Details of maintenance work to be carried out	Frequency	Remarks
1.	Checking for any breakage/cracks in HV bushing	Monthly	
2.	Checking of oil leaks	Monthly	
3.	Spark gap cleaning	Monthly	
4.	Cleaning and tightness of connections	Yearly	
5.	Capacitance and Tan (Delta) measurement	Yearly	
6.	Measurement of secondary winding resistance	Yearly	
7.	Measurement of voltage at metering points and at CVT terminals	Yearly	
8.	Testing of oil for breakdown voltage, Tan (Delta) and moisture	Half yearly	
9.	Capacitance and Tan (Delta) measurement of each capacitor stack	S.O.S.	
10.	Cleaning of marshalling box and junction boxes	Yearly	

11.4 PREVENTIVE MAINTENANCE SCHEDULE FOR LIGHTNING ARRESTORS

S/N	Details of maintenance work to be carried out	Frequency	Remarks
1.	Checking of leakage current by current analyzer, Isolators resistance measurement, testing of connector, cleaning of housing	Yearly	
2.	Checking/Tightness and MV drop test of all the connector	3-yearly	

11.5 PREVENTIVE MAINTENANCE SCHEDULE FOR CIRCUIT BREAKER

Circuit breakers, the mechanical switching devices, are designed to automatically open a circuit under the specified abnormal conditions without any damage to the system. Circuit breakers should be well qualified to protect itself even under abnormal conditions. The frequency of tripping is the main concern of its maintenance. Detail of the preventive maintenance schedule is given hereunder.

SN	Details of maintenance work to be carried out	Frequency	Remarks
SF-6 CIRCUIT BREAKER			
1.	SF-6 gas leakage lest	Yearly	
2.	Humidity (Dew point) measurement in SF-6 gas	3-yearly	Evacuate gas & change filter
3.	Air content measurement of SF-6 gas	3-yearly	
4.	SF-6 gas acidity measurement	3-yearly	
5.	Hydraulic Operating Mechanism		
	(a) Checking of oil level and topping up if necessary	Monthly	
	(b) Checking of oil pressure and oil leakage	Monthly	
	(c) Checking of oil pressure drop during duty cycle operation check	Monthly	
	(d) Checking of auto starting/stopping of oil pump	Monthly	
	(e) Checking of pressure switch setting	Monthly	

Table Contd.

SN	Details of maintenance work to be carried out	Frequency	Remarks
6.	Pneumatic Operating Mechanism		
	(a) Functional checking of auto starter of air compressors and relays	Half yearly	
	(b) Checking of air pressure drop during duty cycle operation	Yearly	
	(c) Maintenance of air dryer	Half yearly	
	BREAKER OPERATION CHECKS AND MEASUREMENT		
1.	Checking of circuit breaker timing setting	Yearly	
2.	Checking of pole discrepancy relay	Yearly	
3.	Functional checks, duty cycle operation including rapid reclosing	Yearly	
4.	Checking of all operational lock outs and interlocks	Yearly	
5.	Checking of pressure settings and calibrations of switches	Yearly	
6.	Checking of close/trip coils, measurement of resistance, pick up voltage of trip coils	Yearly	
7.	Checking of breaker poles, support insulator, PIR and grading capacitor	Yearly	
8.	Checking of healthiness of counters, measurement of contact resistance, tan(delta) measurement of grading capacitors	3-yearly	

11.6 PREVENTIVE MAINTENANCE SCHEDULE FOR CURRENT TRANSFORMERS

SN	Details of maintenance work to be carried out	Frequency	Remarks
1.	Insulation-resistance measurement	Yearly	Only when test tap is not provided
2.	Measurement of Tan(Delata) and capacitance	Yearly	
3.	Checking the tightness of all connections	3-yearly	
4.	Measurement of CT secondary resistance	S.O.S.	
5.	Magnetization characteristics checking	S.O.S.	
6.	CT ratio test	S.O.S.	
7.	Checking of space heater	Quarterly	
8.	Cleaning of marshalling box and junction box	Yearly	

11.7 PREVENTIVE MAINTENANCE SCHEDULE FOR ISOLATORS AND EARTH SWITCHES

Isolators are used for isolating the circuits. Isolation must be visible. Hence, its frequency of operation is main consideration for routine maintenance of the circuit breaker and proper maintenance is essential for better performance and reliable operations of the isolators. Detailed schedule of periodical maintenance of circuit breaker is given below.

S/N	Detail of maintenance work to be carried out	Frequency	Remarks
	OPERATING MECHANISM		
1.	Maintenance of linkage including transmission gears, stopper bolts, clearing of auxiliary switch contacts, checking of electrical/ mechanical interlocks	Yearly	
2.	Lubricating of operating mechanism, hinges, locks, joints on levers, bearings and checking of all mounting bolts for tightness	Yearly	
MAIN CONTACTS			
1.	Clearing and lubrication of main contacts, alignment, main contact resistance measurement, tightness of bolts and nuts, pins, etc. checking of support insulators and look for cracks	Yearly	
2.	M.V. drop test across terminal connection	3-yearly	
	EARTH SWITCH		
	Checking and alignment of earthing blades, operations of earthing switch, clearing of contacts, checking of flexible breading, checking of earth connectors of structures and MOM box.	Yearly	

11.8 PREVENTIVE MAINTENANCE SCHEDULE FOR BUSBARS, POST INSULATORS, JUMPERS, CONNECTORS, CLAMPS, STRUCTURES

S/N	Detail of maintenance work to be carried out	Frequency	Remarks
1.	Measurement of station earth resistance	Yearly	
2.	Thermo-vision scanning, checking of insulation for cracks	Do	
3.	Removal of hot spots, de-weeding of switchyard, repainting, rust removal of all structures, equipment, etc.	S.O.S.	

11.9 PREVENTIVE MAINTENANCE SCHEDULE FOR PROTECTION SYSTEMS

S/N	Detail of maintenance work to be carried out	Frequency	Remarks
1.	Calibration of tariff-energy meter, panel meters and non-tariff meters	Yearly	
2.	Secondary injection test of individual protection schemes	Yearly	
3.	Checking of voltage for relays, DC drain of static relays, DC logic circuit testing	Yearly	

11.10 PREVENTIVE MAINTENANCE SCHEDULE FOR BATTERY BANK AND BATTERY CHARGER

SN	Detail of maintenance work to be carried out	Frequency	Remarks
1.	Checking electrolytic level and topping up with distilled water of all cells	Monthly	
2.	Checking of emergency DC lighting to control room	Monthly	
3.	Checking of electrical connection of charge panel for tightness and cleanliness and other electrical connection in DCDB	Half yearly	
4.	Battery impedance testing	Yearly	
5.	Curative discharge of battery for capacity test	3-yearly	

Battery banks installed in the substation play vital role in providing the required supply in the absence of AC supply. It provides the backup energy during emergencies. Details of maintenance schedule are given below.

11.11 PREVENTIVE MAINTENANCE SCHEDULE FOR TRANSMISSION LINES

Transmission line is termed as conveyance of electrical power at high/extra high voltages from generating station to the grid substation. Due to economic reasons, overhead lines are preferred for electrical power transmission. But we should not forget that it is prone to disturbances from weather elements, lighting strokes and other unavoidable elements. So, the preventive maintenance of the transmission lines is essential for smooth and regular supply of power from one place to other. Details maintenance schedule is given as under:

SN	Detail of maintenance work to be carried out	Frequency	Remarks
1.	Patrolling of the lines	Quarterly	
2.	Tree cutting and loping of branches	S.O.S.	
3.	Back filling of foundation	S.O.S.	
4.	Tightening of jumpers	S.O.S.	
5.	Adjustment/replacement of vibration damper	S.O.S.	
6.	Replacement of broken insulator	S.O.S.	Hot-line maintenance can be done

EXERCISES

1. Describe necessity of preventive maintenance of the equipment.
2. Write systematically, preventive maintenance schedule for high voltage transformers.
3. Write a note on how to look after hydraulic and pneumatic operating mechanism of a circuit breaker.
4. What do you understand by hot-line maintenance of transmission line?
5. Explain how important is tree pruning and grass cutting inside the electrical substation premises.
6. Write short notes on:
   (a) Preventive maintenance of CTs.
   (b) Preventive maintenance of PTs.
   (c) Preventive maintenance of SF_6 circuit breakers.
   (d) Preventive maintenance of isolators.
   (e) Preventive maintenance of protective systems.

12

Automobile Electrification

12.1 INTRODUCTION

Vehicles, the mechanical devices run day and night in normal as well as extreme weather conditions. Halogen lamps have ability to penetrate fog, drizzle, mist, etc. Bulbs and indicators are used to provide ordinary light. In order to make able the bulbs, indicators and halogen lamps to glow, an electric circuit with power source such as DC batteries is needed in the automobiles.

Various electrical circuitry with necessary electrical apparatus used in the automobiles are described hereunder:

- Lighting circuit
- Battery charging circuit
- Starting circuit
- Ignition circuit
- Instrument circuit

12.2 LIGHTING CIRCUIT

The lighting circuit consists of the following sub-circuits:

- Headlight
- Sidelight
- Park light
- Tail light
- Stop light
- Doom-light
- Door-light
- Dash light

Head-light system has double filament lamps, one of 30 watts and the other of 5 watts. Lamps are controlled by two-way switch. A common circuit is shown in the Figure 12.1.

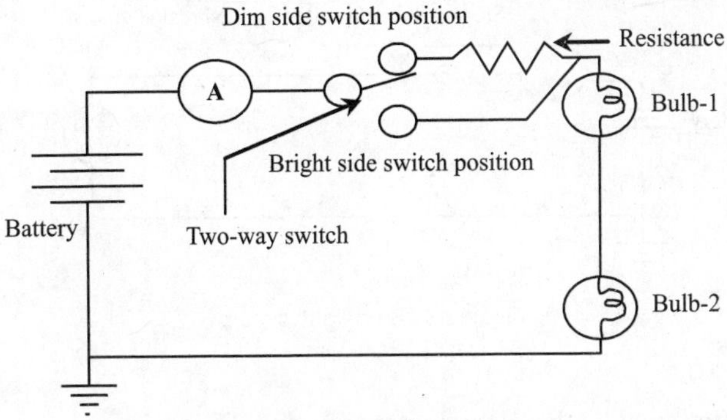

Fig. 12.1. Head-light circuit.

This light is required to identify the route for the vehicle during poor visibility in the night or low light duration. If the brightness of the headlights of two vehicles coming from opposite directions remains on their peak, there is sharp glare on the driver's face, which creates problems of recognizing the exact position of the vehicle coming from opposite direction. So to avoid the glare on the driver's face, the dim light is used when any other vehicle is coming from the opposite side.

Sidelights are to indicate the breadth of the vehicle. They are, therefore, fitted on the front sides. The wattage of the bulbs is about 10 watts.

Park-lights are fitted near the sidelights and also at the back of the vehicle. They are used at the time of parking. Wattage is about 4 watts. The circuit is shown in the Figure 12.2.

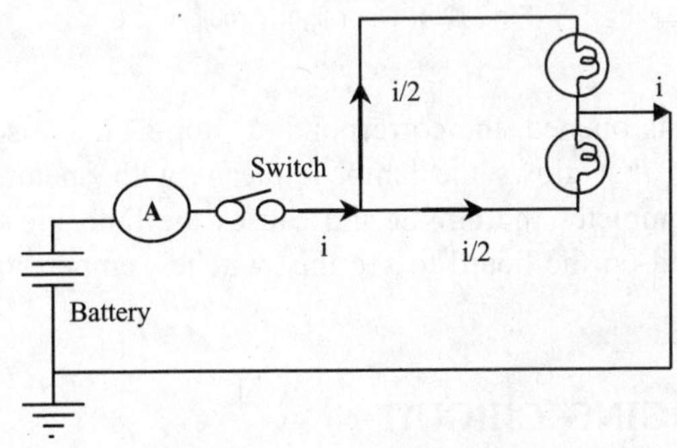

Fig. 12.2. Park-light circuit.

Tail lights are coloured lights and bulbs are fitted at the backside of the vehicle. They indicate the presence of the vehicle to the coming vehicles, animals, bodies and persons. The wattage of the lights is in the range of 5 watts.

Stoplight is also fitted at the backside of the vehicle. This light is controlled with break system. When driver tends to apply the break, stoplights glow and persons around the vehicle understand that vehicle is to stop. The wattage of the bulbs is about 3 watts. The circuit is shown in the Figure 12.3.

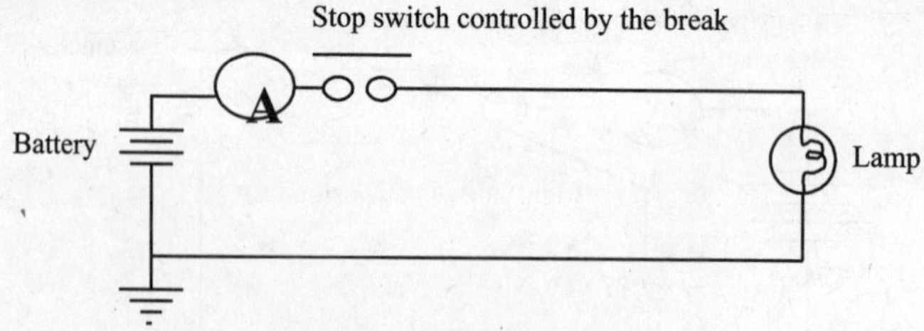

Fig. 12.3. Stop-light circuit.

Doom-light is fitted inside the vehicle for lighting. The circuit is similar to that of the sidelights. Wattage of the lamp is kept in the range of 20 watts.

Door-light helps the driver to identify the door position whether it is open or closed. The door opening system controls the switches. The wattage of the lamps is about 3 to 5 watts. The circuit is shown in the Figure 12.4.

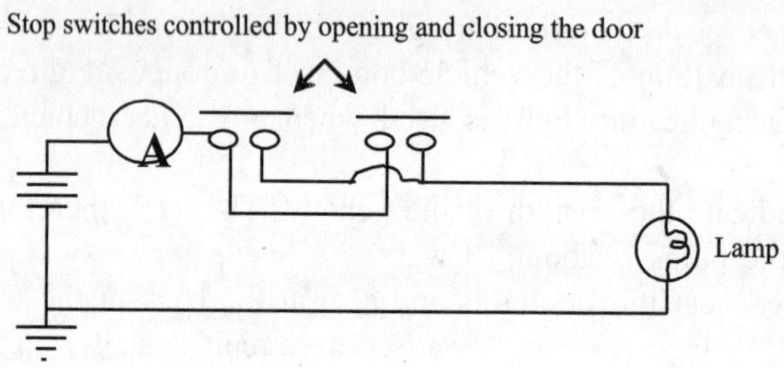

Fig. 12.4. Door-light circuit.

As any of the doors gets opened, the corresponding stop-switch also known as door switch is pushed to 'ON' position that glows the lamp. It means with opening of any one door, the switch gets 'ON' which completes the circuit and causes the bulb glow.

Dashboard light is used on the board to see the switches, ampere meters and other meters. Wattage is about 10 watts.

12.3 BATTERY-CHARGING CIRCUIT

Battery charging is required in a vehicle to keep ready charged the battery of the vehicle. A shunt generator is mounted near the engine. It is driven through a 'V' belt. It gives supply for charging the battery. The same belt derives the radiator fan also. To protect the battery charging unit, a reverse current relay is connected between the battery and the generator. The relay consists of two coils wound over a magnetic frame.

One coil is of fine wire connected across the armature while the other is of heavy wire with a few turns connected in series with the generator's armature, battery and the contacts. Contacts open and close by the armature of the relay. When the generated voltage reaches sufficient to charge the battery, the shunt coil of the relay closes the switch connecting the battery to the generator. If the generated voltage falls below the preset value, a current flows from the battery to the generator and thus the current direction in the series coil gets reversed. The flux due to this opposes the flux due to the shunt coil and opens the switch, which disconnects the battery.

Two types of the shunt generators are in use for charging the batteries:

(i) Two-brush Dynamo

It consists of two brushes. One is earthed and the other gives supply to the load and battery. Separate regulators are used to regulate the currents and voltages. Current regulator (CR) is a series coil of thick wire with a few turns and when there is short circuit in the external circuit, heavy current flows through the series coil. Heavy short circuit current attracts the movable arm of the relay system below, which opens the circuit. Now current flows through the resistance and current value decreases. In this way the current regulator regulates the current flow and protects the circuit from the dangerous current. The circuit is shown in the Figure 12.5. A relay is also used in the circuit for protection purpose.

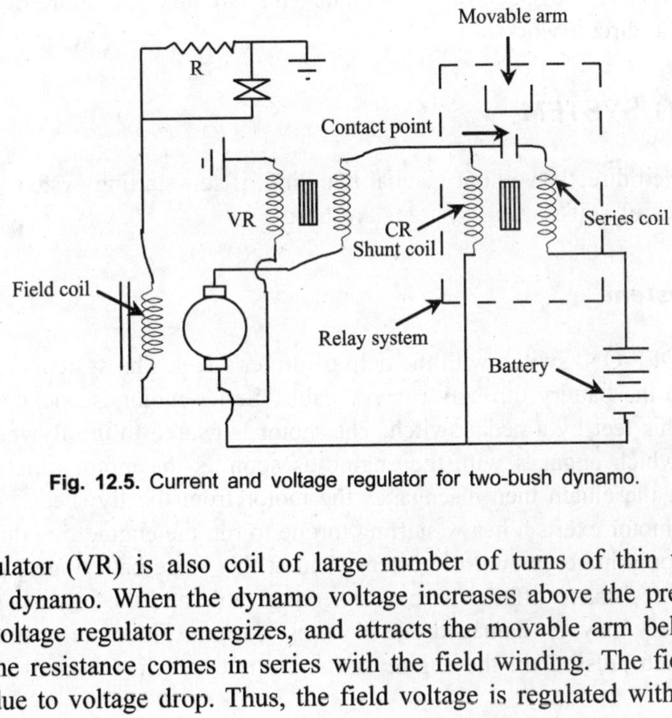

Fig. 12.5. Current and voltage regulator for two-bush dynamo.

Voltage regulator (VR) is also coil of large number of turns of thin wire connected in parallel with the dynamo. When the dynamo voltage increases above the prescribed value, the coil called the voltage regulator energizes, and attracts the movable arm below and opens the contact point. The resistance comes in series with the field winding. The field of the dynamo becomes weak due to voltage drop. Thus, the field voltage is regulated without any difficulty.

(ii) Three-brush Dynamo

This system is another type of battery charger. This does not require current and voltage regulators. Here, current and voltages are regulated by adjusting the third brush of the dynamo. In the Figure 12.6, three-bush dynamo circuit is shown. Reverse current relay prevents the battery

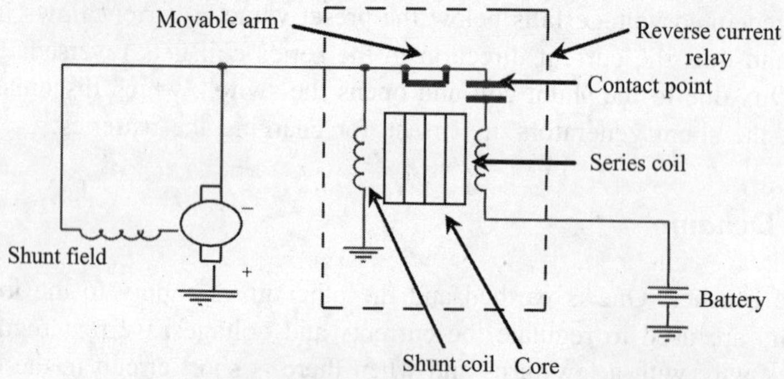

Fig. 12.6. Tree bush dynamo.

from being discharged. Since the voltage is proportional to the speed of the vehicle, this type of dynamo needs great care in adjusting the third brush when the voltage and current rise take place suddenly by raising the speed of the vehicle. This circuit works more or less similar way as two-bush dynamo circuit works.

12.4 STARTING SYSTEM

Vehicle can be started directly without manual handling, if self-starting system is provided in it. Generally, two types of self-starting systems are used.

(i) Pedomax System

This controls the 'OFF-ON' switch with the help of driver's feet. The system consists of a series motor connected to the battery through a heavy cable. Series motor is operated and controlled by the driver with his feet by a pedal switch. The motor is geared to the flywheel of the engine through a clutch, which engages with the engine as soon as the motor turns 'ON'. After the engine gets started, the clutch then disengages the motor from the flywheel.

The DC series motor exerts a heavy starting torque to run the engine. So, during the starting, battery needs to supply huge amount of current. And at this stage battery gets practically short-circuited. The only resistance offered is the connecting cable and the internal resistance of the battery. So, the battery may get damaged due to excess current. Therefore, it is recommended to keep the pedal switch pressed till the engine starts. Do not switch 'ON and 'OFF' frequently.

(ii) Relay Coil System

Relay coil system of vehicle starting has become prominently popular. Here, relay coil works itself to make the starting switch 'ON' and putting only the relay switch in the 'ON' position is required. High current rated cable is required for connecting the circuit because of the large starting current. In the Figure 12.7, relay coil self starter circuit is shown.

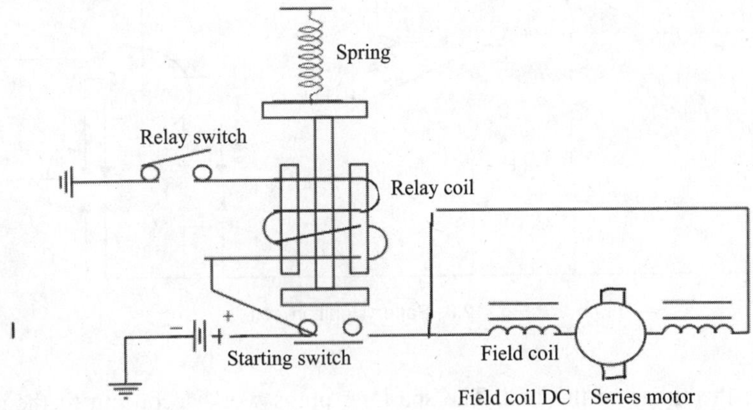

Fig. 12.7. Relay coil self-starter.

12.5 IGNITION SYSTEM

Ignition system gives flame to the fuel to make the engine start. Impulse voltage (up to 10 kV of very short duration) is required for igniting the spark plug. Magneto generator or a battery operated ignition system is used. Following are the ignition systems in use whose purpose is to give flame to the fuel.

(i) Magneto-ignition System

It is an alternating current generator in which the field is provided by a permanent magnet that makes magnetization system. The armature carries two windings, namely, primary and secondary windings. When armature geared to the engine, it rotates and heavy current circulation takes place in the primary winding. The effect of this current is the production of large amount of flux. At a certain position of the armature, a cam fitted to the magneto shaft opens the primary circuit mechanically. It induces e.m.f. in the secondary winding. There is also a device to advance the timing of the spark with the speed.

(ii) Battery Ignition Coil

It is very popular system for giving flame to the fuel. This system consists of an induction coil and a contact breaker. The battery current is given to the ignition coil (induction coil). Here, low DC voltage is converted into the AC voltage by the making and breaking system. A huge voltage is induced in the secondary, which is connected in turn to each spark plug of the cylinder of the engine through distributor bus. The ignition coil is a primary coil having low voltage and low current. Making and breaking is performed with the help of a mechanical device. Induced secondary low voltage is supplied to the rotor, which transfers it to the distributor bus. The distributor distributes the impulse voltages to the sparking plugs in turn of each engine

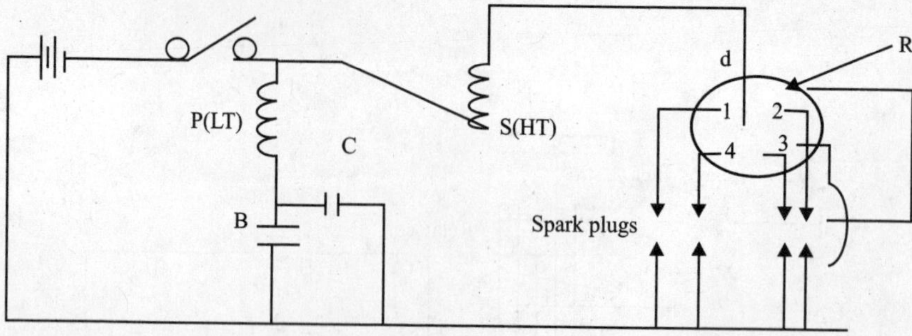

Fig. 12.8. Battery ignition coil.

cylinder to give the flame to the fuel. The sparking plugs work according to their firing order. Engine through gearing drives the distributor. The distance between the contacts ranges from 0.025 to 0.03 cm. The distance between the sparking plugs varies from 0.0625 to 0.125 cm.

Here

P = Primary of the ignition coil (L.T. side)
S = Secondary of the ignition coil having (H.T. side)
C = Contacts
B = Decoupling condenser (Bypass condenser)
d = Distributor distributing secondary of the impulse voltage to the four cylinders in turn
1, 2, 3, 4 = Spark plugs and distributor contacts

(iii) Sparking Plugs

These are made up of steel shells carrying porcelain bushes into which pointed electrodes are fitted. The pointed electrodes are made up of nickel alloy. The shell is screwed to the cylinder. Spark jumps between the two electrodes at the moment of the ignition. Generally, positive of the battery is earthed to the frame and the direction of the voltage induced in the secondary is sufficient to produce the required amount of intensity of the spark. If the negative is earthed, a higher voltage is required to produce a required intensity of the spark.

12.6 INSTRUMENT CIRCUITS

The following instruments are used in automobiles:

- Electric petrol gauge
- Electric petrol pump
- Wiper
- Horn
- Side indicator

1. Electric Petrol Gauge

The quantity of the petrol in the tank is measured by the petrol gauge. It consists of two coils, a current coil connected in series and a shunt coil also known as pressure coil connected in parallel. A pointer controlled by field produced by these two coils moves over the indicating scale and indicates whether the tank is empty, half or full of petrol. A variable resistance with gears adjustment and float is also connected in the circuit. The float floats in the petrol tank and is attached with the gears. This instrument is fitted on the dashboard. The series and shunt coils are connected with the battery through an ammeter.

2. Electric Petrol Pump

This fills petrol in the carburettors of the vehicle from the main petrol tank. Two valves, one inlet valve and the other outlet valve are fitted in the electric petrol pump. The petrol enters the pump tube through the inlet valve and goes out of it through the outlet valve. A diaphragm is attached with the pump-tube that works by an electric system. The electric system has a coil wound over the iron cores. A contact opens and closes by a diaphragm rod, which moves between the cores. A condenser connected across the contacts limits sparking. The coil circuit is completed through ignition switch, ampere-meter and battery as shown in Figure 12.9.

3. Screen Wiper

Screen wiper is used to clean/wipe the glass-screen situated at the front of the driver. It consists of a series motor. Motor's rotor is attached with a movable arm. Armature (rotor) of the motor causes movement in the movable arm carrying the wiper. Wiper moves to and fro on the screen and cleans it especially in the rain.

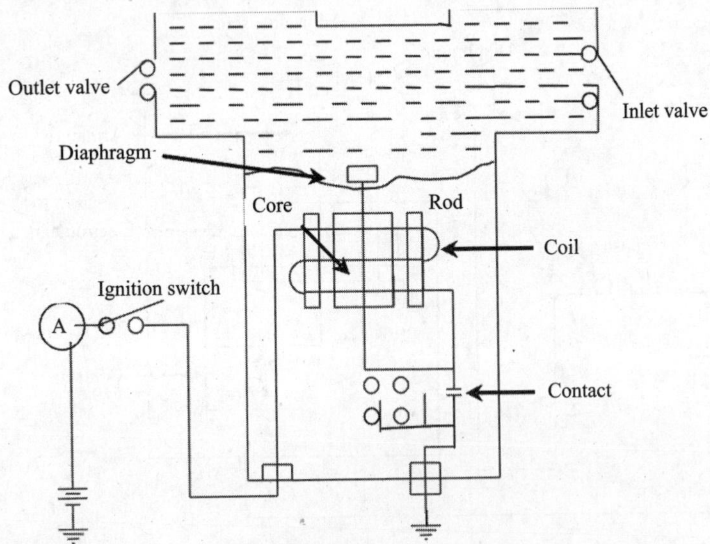

Fig. 12.9. Electrical petrol pump.

4. Horn

It is used to produce sound. It consists of a diaphragm, which is fitted with a movable armature between the electromagnets. A battery supplies current to the coil. The circuit makes and breaks continuously by a contact inserted in the circuit of the electromagnet coil. Diaphragm with an adjustable screw controls the sound. Horns are of two types, the magnet type and motor type.

Magnet type horn circuit consists of an electromagnetic coil, which is attached with the diaphragm. Contacts are normally closed. With the application of the current, the coil gets magnetized and attracts the strip fitted with armature. The contacts breakup and the coil become demagnetized. As a result of it, the armature falls down and circuit once again becomes complete. Rapid breaks and makes vibrate the diaphragm and sound gets produced.

Motor type horn consists of a series motor, which moves the threaded plate fixed upon the motor shaft on the other side of the commutator. Another threaded plate attached with the diaphragm is also fixed. The armature movements are thus communicated to the diaphragm, which vibrates and produces sound.

5. Side Indicator

This indicates the side to which the movable vehicle has to turn. A coil, inside which an arm is fitted, is the heart of the side indicator. A three watt bulb is fitted in the arm which acts as indicator. Two indicators are fitted on the two sides of the vehicle and a two-way switch controls it. As current passes through the coil, it gets energized and attracts the movable arm, which indicates the direction. As switch gets off, the coil is demagnetized and the arm goes down to its original position. The bulb also glows as it is connected in parallel to the coil. A battery is used as supply source to the indicators

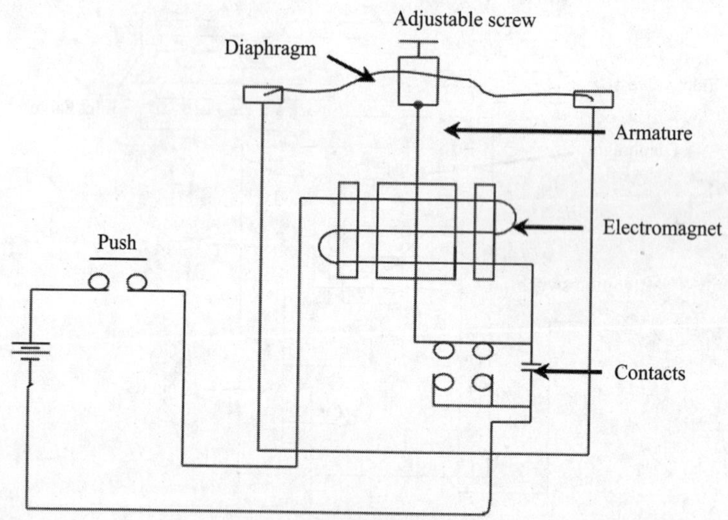

Fig. 12.10. Magnet type horn.

EXERCISES

1. Name the various electrical circuits used in a vehicle and state briefly the purpose of each.
2. Give a brief description of two-brush dynamo used in the vehicle for charging battery.
3. Write short notes on the following:
   (a) Electrical petrol gauge.
   (b) Electric petrol pump.
   (c) Magnet type horn.
4. Describe self-starting circuit.
5. Explain in detail the lighting circuit of the vehicle.
6. Explain in detail a battery charging system.
7. Describe ignition system of a vehicle.
8. Write in brief about
   (a) Side indicator of a vehicle
   (b) Screen wiper for a vehicle
   (c) Sparking plug

13

Electrical Wiring Systems

13.1 ILLUMINATION

Use of right fixtures and technology enhances the quality of light and helps in conserving the electricity. Technically proper selection of the light source is economical as well as releases less heat thereby puts less load on air conditioners. The following aspects should be taken into mind while planning illumination system in houses, hospitals, schools, stadiums, auditoriums, factories, etc.

- Use of occupancy sensors at correct locations
- Plans for the maximum utilization of sunlight and moonlight
- Placing of the fixtures at right place and height
- Selection of the right fixtures depending on the requirement of the illumination level
- Use of the dimmers for the luminance control, which help in giving the needed illumination with lesser electricity usage
- Proper design of light shelves, windowsills, light reflectors, louvres, walls, blinds, etc helps in maximum utilization of sunlight
- Design of windows and its glasses should be proper

13.2 HOUSE WIRING

For house wiring, following two systems of wiring are used:

13.2.1 Tree System of Wiring

In this system, branch circuits are tapped from the main circuit at convenient and required points. This system is now not in use in electrical purposes because of following disadvantages.

Disadvantages

- Appearance is not good.
- Due to large number of joints, faults occur very often at joints.
- Locating the fault is difficult.
- Considerable voltage-drops affect the entire system.
- The required fuses are placed here and there.

Advantage

It is cheap.

13.2.2 Distribution System of Wiring

In this system, mains are taken to one or more convenient distribution boards. Separate connections for each branch circuit are taken from the distribution boards. Because of the following merits, this system of wiring is nowadays widely used for interior wiring in buildings.

Advantages

- Good appearance.
- No joints are required.
- Each circuit can be readily disconnected from the mains without affecting the other circuits.
- Locating the fault is easy.
- All the points remain, almost at the same voltage level.
- A single fuse protects each circuit independently.
- System is simple, and capable of further extension, if required.
- Blown fuse replacement is easy.

Disadvantage

Its cost is high in comparison to tree system of wiring, because of the requirement of more wires.

13.3 DIFFERENT TYPES OF WIRING USED FOR DOMESTIC PREMISES

Selection of suitable type of wiring depends on the following:

- Position of light/fan points
- Safety
- Durability
- Initial cost
- Appearance

Different types of wiring used for domestic premises are described hereunder.

13.3.1 TRS/PVC Sheathed Batten Wiring

Tough rubber sheath (TRS) or PVC (Polyvinyl Chloride) insulated wires are used in these wiring. The wiring is carried out on wooden batten fixed on the wall or ceiling. The wires/cables are laid down on the fixed wood battens on walls or ceilings on wooden plugs. Distance between two wooden plugs on the wooden batten should not be more than 75 cm.

In order to fasten the wires to the batten, tinned brass/iron link clips fixed with small nails are used. The clips are spaced on the wooden batten at about 10 cm intervals. Flexible pipes are used for passing cables on walls and floors. For light, and fan points, a twin core cable is used. While for 3-pin socket outlet, a twin core cable with earth conductor is used.

Advantage

- A very cheap method for small and medium residential premises.
- It is quick to do batten wiring.
- Easy to locate fault and redress it.

Limitations

- This wiring method is suitable only for dry conditions and not suitable for locations exposed to sun or excessive dampness. Hence, it is not suitable for outdoor use.
- This is also not recommendable where acids and alkalis are to be present.
- Wood batten may catch fire and there is risk of fire in case of short circuit.
- Appearance is not so good as conduit wiring.

Note: This type of wiring has become obsolete nowadays and is not preferred.

Precautions

- Well seasoned, teakwood batten of 13 mm thickness should be used.
- The batten wood must be perfect straight.
- The batten should be well varnished.
- While passing the wires from walls or ceilings, the conduit or flexible pipes must be used.

13.3.2 Wooden Casing and Capping Wiring

This wiring was used for indoor and domestic installations. In this type of wiring, VIR (Vulcanized India Rubber) or PVC (Polyvinyl Chloride) insulated wires are laid in two or three channels of wooden casings. Wooden casings are fixed on walls or ceilings with the help of wooden plugs.

Wooden casings consist of 'U' shaped grooves usually two in number, into which, the wires are laid down in a manner that wires of opposite polarity are laid down in different grooves. After laying down the wires, the casing is covered with varnished clipping with the help of small brass or iron nails or screws. The cables are passed through conduit while crossing wall or ceiling. Positive and negative wires should be taken in different channels/grooves. The capping is fitted to the casing with the help of wood screws of 12.7 mm length. The casing and capping are made up of seasoned teakwood. All sides of casing should be of smooth finish and well varnished. The size of casing and capping are selected according to the number and size of wires to be accommodated.

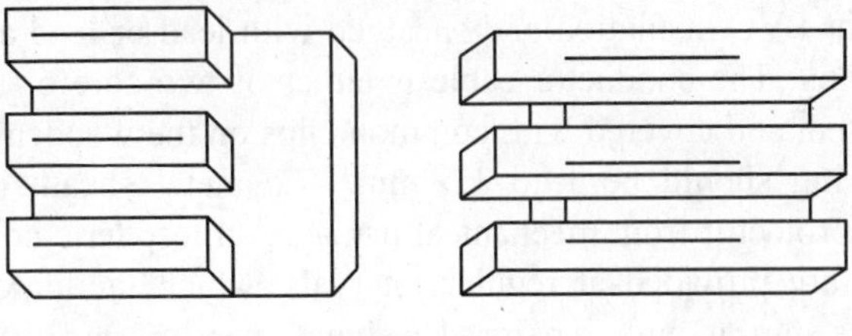

Two casing for jointing

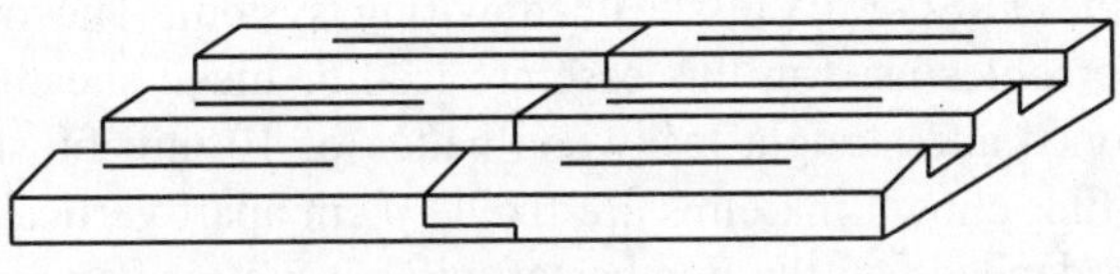

A finished joint

Advantages

- This wiring has longer lifespan in comparison to the TRS/PVC sheathed batten type wiring.
- Capping cover protects the wires by exposing to the weather.
- Phase conductor and neutral conductor are placed separately so there is less chance of short circuit.
- Inspection of wiring is easy and can be repaired by just opening the capping.

Limitations

- There is danger of fire hazards because wooden casings and clippings are inflammable.
- The wirings are not perfectly damp proof.
- The method is costlier than TRS/PVC sheathed batten wiring.
- Appearance is not good.

Note: This is the earliest method of wiring and has become obsolete nowadays.

PVC Casing and Capping Wiring

Nowadays PVC makes casings are fixed on the walls or ceilings with the help of plugs and countersunk screws with an interval not more than three feet. The PVC casing capping wiring gives better look than wooden casing capping wiring system. Inspection of fault and wire conditions is easier which is done just opening the capping. The other features, advantages and limitations are same as wooden casing and capping wiring.

13.3.3 Metal/Lead Sheathed Wiring System

Wires/cables with VIR or PVC insulation and sheathed with lead or lead alloy or aluminium are used in this type of wiring. The conductor cable is either of two-core or three-core types. Each core is separately insulated, and covered with common clips on the wooden batten. The maximum thickness of lead covering should be 1 to 1.5 mm. The metal sheath provides covering the wires/cables in order to protect it from mechanical damage, atmospheric corrosion and dampness. Usually junction boxes are provided at regular intervals, which facilitate for checking of any broken wire connections during fault repairs. The junction boxes are provided for connecting the earthing terminals.

This is similar to the TRS or PVC sheathed wiring system. The only difference in this wiring system is that the cables used in this case are lead or metal sheathed. The lead sheathed cables run on well-seasoned and straight teakwood batten of 10 mm finished smooth thickness. The cables are fixed by link-clips. Link clips are fixed 10 cm apart vertical and 15 cm horizontal run. The width of batten depends on the number of wires it may have to carry. An earth wire is solidly connected with the sheath to prevent the lead sheath becoming electrically alive and giving electric shock.

Advantages

- It is suitable to use even in the places exposed to sun or rain. Joints should be avoided if wiring is exposed to the sun or rain.
- This can be used in the damp situations by covering the ends.
- It has comparatively longer life.
- It gives comparatively better look.

Note: This type of wiring can easily be seen in the workshops and industrial setups.

Limitations

- In the event of improper earthing or damaged insulation, the metal sheath becomes alive.
- It is not preferable where chemical erosion is possible.
- It is comparatively costly.

Precautions

- Supports should be chemically inert with respect to the sheath.
- Sharp bends should be avoided at the corners.
- The electrical continuity should be maintained and earthing should be provided at suitable places preferably at the entry and joints.

13.3.4 Cleat Wiring

This type of wiring is preferred for farmhouses and temporary shelters. In this wiring cleats are made up of two halves base or bottom and top. Cables are supported and gripped between porcelain or seasoned teak wood cleats 6 mm above the wall or ceiling. The base of the cleat is grooved to accommodate the cable on it and top in the axial direction. Cleats are fixed with the help of the shades on the wooden plugs fitted on the wall at regular intervals. Gap between the two cleats must be maintained within 30 cm to 60 cm. The distance between two cables should be 2.5 cm apart centre to centre for the branch load of the single core cables. It should be 4 cm for sub-mains and service connections. VIR, PVC wires are generally used in these types of wirings.

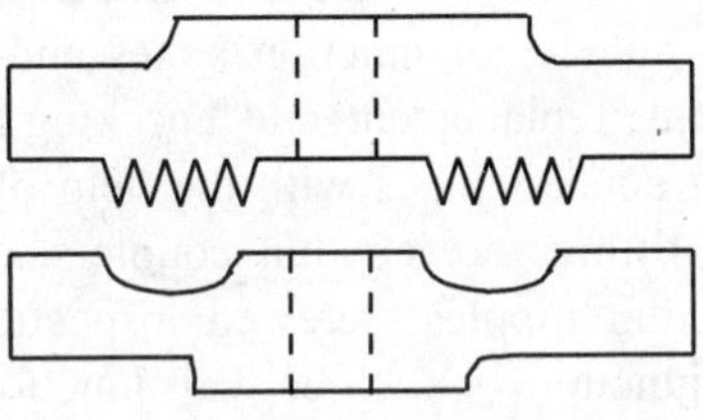

Cleat for two wires

Advantages

- Cheaper
- Easy to work

Disadvantages

- Unsuitable for damp and wet area
- Poor appearance
- Mechanically unprotected
- Little safe

Precautions

- These may cause mechanical injury. Tubing across the wires should be done.

- Whenever wires cross walls and pipelines, these should be drawn in porcelain tubes or conduits.
- Wet walls should be avoided.
- Whenever cleated conductors cross each other, they should be separated by an insulating bridge piece (1.3 cm between conductors).
- Over or loose tightening of the cleats should be avoided.
- Only one wire should be placed in one groove of the porcelain.
- When cleat wiring is done along any iron joist or any metal, dry and properly varnished wood should be placed between such metal and porcelain cleats.

13.3.5 Conduit Wiring

Surface Conduit Wiring

This type of wiring is very popular and widely in use. This method of wiring adapts the carrying VIR or PVC insulated cables inside the enclosed steel, iron or PVC conduit pipes. The clips are screwed to wooden/plastic plugs. Plugs are fitted into the drilled holes in the wall or concrete surface. For better look, it is option to keep the conduits inside the walls. Conduits are sunk into the surface and then plastered so that conduits are hidden inside the wall/roof. In order to facilitate the drawing of wires, a number of junction boxes and other accessories are required. Junction boxes help in withdrawal, replacement and checking of wires in case of any fault. Usually, wires are drawn through conduit pipes with the help of fish wire (a steel wire).

The conduit accessories and fittings are conduit couplers, conduit bends, conduit elbows, conduit bushings, lock nuts, conduit nipples, screwed inspection bend or inspection elbows, terminal junction box, two-way junction box, three-way junction box, four-way junction box, back-outlay junction box, saddle, clip, hook, outlet box, flexible pipes etc. Some of the conduit fittings being used in this type of wiring are shown in the Figure 13.1.

Uses: The uses of this wiring are:

- This wiring system is most suitable in damp situations.
- This wiring system is recommended for workshops and factories.
- For houses, offices and hotels, it is the best.
- It is best suitable in the fire prone zones like paint, varnish and explosive making factories.

Advantages

It is the best and most desirable system of wiring.

- It gives very good appearance, when conduits and junction boxes, are concealed in the walls, and concrete.
- Buried conduits give protection from mechanical injury and fire risks to the buried wires.
- Flexible PVC pipes can be bent easily to suite the requirement, and hence wire joints are avoided.
- It has long life, if properly executed.
- It provides protection against moisture because conduits can be made watertight.

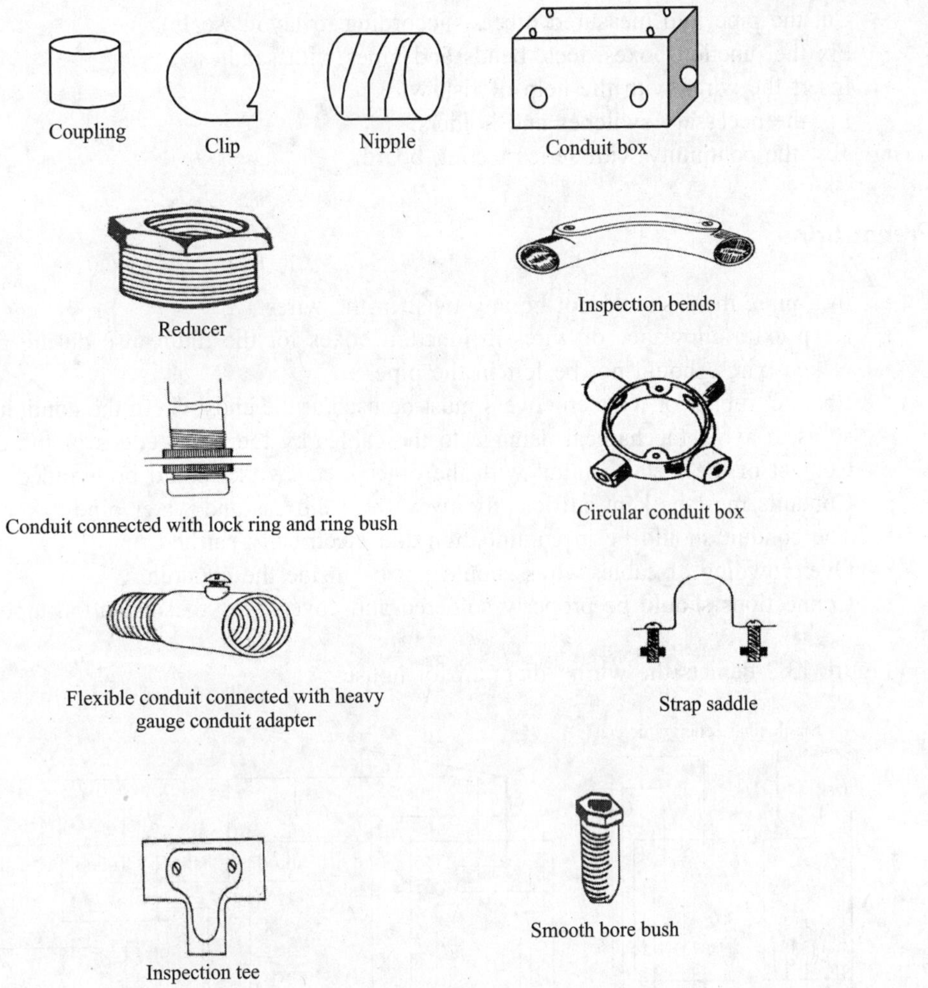

Fig. 13.1. Elements of conduit fittings.

Limitations

- It is costly.
- Replacement of the defective wire in the conduit is not easy.
- It takes longer duration to do wiring.
- Skilled workforce is required.

Procedure

Steps

- Make the layout diagram on the board with the help of chalk and scale as shown in the Figure 13.2.

- Cut the pipe into measured pieces (according to layout scale).
- Fix the junction boxes, tees, bends and pipes with saddles.
- Insert the wires with the help of fish wire.
- Fix the necessary switches and holders.
- Test the continuity with series testing board.

Precautions

- In a pipe, there should not be any twist in the wires.
- Keep extra allowance of wires in junction boxes for the future use during repair.
- Naked wires should not be left in the pipe.
- Ebonite, rubber or wooden covers must be used at the ends where the conduit terminates so as to avoid mechanical damage to the cables by the sharp edges of the conduit.
- Contact of the metal conduit with the other metal work should be avoided.
- Conduits must be kept sufficiently away from the gas and water pipes.
- The conduit should be in continuation and electrically earthed.
- Overcrowding of cables/wires should not be inside the conduits.
- Connections should be properly soldered and covered by the insulation tapes.

Figure 13.2 depicts the wiring diagram for houses.

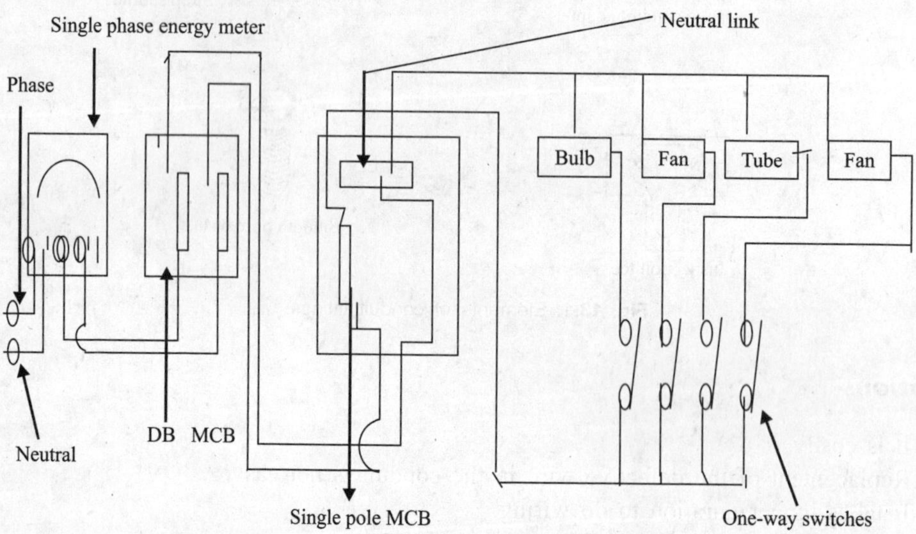

Fig. 13.2. House wiring connection diagram.

Concealed Conduit Wiring

Mechanically continuous conduits are embedded inside the walls or ceiling during the time of construction. Care should be taken not to place the wiring conduits near the water pipes and absolute protection against moisture must be ensured. Conduits are also required to be coated from outside for protection against corrosion. The preparation of the inner surface conduits should be such that the wires can be easily pulled into it without damage to the insulation or outer braid. The PVC wires or VIR wires are drawn into the conduits with the help of GI wires of 18 SWG.

The other similar wiring elements of surface conduit wiring are also required in this wiring. This type of wiring looks better as compared to other types of wiring but this is costlier than the others.

Flexible Conduit Wiring

This system of wiring is not useful for general electrical wiring. Flexible conduit wiring is used to connect the machine terminal boxes and the meters with main switchboards. Flexible pipes are also used to carry main wires while crossing the walls and turning at the end of the conduits. The PVC wires or VIR wires are drawn into the flexible pipes before connecting at the terminals.

13.4 STAIRCASE WIRING

In order to control one light point independently from two switches or more placed at different places on the staircases or in room corridors, this type of wiring is required. Wiring circuit diagram using two switches both of two-way type is shown in the Figure 13.4 whose schematic diagram is shown in the Figure 13.3. The phase wire is required to be connected to the common terminal C_1 of the first two-way switch. According to the requirement, moving the dolly of the switches, the connections can be made either to the terminal B_1 or A_1 and B_2 or A_2. Similarly, common terminal of the second two-way switch C_2 is connected with the neutral point (N) through a lamp. Two wires connect electrical wiring systems A_1 & A_2 and B_1 & B_2. One two-way switch is suitably installed near the first step of the stair and the other two-way switch is installed at the end of the stairs at first floor.

By making movement of the dolly of the two-way switches makes the circuit complete, and the lamp becomes 'ON'. Thus, either in position C_1-A_1-A_2-C_2 or C_1-B_1-B_2-C_2 of the

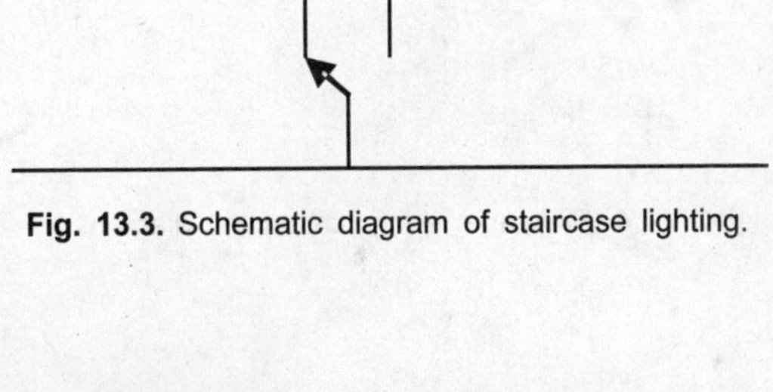

Fig. 13.3. Schematic diagram of staircase lighting.

switches, the lamp gets phase current, and hence, becomes 'ON'. On the other hand, in other alternate positions of the switches, the lamp does not get phase current, and hence, remains 'OFF'. So it is clear that the lamp can be made 'ON' or 'OFF' from any one of the two switches.

Suppose a person reaches near stair, where first switch is present and moves its dolly so that position C_1-A_1-A_2-C_2 or C_1-B_1-B_2-C_2 completes the circuit, and the lamp becomes 'ON' and hence glows.

After reaching the upper floor, where second switch is situated, the person moves dolly of the second switch so as circuit gets open, and then the lamp become 'OFF'. Now on returning, he moves dolly of the second switch so that the circuit is again complete, and lamp becomes 'ON'. After reaching downstairs, he moves the dolly of the first switch and the lamp gets 'OFF'. So, it is clear that when light is required on stairs during movement from down floor to the upper floor, the person switch 'ON' the lamp from entering the down floor and after reaching the upper floor he does not want light on stairs, he switches 'OFF' the lamp. The same can be repeated during movement in the reverse direction.

Procedure

- Connect the circuit and complete the wiring as per circuit diagram shown in Figure 13.4.
- Get checked the circuit connection properly.
- Do the continuity test before switching ON.

Precautions

- Connections should be tight and correct.
- Do not touch live conductor.
- Check all the switches and lamp before connecting it.
- Test the continuity after completing the circuit.
- Do not switch 'OFF' or switch 'ON' the lamp during testing the circuit.

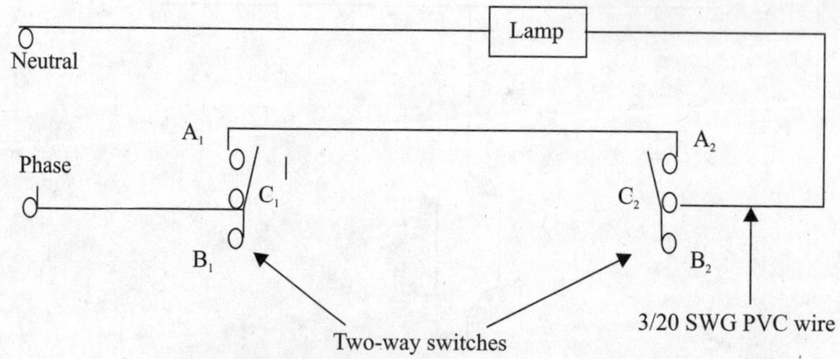

Fig. 13.4. Staircase wiring connection diagram.

13.5 PROTECTION FOR WIRING AND APPARATUSES

Fuses, relays and circuit breakers are used in the electrical circuitry to protect the apparatus, equipment and circuit from overloads or short circuit. The causes of the overloading or short-circuiting may be:

- Connecting more than the specified number of appliances to any one socket with the help of multi-plugs.
- Providing several electrical points to one sub-circuit without using distribution board.
- Accidental contact between phase with phase, phase with neutral or phase with earth wire.
- Due to aging or poor quality, sometimes insulation fails. Insulation failure causes short circuit.

13.5.1 Fuse

A fuse is a device, which protects the electrical wires, cables and other electrical equipment against overloading and occurrence of the short circuit. It contains a fuse element; generally thin wire or strip mounted on insulated base. Fuse member gets heated and destroyed while passing excessive current through it. It breaks the circuit by melting the fuse element when the current flowing in the circuit exceeds a certain predetermined value. The fuse is placed in series with the equipment or its part to be protected.

Fuse elements are made up of generally tin, lead, silver, zinc, aluminium and copper, etc. For small current applications, an alloy of lead and tin in the ratio of 37% and 63% respectively are used. For currents above 15 amperes, this alloy is not used because its size increases and after melting metal releases excessive volume of it. For the 15 A, and above, material for fuse element, is silver but it is costlier. The advantage of silver using as fuse material is that it is not subjected to the oxidation. Its oxide is unstable. The melting point of silver is 960°C and tin is 230°C.

Types

1. Re-wirable Fuse

This type of fuse is re-wirable, i.e., the blown out fuse element can be replaced by a new element. The fuse element can be either open or semi-enclosed. Open type is rarely used. Semi-enclosed type is generally used for low voltage applications in houses and installations for up to 440 V. These are the cheapest and simplest forms of the protection.

Re-wirable fuse is a set of two units. One is known as Base-unit and the second is known as Fuse-carrier. Fuse element can be fixed, removed and re-fixed in the Fuse-carrier. Fuse-carrier is also known as removable unit. The Base-unit is permanently fixed unit with switchboard.

Both the units are fixed on porcelain chassis. Incoming wire is connected with the metal clip through one side of porcelain base. And outgoing wire is connected with the other terminal of the porcelain base. These two terminals situated inside the base of the fuse-carrier are interconnected by fuse wire.

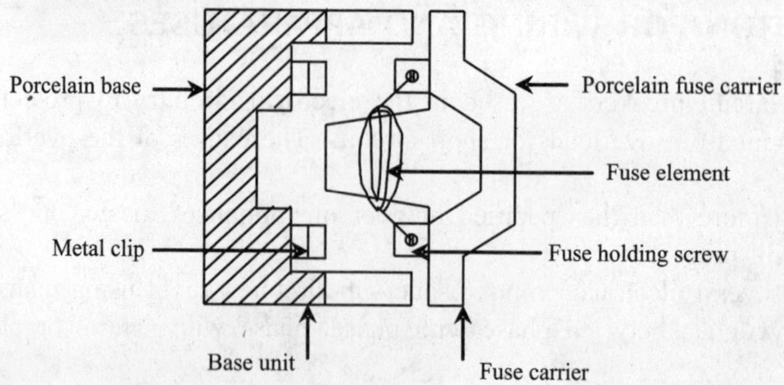

Fig. 13.5. Constructional diagram of re-wirable fuse.

This fuse wire is the weakest part of the circuit and get melts in the event of the overloading or short circuit. After that a new fuse wire gets connected between the terminals. A constructional diagram of re-wirable fuse is shown in the Figure 13.5. Table 13.1 below gives the values of rated current and approximate fusing current for tinned copper fuse wires of different SWG and diameter.

Table 13.1

SWG	Diameter (cm)	Rated current (Ampere)	Approximate fusing current (A)
20	0.09	34	70
25	0.05	15	30
30	0.031	8.5	13
35	0.021	5.0	8
40	0.012	1.5	3

Fusing factor = Minimum fusing current/fuse rating (current)

2. *Totally Enclosed or Cartridge Fuse*

The fuse element of this is enclosed in a totally enclosed insulated container and is provided with metal contacts on both sides. These are of two types.

- D-type cartridge fuse
- Link type cartridge fuse or high rupturing capacity (HRC) cartridge fuse.

D-type cartridge fuse unit is totally enclosed in an insulating container of tube shape. The container is filled up with power of arc quenching property, which is sealed at its ends with metallic cap. Fuse wire is fixed inside the container and can not be removed or re-fixed. Whole unit is required to be replaced when fuse blows. These are used for about rating up to 800 A and voltage up to 660 volts. A constructional diagram of D-type cartridge fuse is shown in the Figure 13.6.

HRC (High Rupturing Capacity) fuse unit consists of a ceramic body usually of steatite, silver make fuse element, clean silica quartz, asbestos washers, porcelain plugs, brass end caps and copper tags. The space around the fuse element within the main body is filled with power,

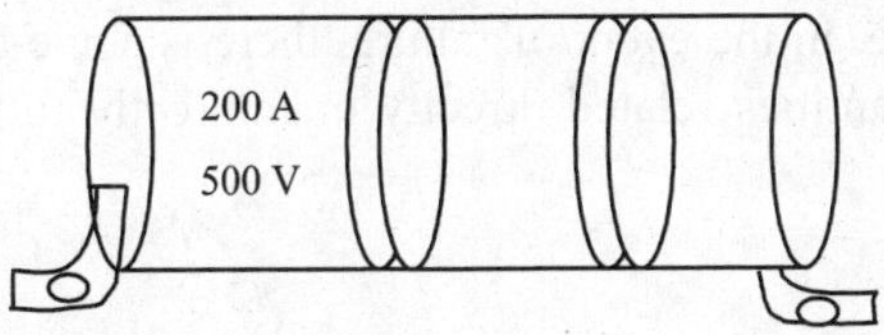

Fig. 13.6. Constructional diagram of D-type cartridge fuse.

which has to promptly extinguish the arc produced during melting of the fuse element. The brass end caps and copper tags are generally electro-tinned. Special forged screw tightens the metal end caps with the ceramic body. End caps are welded with the contacts.

HRC fuses are uneconomical because whole unit is required to be replaced after the fuse blows. Another problem with HRC fuse is to identify the blown fuse among the number of similar fuses applied in the same board or feeder. However, by checking one by one the blown one can be identified. But HRC fuses are most reliable to withstand heavy stresses in clearing the fault. There is less temperature rise at even full load. Since HRC fuses are enclosed type, therefore there is no chance to damage nearby area. A constructional diagram of HRC cartridge fuse is shown in the Figure 13.7.

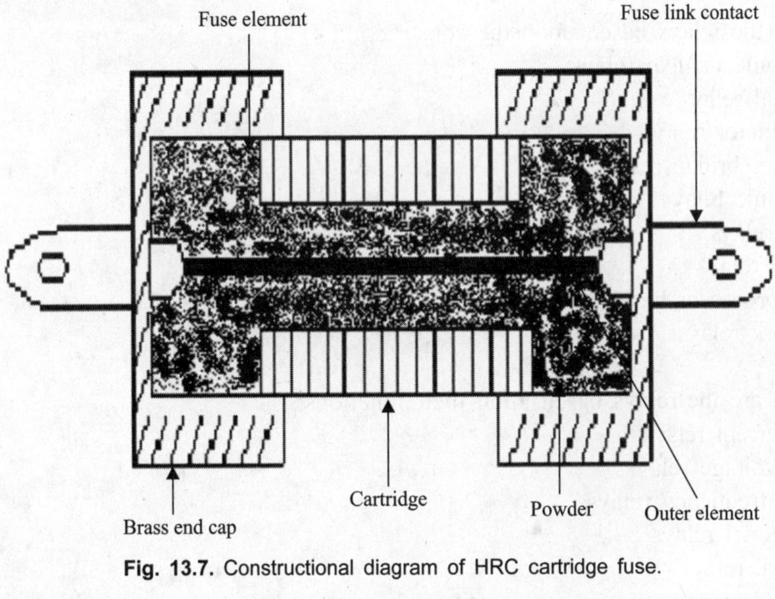

Fig. 13.7. Constructional diagram of HRC cartridge fuse.

13.5.2 Relays

A relay is an automatic device, which senses an abnormal condition in an electric circuit and closes its contacts. These contacts in turn close the circuit breaker trip coil circuit, thereby it opens the circuit breaker and the faulty part of the electric circuit is disconnected from the rest of the healthy circuit. Question arises, how protective relays operate?

For any location and type, in the event of failure, there is some distinctive difference in one or more of the following quantities related directly or indirectly to electricity:

- Magnitude
- Frequency
- Duration
- Phase angle
- Rate of change
- Order of change
- Wave shape

So, the relays may be of any type available simply recognizing a particular difference in the above quantities and operates in response to the level of difference. For example, as the current in primary of current transformer increases, secondary current also increases as a result; relay operates if the current exceeds the predetermined value. The coil of definite relay is energized by the fault/overload current. After certain time delay, the contacts of definite time relays operate and the trip coil gets energized and the breaker opens.

Types

Following are the relays based upon the construction:

- Electromechanical relays
- Thermal relays
- Transductor relays
- Rectifier bridge relays
- Electronic relays
- Static relays
- Digital relays
- Microprocessor based relays
- Network relays

Following are the relays based upon their functions:

- Overcurrent relays
- Undervoltage relays
- Under frequency relays
- Directional relays
- Thermal relays
- Phase sequence relays, i.e., Negative sequence and Zero-sequence relays
- Differential relays and Percentage biased relays
- Distance relays, i.e., Impedance relays, Reactance relays, Mho relays, Offset and Restricted relays
- Pilot relays, i.e., Wire pilot relays, Carrier channel pilot relays, Fibre optic based relays and Microwave pilot relays
- Adaptive relays

- Neural-Network based relays
- Wavelet based relays
- Wide area based relays
- Hybrid based relays

These relays are used for apparatus protection such as busbars, transformers, generators, motors and transmission lines protection, etc. Detail of the different types of relays and relaying schemes do not come under the scope of this book. However, a brief introduction is being presented hereunder.

Protective gear or equipment

It includes protective relays, potential transformers, current transformers and ancillary equipment to be used in a protective system.

Protective system

Protective system consists of protective gear equipment to secure isolation of the fault element under predetermined conditions, usually abnormal or to give an alarm signal or both.

Protective scheme

A scheme, which may consist of several protective systems and is designed to protect one or more elements of a power system, is known as protective scheme.

Protective zone

A power system network divided into a number of zones from the protection point of view has separate protective scheme for its protection. A protective relay has its own zone of protection.

Operating force/torque

A force/torque generated by the relay coils, which causes to close the contacts of the relay.

Restraining force and torque

A force and torque which opposes the operating force and torque.

Pick-up

The minimum value of the actuating quantity (current, voltage, frequency, etc.) above, which the relay initiates to operate.

Reset time

It is the time, which elapses from the moment the actuating quantity, falls below its reset value to the instant when the relay comes back to its normal (initial) position.

Operating time

It is the time, which elapses from the instant at which the actuating quantity exceeds the relay pick-up value to the instant at which the relay closes its contacts.

Setting

The value of the actuating quantity at which the relay is set to operate.

Back-up relay

A back-up relay operates after a slight delay, if main relay fails to operate.

Back-up protection

The back-up protection acts as second line of defence and is designed to clear the fault if the primary protection fails.

Primary protection

Primary protection acts as first line of defence and is designed to clear the fault without any delay.

Flag indicator

Flag is a device, which gives visual indication in the control room, whether a relay has operated, or not.

Electromagnetic relays

These relays operate on the electromagnetic principle. Moving iron, moving coil, attracted armature, induction disc and induction cup type relays are of the category of electromagnetic relays.

Static relays

These are solid-state relays and employ semiconductor diodes, transistors, thyristors, logic gates ICs, etc. In these relays, there are no moving parts.

Microprocessor-based relay

To perform the measurement of the electrical quantities or make comparisons, performing computations, and sending tripping signals, microprocessors can be used. So, the application of the microprocessors in realizing all sorts of relaying characteristics is recognized as microprocessor based relaying schemes.

Over current relay

A relay, which operates when the actuating current exceeds a certain, preset value, is known as over current relay.

Under voltage relay

A relay, which operates when the system voltage falls below a certain, preset value.

Directional or reverse power relay

A directional relay senses the direction of the flow of the power and detects whether the point of fault lies in the forward or reverse direction with respect to the relay location.

Time-lag relay

A time-lag relay operates after a certain preset time lag. The time lag may be due to the presence of a time-delay component or its inherent design feature. These relays make eligible the protection schemes for time discrimination and are useful for control and alarm circuits.

Instantaneous relay

These relays work with no intentional time delay in its operation. It operates in 0.1 second or sometimes less than this.

Inverse time relay

For this relay operating time is inversely proportional to the magnitude of the operating current.

Definite time relay

For this relay operating time is independent of the magnitude of the actuating wiring current.

Inverse definite minimum time (IDMT) relay

These types of relays give an inverse time characteristic at lower values of the operating current and definite time characteristic at higher values of the operating current.

Induction relay

Induction relays like induction disc relays, induction cup relays, etc. operate on the principle of induction.

Moving coil relay

Moving coil relay also called a permanent magnet DC moving coil relay has a permanent magnet and moving coil. The moving coil carries the actuating current.

Moving iron relay

This is a DC polarized, moving iron type relay. Its construction comprises electromagnet, permanent magnet and a moving armature in its construction.

Comparator

It is another name of the relay.

Phase comparator

These relay compare only the phase angle between input quantities.

Amplitude comparator

These relay compare only the magnitude between input quantities.

Instantaneous comparator

These relay compare only the instantaneous values of input signals.

Dual input comparator

Dual input comparators have two input signals.

Thermal relay

Electrothermal effect of the actuating current is used for the operation of thermal relays.

Frequency sensitive relay

These relays operate at a predetermined value of the system frequency. It may be an under-frequency relay or an over-frequency relay. An under-frequency relay operates if the system frequency falls below a certain value. An over-frequency relay operates if the system frequency exceeds below a certain preset value of the frequency.

Differential relay

Differential relays operate in response to the difference of two actuating quantities.

Earth fault relay

Relays used for the protection of an element of a power system against earth faults are known as an earth fault relays.

Phase fault relay

Relays used for the protection of an element of a power system against phase faults are known as a phase fault relay.

Starting relay or fault detector

This relay detects abnormal conditions and initiates the operation of other elements of the protective scheme.

Sensitivity

Sensitivity is the measure of the degree of the effectiveness of a protective relay. A relay should be sensitive enough to operate when the magnitude of the actuating quantity exceeds its pick-up value.

Selectivity or discrimination

The ability of a relay to discriminate between faulty condition and normal conditions or between a fault within the protected section and outside the protected section is described as selectivity or discrimination.

Stability

Stability is the ability of the protective system to remain inoperative under all normal load conditions, and also in case of faults outside its zone. The relay should remain stable when heavy current due to an external fault flows through it.

Blocking

Blocking is the prevention of tripping of the relay, which may be due to the operation of an additional relay or due to its own characteristic.

Blinders

A blinder makes the relay immune for certain faults.

Burden

The power consumed by the relay circuitry at the rated current is known as its burden.

Unit system of operation

This is the system, which is able to detect and respond to fault occurring only within its own zone of protection having absolute discrimination. Its zone of protection is well defined. Unit system of protection is differential protection of alternators, transformers or busbars, frame leakage protection, pilot wire and carrier current protection.

Non-unit system of protection

Non-unit system of protection schemes does not have absolute discrimination or selectivity. In this system, all relays may respond to a given fault, for example, non-unit system of protection are distance protection and time graded, current graded or both time and current graded protection.

Restricted earth fault protection

In the restricted earth fault protection, zone of protection is restricted only to the winding of the alternator or transformer. The scheme responds to the faults occurring within its zone of protection.

Transactor

This is an air-cored coil, which converts a current signal to voltage.

Transducer

Transducers produce the proper relay signals from the power system.

Threshold operation

This operation is linked to the relay about to operate means the fault has occurred at the balance point such as at the reach of the relay.

13.5.3 Contactors

Contactor is a mechanical switching device capable of making, carrying, and breaking electric current under normal circuit conditions including operating overload conditions.

Types

1. Push Button type Contactor

This provides control of an equipment make or break of an electric circuit by pressing a button. As long as the push button is held pressed, it remains pressed. Spring inside the push button brings the contact back its initial position as and when the pressure on the push button is released.

Two sets of contacts, one 'NO' (Normally Open) and second 'NC' (Normally Closed) are generally used. Both sets are used in the circuit when one is kept 'NO' state the other goes to 'NC' state.

For alternating current, push buttons are rated for 2 to 3 A, 500 V. The contact tips are made up of silver that can withstand several switching operations. These are used for starting, stopping, reversing, getting multi-speed of motors and other appliances. Figure 13.8 shows the combination of two sets of push buttons.

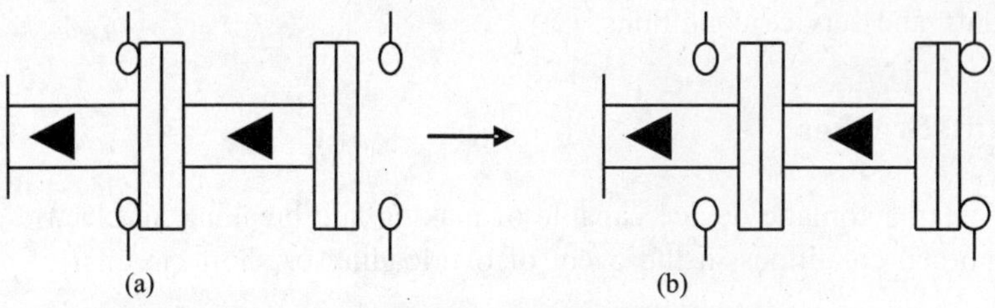

Fig. 13.8. (a) Original position of one NO and other NC push button.
(b) Push button in pressed condition.

2. Electromagnetic type Contactor

This is made to operate by electromagnetic force rather than by hand pressure for making and breaking an electric circuit and its operating speed does not depend on the operator. In this contactor, a coil is connected across a voltage source as per rating of the contactor. For making and breaker of three-phase supply, three main contacts one for each phase are used. Supply source energizes the coil. Main working component, the contactor coil is available for different voltage ratings of AC, DC, or combination of both. Ratings are selected as per requirement. Figure 13.9 shows contactor with a core.

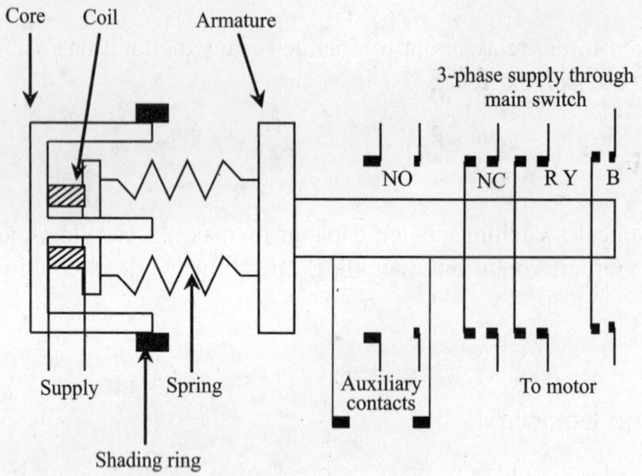

Fig. 13.9. Electromagnetic contactor.

Ratings

- Rated voltage
- Rated operational voltage
- Rated insulation voltage
- Rated current
- Rated thermal current
- Rated operational current
- Rated duty and service conditions.

13.5.4 Circuit Breaker

Circuit breaker is an automatic device capable of making and breaking an electric circuit under normal and abnormal conditions in the event of overloading or short circuits.

Main Parts

- Poles with interrupter, porcelain support, arc quenching medium, etc.
- Operating mechanism
- Support structure and
- Control cabinet

Main Types

- Air circuit breaker
- Oil circuit breaker
- Minimum oil circuit breaker
- Air blast circuit breaker
- Sulphur hexa fluoride (SF_6) circuit breaker
- Vacuum circuit breaker

Rating of AC Circuit Breaker

Each AC circuit breaker should have the following specified values:

1.	Rated voltage	240/415 V, AC, 220 V, DC
2.	Rated normal current	0.5 A to 63 A
3.	Protection class	P 30 as per IS 2147
4.	Rated frequency	50 to 60 Hz
5.	No. of poles	Three pole
6.	Rated breaking current	9 kA at 240/415 V
7.	Permissible ambient temperature	5 + 55°C
8.	Connecting conductor size	as per requirement, say up to 35 mm^2
9.	Electrical endurance	More than 50,000 operating cycles
10.	Mechanical endurance	More than one lac operating cycles

13.5.5 MCB (Miniature Circuit Breaker)

MCB provides overload and short-circuit protection and works as a circuit breaker. It also serves the purpose of fuse units during normal operating conditions. But it does not require be servicing or rewiring as in the case of fusing units. MCB gives better service of protection because this operates just on 5% to 15% excessive overloads. Fuse wire of the fuse units melt on 50% to 100% load or on even more. MCB is much faster than fuses. So, MCB is more reliable and easy to handle.

Miniature circuit breakers are basically electromechanical device. They operate (disconnect) the circuit if circuit current reaches the predetermined value. MCBs are the better choice to be fitted on the consumer's distribution board. MCBs operate accurately and efficiently during overloading and short circuit conditions both. Fuses operate in mostly short circuit conditions and in excessive overloading conditions. Looking on the operating knob that moves automatically to the 'OFF' position can identify the tripped MCB. A single pole MCB unit is shown in the Figure 13.10 and cross-sectional view of a MCB is shown in the Figure 13.11. Arc produced due to the current interruption is broken down into small pockets of arcs in the arc chute stack in order to reduce its impact.

The arc produced at the time of separation of contacts actually moves very fast under the influence of magnetic field and is directed towards arc-chute stack. Increase in the number of arc-chute plates reduces the plate voltage and quickly extinguishes the arc. A non-melting material

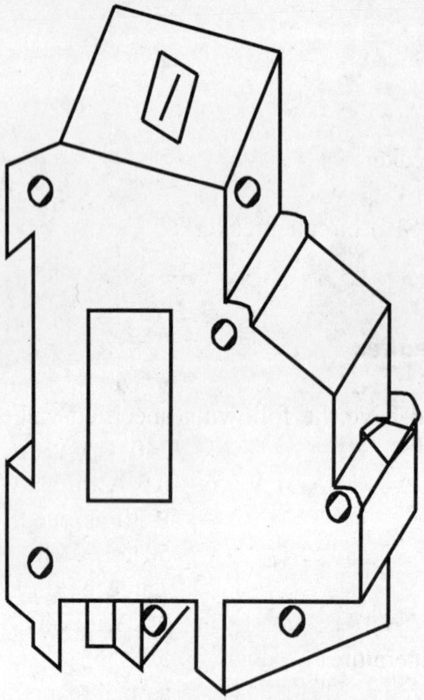

Fig. 13.10. A single pole MCB unit.

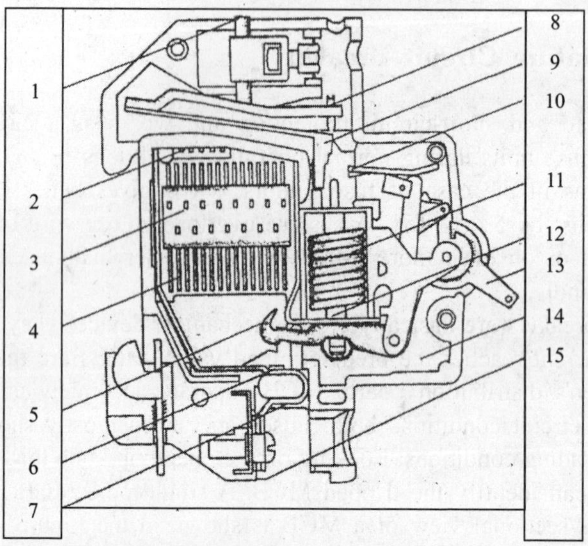

1. Outgoing terminal; 2. Angular vents; 3. Arc chutes holder; 4. Arc chutes; 5. Moving contact; 6. Silver graphite contact tip; 7. Fixed contact; 8. Bimetal carrier; 9. Bimetal strip; 10. Trip lever; 11. Solenoid; 12. Plunger; 13. Operating knob; 14. Toggle; 15. Incoming terminal

Fig. 13.11. Cross-sectional view of the MCB.

(silver graphite) is welded with contact, which makes the contact tips. Property of the housing material of the MCB should be flame retardant, high melting temperature, water non-absorbent, high dielectric strength and low coefficient of linear thermal expansion. So, high strength plastic is used as housing material.

Angular vents are provided to prevent any ingress dust and to make the electrical contact dust free.

The moving parts are encased in a sheet metal casing. Plastic casing cannot withstand several operations of the MCB.

Uses

MCBs have become popular and being used extensively. These are replacing the conventional re-wirable fuses. These are required in homes, shops distribution boards, offices, power loads, etc.

Types

These are categorized into two series.

(i) *L-Series MCB*: These are suitable for low and steady currents like geysers, bulbs, electric iron, oven, heaters, meters, cables, wires, etc.

(ii) *G-Series MCB:* These are suitable for high inrush current peaks. These are applied for protecting the air-conditioners, motors, halogen lamps, machine tools, etc. Other type of circuit breaker is RCCB (Residual Current Circuit Breaker). This is also known as ELCB (Earth Leakage Circuit Breaker).

RCCB or ELCB

Two wires one for phase and other for neutral are used for carrying current in electrical appliances. Leakage current flows if metallic parts of the appliances come in direct contact with naked part of the phase wire that may be due to the insulation failure or accident. In the case if metallic part of the appliance comes in touch with an earthed object leakage current passes to the ground. Absence of proper earthing and shielding of this path, may cause fire hazards and undesirable overloading. The earth leakage is the main cause of electric shock. So, this type of leakage and its after effects may be avoided by using RCCB (Residual Current Circuit Breaker) popularly known as ELCB (Earth Leakage Circuit Breaker). Its construction is more or less same as described earlier and is useful in protecting the following:

- Protection against fire
- Protection against indirect contact
- Protection against electrocution to human and domestic cattle.
- It also protects energy loss otherwise undetected leakage current causes continuous energy loss that may be in small amount but longer duration energy loss becomes considerable enough.

EXERCISES

1. What are the wiring systems? Define them.
2. List out the disadvantages of the batten wiring.
3. List out the accessories required for conduit wiring.
4. What precautions should be taken for conduit wiring?
5. Compare the different types of wirings.
6. Distinguish between surface and concealed conduit wiring.
7. Explain cleat wiring and lead sheathed wiring systems.
8. Write two uses of the staircase wiring.
9. Which type of switch is used in staircase wiring and why?
10. What is the difference between schematic diagram and wiring diagram?
11. How do you perform continuity test?
12. Why is conduit pipe wiring preferred?
13. What do you mean by saddle, bend and tee?
14. Draw wiring diagram of a circuit consisting of a light point, plug point and fan point?
15. What is the role of a fuse in an electrical circuit?
16. Which material is used as fuse element in a re-wirable fuse?
17. Which material is used as fuse element in a HRC fuse and why?
18. List out the properties of filling powder of cartridge fuse.
19. Make comparison between:
    (i) Re-wirable fuse and D-type cartridge fuse.
    (ii) D-type cartridge fuse and HRC fuse
20. Write, in brief, the importance of relays in electrical applications.
21. Describe, in brief, different types of relays.
22. What do you mean by threshold operation of the relays?
23. Describe difference between back-up protection and primary protection.
24. Make a list of the relays based upon their functions.
25. Make a list of the relays based upon their constructions.
26. Why contactors are required?
27. What are the types of contactor?
28. Where MCBs are preferred?
29. Draw a cross-sectional view of MCB and explain its working.
30. How miniature circuit breakers are rated?
31. What is ELCB? List out its uses.

14

Indian Electricity Rules

14.1 INTRODUCTION

Before, actually, studying Indian Electricity Rules and other precautions, we should discuss, why these rules and regulations have been framed?

These rules and regulations have been framed by the Institution of Electrical Engineers (India) to:

- Safeguard consumers of electrical energy from shock
- Minimize fire risk and
- Ensure, as far as possible, satisfactory operation of equipment and apparatus used.

These rules are known as the Indian Electricity Rules, 1956.

14.2 OBJECTIVES

- To quote the relevant Indian Electricity Rules for a given electrical installation.
- Interpret the relevant Indian Electricity Rules (1956) in respect of lighting and power installation.

Some relevant rules are given hereunder.

CHAPTER-I

14.3 PRELIMINARY

Rule 1. Short Title and Commencement

1. These rules may be called the Indian Electricity Rules, 1956.
2. They shall come into force at once.

Rule 2. Definitions

(a) "the Act" means the Indian Electricity Act, 1910;

(b) "accessible" means within physical reach without the use of any appliance or special effort;

(c) *"ampere"* means a unit of electric current and is the unvarying electric current which when passed through a solution of nitrate of silver in water, in accordance with the specifications set out in Annexure-I, deposits silver at the rate of 0.001118 of a gramme per second.

(d) "annexure" means an annexure to these rules;

(e) *"apparatus"* means electric apparatus and includes, all machines, fittings, accessories and appliances in which conductors are used;

(ee) "Authorized person" means a person authorized under rule 3;

(f) *"bare"* means not covered with insulating material;

(g) *"cable"* means a length of insulated single conductor or of two or more such conductors, each provided with its own insulation, which are laid up together. Such insulated conductor or conductors may or may not be provided with an overall mechanical protective covering;

(gg) "flexible cable" means a cable consisting of one or more cores each formed of a group of wires, the diameter and the physical properties of the wires and the insulating material being such to afford flexibility;

(h) *"circuit"* means an arrangement of conductor or conductors for the purpose of conveying energy and forming a system or a branch of a system;

(i) *"circuit breaker"* means a device, capable of making and breaking the circuit under all conditions, and unless otherwise specified, so designed as to break the current automatically under abnormal conditions;

(j) *"concentric cable"* means a composite cable comprising an inner conductor, which is insulated and one or more outer conductors, which are insulated from one another and are disposed over the insulation of, and more or less around the inner conductor;

(k) *"conductor"* means any wire, cable, bar, tube, rail or plate used for conducting energy and so arranged as to be electrically connected to a system;

(l) *"conduit"* means rigid or flexible metallic tubing or mechanically strong and fire resisting nonmetallic tubing in which a cable or cables may be drawn for the purpose of affording it or them mechanical protection;

(m) *"covered with insulating material"* means adequately covered with insulating material of such quality and thickness as to prevent danger;

(n) *"cut-out"* means any appliance for automatically interrupting the transmission of energy through any conductor when the current rises above a predetermined amount, and shall also include fusible cut-out;

(o) *"danger"* means danger to health or danger to life or any part of body from shock, burn or other injury to persons, or property, or from fire or explosion, attendant upon the generation, transmission, transformation, conversion, distribution or use of energy;

(p) *"dead"* means at or about earth potential and disconnected from any live system: PROVIDED that apparatus separated from a live conductor by a spark gap shall not be deemed to be 'dead'.

Note: The term "dead" is used only with reference to current carrying parts when these parts are not live.

(q) *"earthed" or "connected with earth"* means connected with the general mass of earth in such manner as to ensure at all times an immediate discharge of energy without danger;

(r) *"earthing system"* means an electrical system in which all conductors are earthed;

(s) *"enclosed substation"* means any premises or enclosure or part thereof, being large enough to admit the entrance of a person after the apparatus therein is in position, containing apparatus for transforming or converting energy to or from a voltage at or above medium voltage (other than transforming or converting solely for the operation of switchgear or instruments) with or without any other apparatus for switching, controlling or otherwise regulating the energy, and includes the apparatus therein;

(t) *"enclosed switch-station"* means any premises or enclosure or part thereof, being large enough to admit the entrance of a person after the apparatus therein is in position, containing apparatus for switching, controlling or otherwise regulating energy at or above medium voltage but not for transforming or converting energy (other than for transforming or converting solely for the operation of switchgear or instruments), and includes the apparatus therein;

(tt) *"Flameproof enclosure"* means an enclosure for electrical machinery or apparatus that will withstand, when the covers or other access doors are properly secured, an internal explosion of the inflammable gas or vapour which may enter or originate inside the enclosure, without suffering damage and without communicating the internal flammation (or explosion) to the external inflammable gas or vapour in which it is designed to be used, through any joints or other structural openings in the closure;

(u) *"guarded"* means covered, shielded, fenced or otherwise protected by means of suitable casing, barrier, rails or metal screens to remove the possibility of dangerous contact or approach by persons or objects to a point of danger;

(uu) *"hand-held portable apparatus"* means an apparatus which is so designed as to be capable of being held in the hands and moved while connected to a supply of electricity;

(v) *"Inspector"* means an electrical inspector appointed under Section 36;

(w) *"Inspector of Mines"* means an Inspector appointed under the Mines Act, 1952 (35 of 1952);

(x) *"installation"* means any composite electrical unit used for the purpose of generating, transforming, transmitting, converting, distributing or utilizing energy.

(y) *"intrinsically safe"* as applied to apparatus or associated circuits shall denote that any sparking that may occur in normal working is incapable of causing explosion of inflammable gas or vapour;

(yy) increased safety type 'e' means a method of protection by which additional measures are applied so as to give increased security against the possibility of excessive temperatures and of occurrence of arcs or sparks in normal service;

(z) *"lightning arrestor"* means a device, which has the property of diverting to earth, any electrical surge of excessively high amplitude applied to its terminals and is capable of interrupting follow current if present and restoring itself thereafter to its original operating conditions;

(zz) *"linked switch"* means a switch with all the poles mechanically linked so as to operate simultaneously;

(aa) *"live"* means electrically charged;

(ab) *"metallic covering"* means mechanically strong metal covering surrounding one or more conductors;

(aba) *"meter"* means a set of integrating instruments used to measure the amount of electrical energy supplied or the quantity of electrical energy contained in the supply, in a given time, which include whole current meter and metering equipment such as current transformer, capacitor voltage transformer or potential or voltage transformer with necessary wiring and accessories;

(ac) *"neutral conductor"* means that conductor of a multi-wire system, the voltage of which is normally intermediate between the voltages and other conductors of the system and shall also include return wire of the single-phase system;

(ad) *"occupier"* means the owner or person in occupation of the premises where energy is being used or proposed to be used;

(add) "officer appointed to assist the Inspector" means an officer appointed under rule 4A;

(ae) *"ohm"* means a unit of electrical resistance and is the resistance offered to an unvarying electrical current by a column of mercury at the temperature of melting ice 14.4521 grams in mass of a uniform cross-sectional area and of length of 106.3 centimetres; The aforesaid unit is represented by the resistance between the terminals of the instrument marked "Government of India Ohm Standard Verified" to the passage of an electric current when the coil of wire, forming part of the aforesaid instrument and connected to the aforesaid terminals is in all parts at a temperature of 30°C;

(af) *"open sparking"* means sparking which owing to the lack of adequate provisions for preventing the ignition of inflammable gas external to the apparatus would ignite such inflammable gas;

(ag) *"overhead line"* means any electric supply line, which is placed above ground and in the open air but excluding live rails of a traction system;

(ah) *"owner"*, *"agent"* and *"manager"* of a mine have the same meanings as are assigned to them in the Mines Act, 1952 (35 of 1952);

(ai) *"portable apparatus"* means an apparatus, which is so designed as to be capable of being moved while in operation;

(aj) *"portable hand lamp"* means a portable light fitting provided with suitable handle, guard and flexible cord connected to a plug;

(ak) *"section"* means a section of the Act;

(al) *"span"* means the horizontal distance between two adjacent supporting points of an overhead conductor;

(am) *"street box"* means a totally enclosed structure, either above or below ground containing apparatus for transforming, switching, controlling or otherwise regulating energy;

(an) *"supplier"* means a licensee, a non-licensee or any other supplier of energy including the government;

(ao) *"switch"* means a manually operated device for opening and closing or for changing the connections of a circuit;

(aoa) "switchboard" means an assembly including the switchgear for the control of electrical circuits, electric connections and the supporting frame;

(ap) *"switchgear"* shall denote switches, circuit breakers, cutouts and other apparatus used for operation, regulation and control of circuits;

(aq) *"system"* means an electrical system in which all the conductors and apparatus are electrically connected to a common source of electric supply;

(ar) *"transportable apparatus"* means apparatus which is operated in a fixed position but which is so designed as to be capable of being moved readily from one place to another;

(as) *"volt"* means a unit of electromotive force and is the electric pressure, which when steadily applied to a conductor, the resistance of which is one ohm will produce a current of one ampere;

(at) *"voltage"* means the difference of electrical potential measured in volts, between any two conductors or between any part of either conductor and the earth as measured by a suitable voltmeter and is said to be:

"low", where the voltage does not exceed 250 volts under normal conditions, subject, however, to the percentage variation allowed by these rules;

"medium", where the voltage does not exceed 650 volts under normal conditions subject, however, to the percentage variation allowed by these rules;

"high", where the voltage does not exceed 33,000 volts under normal conditions subject, however, to the percentage variation allowed by these rules;

"*extra high*", where the voltage exceeds 33,000 volts under normal conditions subject, however, to the percentage variation allowed by these rules.

14.4 AUTHORIZATION

Rule 3. Authorization

1. A supplier or a consumer, or the owner, agent or manager of a mine, or the agent of any company operating in an oil field or the owner of a drilled well in an oil field or a contractor for the time being under contract with a supplier or a consumer to carry out duties incidental to the generation, transformation, transmission, conversion, distribution or use of energy may authorize

   Sub-rule (2) of rule 36, clause (a) of sub-rule (1) of rule 51, clause (a) of sub-rule (1) [and [clauses (h) and (i) of sub-rule (2) of rule 64], sub-rule (2) of rule 110, sub-rule (1) and (4) of rule 121, sub-rule (4) of rule 123, rule 124 and sub-rule (8) of rule 125.

2. No person shall be authorized under sub-rule (1) unless he is competent to perform the duties assigned to him and possesses either an appropriate certificate of competency or permit to work.

2A. (a) No person shall be authorized to operate or undertake maintenance of any part or whole of generating station of capacity 100 MW and above together with the associated substation unless he is adequately qualified and has successfully undergone the type of

training specified in Annexure XIV. Provided that the provisions contained in this sub-rule shall have effect in respect of the person already authorized to operate or undertake maintenance of any part or whole of generating station as aforesaid from the date to be specified by the appropriate government, but such a date shall not be later than a period of [6 years, 2 months] from the date this rule comes into force.

(b) The appropriate government may, on the recommendations of the owner of such generating station, relax the conditions stipulated in clause (a) of this sub-rule for any engineer and such other persons who have already sufficient experience in the operation and maintenance of a generating station.

(c) The owner of a generating station, in consultation with Central Electricity Authority, may alter the duration and manner of training in respect of those persons who have been already engaged in the operation and maintenance of a generating station or a substation.

2B. The provisions contained in rule 3 (2A) will also be applicable in respect of other substations of 132 kV and above from a date to be specified by the appropriate government but such a date shall not be later than 3 years from which this rule comes into force.

3. No person shall be deemed to be authorized under sub-rule (1) unless his name has been entered in a list maintained at the office or premises of the person authorizing him, and giving the purpose for which such person is authorized and the entry has been attested by the authorized person and the person authorizing him.

4. Every list maintained under sub-rule (3) shall be produced before an Inspector or any officer of a specified rank and class appointed to assist the Inspector when required.

5. An Inspector may cancel or amend, in such manner, as he considers necessary, any authorization made under sub-rule (1).

6. In every registered factory where more than 250 kW of electrical load is connected, there shall be a person authorized by the management of the factory for ensuring the observance of the safety provisions laid under the Act and the rules made there under, who shall periodically inspect such installation, get them tested and keep a record thereof and such records shall be made available to the Inspector or any officer of a specified rank and class appointed to assist the Inspector when required.

7. All suppliers of electricity including generating companies, transmission companies and distribution companies shall appoint a safety officer for proper observance of safety measures in their organization in construction operation and maintenance of power station, substations, and transmission and distribution lines.

CHAPTER-II

14.5 INSPECTORS

Rule 4. Qualifications of Inspectors

No person shall be appointed to be an Inspector unless:

(i) he possesses a degree in electrical engineering or its equivalent from a recognized university or institution; and

(ii) he has been regularly engaged for a period of at least eight years in the practice of electrical engineering, of which not less than two years have been spent in an electrical or mechanical engineering workshop or in generation or transmission or distribution of electricity, or in the administration of the Act and these rules, in a position of responsibility:

Provided that the appropriate government may, for a period of seven years commencing on the date the Indian Electricity (Amendment 4) Rules, 1986 are brought into force, also appoint a person possessing recognized diploma in electrical engineering and having prescribed experience to be an Inspector in respect of low and medium voltage installations only.

The appropriate government may, by notification in their official gazette, appoint as many officers as it thinks fit to assist an Inspector.

1. The appropriate government may, by a separate notification in the official gazette, authorize certain officers appointed under sub-rule (1) for the purposes of sub-rule (4) and sub-rule (6) of rule 3, sub-rule (3) and sub-rule (4) of the rule 5, rule 9. rule 10, clause (c) of sub-rule (1) of rule 32, clause (b) of sub-rule (1) and clause (c) of sub-rule (2) of rule 46, sub-rule (1) of rule 49, sub-rule (3) of rule 51, sub–rule (3) of rule 59, sub-rule (6) of rule 61 and clause (b) of sub-rule (5) of rule 67.

Rule 4B. Qualifications of Officers Appointed to Assist the Inspectors

1. No person shall be appointed as an officer to assist an Inspector, unless:
   (a) (i) He possesses a degree in electrical engineering or its equivalent from a recognized university or institution; and
   (ii) he has been regularly engaged for a period of at least three years in the practice of electrical engineering, of which not less than one year has been spent in an electrical or mechanical engineering workshop or in generation or transmission or distribution of electricity, or in the administration of the Act and these rules, in a position of responsibility:

   Provided that the appropriate government may relax the requirements of experience in case of officers otherwise well qualified.

   (b) (i) he possesses a diploma in electrical engineering or its equivalent from a recognized institution; and
   (ii) he has been regularly engaged for a period of at least six years in the practice of electrical engineering, of which not less than two years have been spent in an electrical or mechanical engineering workshop or in generation or transmission or distribution of electricity or in the administration of the Act and these rules, in a position of responsibility.
2. The persons appointed under rule 4A shall undergo such training as the appropriate government may consider it necessary for the purpose and such training shall be completed to the satisfaction of the appropriate government.
3. The appropriate government may relax requirements of practical experience in case of officers otherwise found suitable but a notification under sub-rule (2) of rule 4A in case

of such officers shall be issued only after they acquired the experience as required by sub-rule (1).

Rule 5. Entry and Inspection

1. Any Inspector or any officer appointed to assist an Inspector may enter, inspect and examine any place, carriage or vessel in which he has reason to believe that there is any appliance or apparatus used in the generation, transmission, transformation, conversion, distribution or use of energy and may carry out tests therein.
2. Every supplier, consumer, owner and occupier shall afford at all reasonable times facilities to any such Inspector or officer to make such examinations and tests as may be necessary to satisfy himself as to the due observance of the provisions of Act, the terms of the license and these rules.
3. Every supplier and every owner of a generating station or of a high or extra-high voltage installation shall, if required so to do by an Inspector or any officer appointed to assist the Inspector and authorized under sub-rule (2) of rule 4A provide reasonable means for carrying out all tests, prescribed by or under the Act, of the appliances or apparatus used for the supply of use of energy by him as the case may be.
4. An Inspector, or any officer appointed to assist an Inspector and authorized under sub-rule (2) of rule 4A may serve an order in the form set out in Annexure IX, upon any supplier, consumer, owner or occupier, calling upon him Annexure IX, upon any supplier, consumer, owner or occupier, calling upon him to comply with any specified rule and the person so served shall thereupon comply with the order within the period named therein, and shall report in writing to the Inspector or the officer serving the order, as the case may be, when the order is complied with.

   Provided that, if within the period specified in the aforesaid order, an appeal is filed against the order, the appellate authority may suspend its operation pending the decision of the appeal.

Rule 6. Appeals

1. An appeal against an order served under these rules shall lie:
   (a) if the order is served by an officer appointed to assist an Inspector and authorized under sub-rule (2) of rule 4A to the Inspector;
   (b) if the order is served by an Inspector, to the Central Government or the State Government, as the case may be.
2. In the case of an order of an Inspector, on an appeal preferred to him under clause (a) of sub-rule (1). A further appeal shall lie to the Central Government or the State Government, as the case may be.
3. Every appeal made under sub-rule (1) shall be in writing, shall be accompanied by a copy of the order appealed against and shall be presented within 3 months of the date on which such order has been served or delivered or is deemed to have been served or delivered as the case may be.

Rule 7. Amount of Fees

1. The fees set out in Annexure II shall be payable in respect of the service therein mentioned where the tests are carried out by comparison with the Government of India Standards referred to in sub-rule (1) of rule 2.
2. The Central Government or the State Government, as the case may be, may levy such fees for testing and inspection and generally for the services of Inspectors (or any officers appointed to assist the Inspectors) as it may from time to time by general or special order, direct; and may, if it thinks fit, remit any fee or any portion thereof.

Rule 8. Incidence of Fees Recoverable in the Cases of Dispute

Where an Inspector is called into decide any difference or dispute and where a fee for such service is recoverable, the Inspector shall decide by whom such fee shall be payable.

Rule 9. Submission of Records

An Inspector or any officer appointed to assist the Inspector and authorized under sub-rule (2) of rule 4A may require a supplier or an owner to submit to him for examination any records of tests made in connection with his works and he shall comply with such requisition. Similarly, a supplier or an owner may require the Inspector or any officer appointed to assist the Inspector and authorized under sub-rule (2) of rule 4A to submit to him, for examination any records of test made by the Inspector or any officer appointed to assist the Inspector and authorized under sub-rule (2) of rule 4A in connection with his works and the Inspector or any officer appointed to assist the Inspector and authorized under sub-rule (2) of rule 4A shall comply with such requisition.

Rule 10. List of Consumers

An Inspector or any officer appointed to assist the Inspector and authorized under sub-rule (2) of rule 4A) may require a supplier to submit to him a list of all persons supplied with energy by him, the addresses at which such energy is supplied, the months of connecting services, the voltage of supply, the connected load, the purpose of supply, and the name of contractor carrying out the installation work and the supplier shall comply with such requisition.

CHAPTER-III

14.6 LICENCE

Rule 11. Application for Licence

1. Every application for a licence shall be signed by or on behalf of the applicant and addressed to such officer as the State Government may designate in this behalf and it shall be accompanied by:

(a) six copies, in print, of the draft licence as proposed by the applicant, with the name and address of the applicant and of his agent (if any) printed on the outside of the draft;

(b) three copies, each signed by the applicant, of maps of the proposed area of supply and of the streets or roads in which the supply of energy is to be compulsory, which shall be so marked or coloured as to define any portion of such area and streets or roads which are under the administrative control of any local authority and shall be on a scale:

(i) of not less than 10 centimetres to a kilometre, or

(ii) if no such maps are available, of not less than that of the largest scale ordinance maps available, or

(iii) on such other scales as may be approved by the State Government;

(c) a list of any local authorities invested with administration of any portion of the area of supply;

(d) an approximate statement describing any lands which the applicant proposes to acquire for the purpose of the licence under the provisions of the Land Acquisition Act, 1894 (1 of 1894).

(e) an approximate statement of the capital proposed to be expended in connection with the undertaking and such other particulars as the State Government may require;

(f) if the applicant is a company which is registered under any of the enactments relating to companies for the time being in force in India or is a corporation by an Act of the Parliament, a copy of the Memorandum and Articles of Association; and

(g) a treasury receipt for such fee not exceeding fifteen hundred rupees, as the State Government may require, paid into a government treasury in the state concerned, unless such fee is remitted, wholly or in part by general or special order of the State Government.

2. If the application for a license is rejected or if a licence is revoked under subsection (2) of Section 4 as to the whole or any part of the area of supply, the State Government may at its discretion refund, wholly or in part, the fee referred to in clause (g) of sub-rule (1).

Rule 12. Copies of Maps and Draft Licence for Public Inspection

The applicant shall deposit at his own office and of his agents (if any) and at the office of every local authority invested with the administration of any portion of the proposed area of supply.

(a) copies of the maps referred to in clause (b) of sub-rule (1) of rule 11 for public inspection; and

(b) sufficient number of copies of the draft licence to be furnished to all persons applying for them at a price not exceeding twenty-five paisa per copy.

Rule 13. Contents of Draft Licence

The draft licence shall contain the following particulars:

(a) a short title descriptive of the proposed undertaking together with the address and description of the applicant, or in the case of a firm, the names of all the directors or partners of the firm;

(b) a statement of the boundaries of the proposed area of supply;

(c) if the generating station is situated or is to be situated outside the area of supply or if any intervening area not included in the area of supply is to be crossed, a list of the streets not included in the area of supply along or across which electric supply lines are to be laid down or placed;

(d) the proposed limits within which and the conditions under which the supply of energy is to be compulsory or permissive, the nature and amount of the supply (if limited) and the like;

(e) a list of the streets (if any) which are repairable neither by the Central or the State Government nor by a local authority and of the railways and tramways (if any) the soil or pavement of which the applicant seeks powers to open or break up, and the names of the persons, or designations of authorities by whom such streets are repairable or who are for the time being entitled to work such railways or tramways;

(f) the proposed period after which the right to purchase is to take effect;

(g) a statement of any special terms of purchase or orders proposed to be made under section 10; and

(h) any proposed modification of the Schedule to the Act to be made under clause (f) of subsection (2) of Section 3.

Rule 14. Form of Draft Licence

The form of draft licence contained in Annexure III may, with such variation as the circumstances of each case require, be used for the purposes of rules 11 and 13 and if used, shall be sufficient.

Rule 15. Advertisement of Application and Contents Thereof

1. The applicant shall, within fourteen days from the submission of the application under rule 11, publish notice of his application by public advertisement, and such advertisement shall publish such particulars as the State Government may specify.
2. The advertisement shall be headed by a short title corresponding to that given at the head of the draft licence and shall give the addresses of the offices at which, under Rule 12, copies of maps therein referred to may be inspected and the copies of draft licence perused or purchased and shall state that every local authority, company or person, desirous of making any representation with reference to the application to the State Government, may do so by letter addressed to such officer as the State Government may designate in this behalf, within three months of the date of issue of the first advertisement.
3. The advertisement shall be inserted by the applicant in at least two successive issues of such newspaper as the State Government, having regard to its circulation among persons likely to be interested, may direct, and in the absence of any such direction, in at least

two successive issues of any newspaper, published within the proposed area of supply or if there is no such newspaper, in any newspaper published within the State.

4. The applicant shall send a copy of each of the two successive issues of the newspaper containing the advertisement to such officer as the State Government may designate in this behalf as soon as the second issue has appeared and the State Government shall publish the advertisement at least once in the official gazette within six weeks from the date of the first advertisement published under sub-rule (3):

   Provided that any failure or delay on the part of the State Government in publishing the advertisement shall not itself preclude the grant of a licence.

Rule 16. Amendment of Draft Licence

Any person who desires to have any amendment made in the draft licence shall deliver a statement of the amendment to the applicant and to such officer as the State Government may designate in this behalf within the time allowed under sub-rule (2) of rule 15 for the submission of representations referring to the application.

Rule 17. Local Inquiries

If any person locally interested objects to the grant of a licence applied for under the Act, the State Government shall, if either the applicant or the objector so desires, cause a local inquiry to be held of which the notice in writing shall be given to both the applicant and the objector:

Provided that, the State Government may refuse such an inquiry if in its opinion the objection is of a trifling or vexatious nature.

Rule 18. Approval of Draft Licence

When the State Government has approved a draft licence, either in its original form or in a modified form, such officer as the State Government may designate in this behalf shall inform the applicant of such approval and of the form in which it is proposed to grant the licence.

Rule 19. Notification of Grant of Licence

On receiving intimation in writing from the applicant that he is willing to accept a licence in the form approved by the State Government, the State Government shall publish the licence within two months by a notification in the official gazette, together with a statement that it has been granted.

Rule 20. Date of Commencement of Licence

The date of a notification under rule 19 shall be deemed to be the date of commencement of the licence.

Rule 21. Deposit of Maps

When a licence has been granted, three sets of maps showing, as regards such licence, the particulars specified in clause (b) of sub-rule (1) of rule 11 shall be signed and dated to correspond with the date of the notification of the grant of the licence by such officer as the State Government may designate in this behalf. One set of such maps shall be retained as the deposited maps by the said officer and of the remaining two sets, one shall be furnished to the State of Electricity Board constituted under Section 5 of the Electricity (Supply) Act, 1948 and the other to the licensee.

Rule 22. Deposit of Printed Copies

1. Every person who is granted a licence, shall within thirty days of the grant thereof:
   (a) have adequate number of copies of the licence printed;
   (b) have adequate number of maps prepared showing the area of supply and the compulsory areas specified in the licence;
   (c) arrange to exhibit a copy of such licence and maps for public inspection at all reasonable times at his head office, his local offices (if any), and at the office of every local authority within the area of supply.
2. Every such licensee shall, within the aforesaid period of thirty days, supply free of charge one copy of the licence and the relevant maps to every local authority within the area of supply and shall also make necessary arrangements for the sale of printed copies of the licence to all persons applying for the same, at a price not exceeding twenty-five paisa per copy.

Rule 23. Application for Written Consent of State Government in Certain Cases

If a licensee desires the written consent of the State Government under subsection (5) of Section 12 to enable him to open or break up the soil or pavement of any street (which is reparable neither by the Central or the State Government nor by a local authority), or any railway or tramway, he shall apply for such consent in writing to such officer as the State Government may designate in this behalf and shall describe accurately the street, railway, or tramway which he seeks power to open or break up and the names of the persons or designations of the authorities by whom such street is reparable or who are for the time being entitled to work such railway or tramway; and the extent to which he proposes to open or break up the same.

Rule 24. Amendment of Licence

1. If a licensee desires that any alterations or amendments should be made in the terms and conditions of his licence under (subsection (1) of Section 4A) he shall submit a written application to the officer designated by the State Government under Rule 11 and shall within fourteen days from the submission of the application publish notice of his application by public advertisement; and the provisions of sub-rules (2), (3) and (4) if rule 15 shall apply to such publication.

2. The State Government shall within six months of the date of submission of the application, either approves of the alterations or amendments in the form proposed by the licensee, or in any other modified form, which he accepts, or rejects them. When the State Government has approved of the alterations, or amendments either in the form proposed by the licensee or in any other modified form, which he accepts, it shall notify the alterations or amendments so approved, in the official gazette.

Rule 25. Sale of Plans

Copies of plans or sections such as are referred to in clause XVI of the schedule to the Act shall be supplied by the licensee to every applicant at a price not exceeding one rupee per 1000 square centimetres.

Rule 26. Preparation and Submission of Accounts

1. Every licensee, unless exempted under Section 11, shall cause the accounts of his undertaking to be made up to the thirty-first day of March each year.
2. Such licensee shall prepare and render an annual statement of his accounts in accordance with the provisions of Section 11 within a period of six months from the aforesaid date, or such extended period as the State Government may authorize after it is satisfied that the time allowed is insufficient owing to any cause beyond the control of the licensee; and the statement shall be rendered in quadruplicate if the State Government so desires.
3. The accounts shall be made up in the prescribed forms set out in Annexures IV and V and shall be rendered in Indian currency.
   All the forms shall be signed by the licensee or his accredited agent or manager.
4. The State Government may, by special or general order direct that in addition to the submission of the annual statements of accounts in the form prescribed in sub-rule (3), a licensee shall submit to the State Government or such other authority as it may appoint in this behalf such additional information as it may require for the purpose.

Rule 27. Model Conditions of Supply

1. Without prejudice to the powers conferred by Section 21 on the State Government in this behalf, the model conditions of supply contained in Annexure VI may, with such variations as the circumstances of each case require, be adopted by the licensee for the purpose of subsection (2) of that section with the previous sanction of the State Government.
2. The licensee shall always keep in his office an adequate number of printed copies of the sanctioned conditions of supply and shall, on demand, sell such copies to any applicant at a price fixed, on cost basis, from time to time.

Rule 28. Forms of Requisitions

Requisitions under subclause (4) of clause V or subclause (5) of clause VI, as the case may be, of the Schedule to the Act shall be made in the form set out in Annexure VII or Annexure VIII.

CHAPTER-IV

14.7 GENERAL SAFETY REQUIREMENTS

Rule 29. Construction, Installation, Protection, Operation and Maintenance of Electric Supply Lines and Apparatus

1. All electric supply lines and apparatus shall be of sufficient ratings for power, insulation and estimated fault current and of sufficient mechanical strength for the work they may be required to perform under the environmental conditions of installation, and shall be constructed, installed, protected, worked, and maintained in such a manner as to ensure safety of human being, animals and property.
2. Save as otherwise provided in these rules, the relevant Code of Practice of the Bureau of Indian Standards, including National Electricity Code, if any, may be followed to carry out the purposes of this rule and in the event of any inconsistency, the provisions of these rules shall prevail.
3. The material and apparatus used shall confirm to the relevant specification of the Bureau of Indian Standards, including National Electricity Code, where such specifications have already been laid down.

Rule 30. Service Lines and Apparatus on Consumer's Premises

1. The supplier shall ensure that all electric supply lines, wires, fittings and apparatus belonging to him or under his control which are on a consumer's premises are in a safe condition and in all respects for supplying energy, and the supplier shall take due precautions to avoid danger arising on such premises from such supply lines, wires, fittings and apparatus.
2. Service line placed by the supplier on the premises of a consumer which are underground or which are accessible shall be so insulated and protected by the supplier as to be secured under all ordinary conditions against electrical, mechanical, chemical or other injury to the insulation.
3. The consumer shall, also as far as circumstances permit, take precautions for the safe custody of the equipment on his premises belonging to the supplier.
4. The consumer shall also ensure that the installation under his control is maintained in a safe condition.

Rule 31. Cutout Consumer's Premises

1. The supplier shall provide a suitable cutout in each conductor of every line other that an earthed or earthed neutral conductor or the earthed external conductor of a concentric cables within a consumer's premises in an accessible position. Such cut out shall be contained within adequately enclosed fireproof receptacle. Where more than one consumer is supplied through a common service line, each such consumer shall be provided with and independent cutout at the point of junction to the common service.
2. Every electric supply line, other than the earthed or earthed neutral conductor of any system or the earthed external conductor of a concentric cable, shall be protected by suitable cutout by its owner.

Rule 32. Identification of Earthed and Earthed Neutral Conductor and Position of Switches and Cut-outs Therein

Where the conductors include an earthed conductor of a two-wire system or an earthed neutral conductor of a multi-wire system or a conductor, which is to be connected thereto, the following conditions shall be complied with:

1. An indication of a permanent nature shall be provided by the owner of the earthed or earthed neutral conductor, or the conductor, which is to be connected thereto, to enable such conductor to be distinguished from any live conductor. Such indication shall be provided:
   (a) Where the earthed or earthed neutral conductor is the property of the supplier, at or near the point of commencement of supply.
   (b) Where a conductor forming part of a consumer's system is to be connected to the supplier's earthed or earthed neutral conductor, at the point where such connection is to be made.
   (c) In all other cases, at a point corresponding to the point of commencement of supply or at such other point as may be approved by an inspector or any officer appointed to assist the Inspector and authorized under sub-rule (2) of rule 4A.
2. No cutout, link or switch other than a linked switch arranged to operate simultaneously on the earthed or earthed neutral conductor and live conductors shall be inserted or remain inserted in any earthed or earthed neutral conductor in a two-wire system or in any earthed or earthed neutral conductor of a multi-wire system or in any conductor connected thereto with the following exceptions.
   (a) A link for testing purposes, or
   (b) A switch for use in controlling a generator or transformer.

Rule 33. Earthed Terminal on Consumer's Premises

1. The supplier shall provide and maintain on the consumer's premises for the consumer's use a suitable earthed terminal in an accessible position at or near the point of commencement of supply as defined under rule 58.

Provided that in the case of medium, high or extra high voltage installation the consumer shall, in addition to the aforementioned earthing arrangement provide his own earthing system with an independent electrode and maintain the same.

Provided further that the supplier may not provide any terminal in the case of installations already connected to his system on or before the date to be specified by the State Government in this behalf if he is satisfied that the consumer's earthing arrangement is efficient.

2. The consumer shall take all reasonable precautions to prevent mechanical damage to the earthed terminal and its lead belonging to the supplier.
3. The supplier may recover from the consumer the cost of installation on the basis of schedule of charges notified in advance, and where such schedule of charges is not notified, the procedure specified in sub-rule (5) of rule 82 will apply.

Rule 34. Accessibility of Bare Conductor

Where bare conductors are used in a building the owner of such conductors shall:

(a) Ensure that they are inaccessible.

(b) Provide in readily accessible position switches for rendering them dead whenever necessary.

(c) Take such other safety measures as are considered necessary by the Inspector.

Rule 35. Danger Notices

The owner of every medium, high and extra high voltage installation shall affix permanently in a conspicuous position a 'Danger' notice in Hindi or English and the local language of the district, with a sign of skull and bones of a design as per the relevant ISS No. 2551 on:

(a) every motor, generator, transformer and other electrical plant and equipment together with apparatus used for controlling or regulating the same;

(b) all supports of high and extra high voltage overhead lines, which can be easily climbed upon without the aid of ladder or special appliances.

Rails, tubular poles, wooden supports, reinforced cement concrete poles without steps, I-sections and channels shall be deemed as supports which cannot be easily climbed upon for the purposes of this clause.

(c) luminous tube light requiring high voltage supply X-ray and similar high frequency installation:

Provided that where it is not possible to affix such notices on any generator, motor, transformer or other apparatus, they shall be affixed be near as possible thereto or the word "danger" and the voltage of the apparatus concerned.

Provided further that where the generator, motor transformer or other apparatus is within an enclosure, one notice affixed to the said enclosure shall be sufficient for the purposes of this rule.

Rule 36. Handling of Electric Supply Lines Apparatus

1. Before any conductor or apparatus is handled, adequate precautions shall be taken by earthing or other suitable means, to discharge electrically such conductor or apparatus and any adjacent conductor or apparatus if there is danger therefrom, and to prevent any conductor or apparatus from being accidentally or inadvertently electrically charged when persons are working thereon:

   Every person who is working on an electric supply line or apparatus or both shall be provided with tools and devices such as gloves, rubber shoes, safety belts, ladders, earthing devices, helmets, line testers, hand lines and the like for protecting him from mechanical and electrical injury. Such tools and devices shall always be maintained in sound and efficient working conditions.
2. No person shall work on any live electric supply line or apparatus and no person shall assist such person on such work, unless he is authorized in that behalf, and takes the safety measures approved by the Inspector.
3. Every telecommunication line on supports carrying a high or extra high voltage line shall, for the purpose of working thereon, be deemed to be a high voltage line.

Rule 37. Supply of Vehicle Cranes

Every person owning a vehicle, travelling crane or the like to which energy is supplied from an external source shall ensure that it is efficiently controlled by a suitable switch enabling all voltages to be cut of in one operation and where such vehicles, travelling crane or the line runs on metal rails, the owner shall ensure that the rails are electrically continuous and earthed.

Rule 38. Cables for Portable or Transportable Apparatus

1. Flexible cables shall not be used for portable or transportable motors, generators, transformers, rectifiers, electric drills, electric sprays, welding sets or any other portable or transportable apparatus unless they are heavily insulated and adequately protected from mechanical inquiry.
2. Where the protection is by means of metallic covering, the covering shall be in metallic connection with the frame of any such apparatus and earth.
3. The cables shall be three-core type and four-core type for portable and transportable apparatus working on single phase and three-phase supply respectively and the wire meant to be used for ground connection shall be easily identifiable.

Rule 39. Cables Protected by Bituminous Materials

1. Where the supplier or the owner has brought into use an electric supply line which is not completely enclosed in a continuous metallic covering connected with earth, and is insulated or protected *in situ* by composition or material of a bituminous character:

(i) Any pipe, conduit or the like into which such electric supply line may, have been drawn or placed shall, unless other arrangements are approved by the Inspector, in any particular case, be effectively sealed at its point of entry into any street box so as to prevent any flow of gas to or from the street box; and

(ii) Such electric supply line shall be periodically inspected and tested where accessible and the supplier or the owner shall duly record result of each such inspection and test.

2. It shall not be permissible for the supplier or the owner after the coming into force of these rules, to bring into use any further electric.
3. Supply line as aforesaid which is insulated or protected *in situ* by any composition or material known to be liable to produce noxious explosive gases on excessive heating.

Rule 40. Street Boxes

1. Street boxes shall not contain gas pipes, and precautions shall be taken to prevent, as far as reasonably possibly, say influx of water or gas.
2. Where electric supply lines forming part of different systems pass through the same street box, they shall be readily distinguishable from one another and all electric supply lines at high or extra high voltage in street boxes shall be adequately supported so as to prevent risk of damage to or danger from adjacent electric supply lines.
3. All street boxes shall be regularly inspected for the purpose of detecting the presence of gas and if any influx or accumulation is discovered, the owner shall give immediate notice to any authority or company who have gas mains in the neighbourhood of the street box and in cases where a street box is a large enough to admit the entrance of a person after the electric supply lines apparatus therein have been placed in position, ample provision shall be made:
   (a) to ensure that any gas which may by accident have obtained access to the box shall escape before a person is allowed to enter, and
   (b) for the prevention of danger from sparking.
4. The owners of street boxes or pillars containing circuits or apparatus shall ensure that their covers and doors are so provided that they are opened only by means of a key or a special appliance.

Rule 41. Distinction of Different Circuits

The owner of every generating station, substation, junction box or pillar in which there are any circuits or apparatus, where intended for operation at different voltages or at the same voltage, shall ensure by means of indication of a permanent nature that the respective circuits are readily distinguishable from one another.

Rule 41A. Distinction of the Installations having More than One Feed

The owner of the every installation includes substation, double pole structure or any other structure having more than one feed, shall ensure by means of indication of a permanent nature, that the installation is readily distinguishable from other installations.

Rule 42. Accidental Charge

The owner of all circuits and apparatus shall so arrange them that there shall be no danger of any part thereof becoming accidentally charged to any voltage beyond the limits of voltage for which they are intended.

Where AC and DC circuits are installed on the same support they shall be so arranged and protected that they shall not come into contact with each other when live.

Rule 43. Provisions Applicable to Protective Equipment

1. Fire buckets filled with clean dry sand and ready for immediate use for extinguishing fires, in addition to fire extinguishers, suitable for dealing with electric fires, shall be conspicuously marked and kept in all generating stations, enclosed substations and enclosed switch stations in convenient situations.

   The fire extinguishers shall be tested for satisfactory operation at least once a year and record of such tests shall be maintained.
2. First aid boxes or cupboards conspicuously marked and equipped with such contents as the state government may specify shall be provided and maintained in every generating station, enclosed substation and enclosed switch station so as to be readily accessible during all working hours. All such boxes and cupboards shall, except in the case of unattended substations and switch-stations be kept in charge of responsible persons who are trained in first aid treatment and one of such persons shall be available during working hours.
3. Two more gas masks shall be provided conspicuously and installed and maintained at accessible places in every generating station with capacity of 5 MW and above and enclosed substation with transformation capacity of 5 MVA and above for use in the event of fire or smoke:

   Provided that where more than one generator with capacity of 5 MW and above is installed in power station, each generator would be provided with at least two separate gas masks in accessible and conspicuous position:

   Provided further that adequate number of gas masks would be provided by the owner of every generating station and enclosed substation with capacity less than 5 MW and 5 MVA respectively, if so desired by the Inspector.

Rule 44. Instructions for Restoration of Persons Suffering from Electric Shock

1. Instructions, in English or Hindi and the local languages of the district and where Hindi is the local language, in English and Hindi for the restoration of person suffering from electric shock, shall be affixed by the owner in a conspicuous place in every generating station, enclosed substation and in every factory as defined in clause (m) of Section 2 of the Factory Act 1948 (LXIII of 1948) in which electricity is used and in such other premises where electricity is used as the Inspector or any officer appointed to assist the Inspector may, by notice in writing serve on the owner, direct.

2. Copies of the instructions shall be supplied on demand by an officer or officers appointed by the Central or State government in this behalf at a price to be fixed by the Central or the State government.
3. The owner of every generating station, enclosed substation, enclosed switch-station, and every factory or other premises to which this rule applies shall ensure that all authorized persons employed by him are acquainted with and are competent to apply the instructions referred to in sub-rule (1).
4. In every manned high voltage or extra high voltage generating station, substation or switch-station, artificial respirator shall be provided and kept in good working condition.

44A. Intimation of Accident

If any accident occurs in connection with the generation, transmission, supply or use of energy in or in connection with, any part of the electric supply lines or other works of any person and the accident results in or is likely to have resulted in loss of human or animal life or in any injury to a human being or an animal, such person or any authorized person of the State Electricity Board/Supplier, not below the rank of a Junior Engineer or equivalent shall send to the Inspector a telegraphic report within 24 hours of the knowledge of the occurrence of the fatal accident and a written report in the form set out in Annexure XIII within 48 hours of the knowledge of occurrence of fatal and all other accidents. Where practicable a telephonic message should also be given to the Inspector immediately, the accident comes to the knowledge of the authorized officer of the State Electricity Board/Supplier or other person concerned.

Rule 45. Precautions to be Adopted by Consumers, Owners, [Occupiers], Electrical Contractors, Electrical Workmen and Suppliers

1. No electrical installation work, including additions, alterations, repairs, and adjustment to existing installations, except such replacement of lamps, fans, fuses, switches, low voltage domestic appliances and fitting as in no way alters its capacity or character, shall be carried out upon the premises of or on behalf of any consumer, supplier, owner or occupier for the purpose of supply to such consumer, owner or occupier except by an electrical contractor licensed in this behalf by the State Government and under the direct supervision of a person holding a certificate of competency and by a person holding a permit issued or recognized by the State Government.

   Provided that in the case of works executed for or on behalf of the Central Government and in the cases of installations in mines, oil fields, and railways, the Central Government and in other cases the State Government may by notification in the Official Gazette, exempt on such conditions as it may impose any such work described therein either generally or in the case of any specified class of consumers, suppliers, owners or occupiers from so much of this sub-rule as requires such work to be carried out by an electrical contractor licensed by the State Government in this behalf.

2. No electrical installation work, which has been carried out in contravention of sub-rule (1) shall be connected with the works of any supplier.

Rule 46. Periodical Inspection and Testing of Installation

1. (a) Where an installation is already connected to the supply system of the supplier, every such installation shall be periodically inspected and tested at intervals not exceeding five years either by the Inspector or any officer appointed to assist the Inspector or by the supplier as may be directed by the State Government in this behalf or in the case of installations belonging to, or under the control of the Central Government, and in case of installations in mines, oil fields and railways, by the Central Government.

(aa) The periodical inspection and testing of high voltage and extra high voltage installations belonging to supplier shall also be carried out at intervals not exceeding five years by the inspector or any officer appointed to assist the inspector.

2. Where the supplier is directed by the Central or the State Government, as the case may be, inspect and test the installation he shall report on the condition of the installation to the consumer concerned in a form approved by the Inspector and shall submit a copy of such report to the Inspector or to any officer appointed to assist the Inspector and authorized under sub-rule (2) of rule 4A.

3. Subject to the approval of the Inspector, the forms of inspection report contained in Annexure IXA may, with such variations as the circumstances of each case require, be used for the purpose of this sub-rule.

2. (a) The fees for such inspection and test shall be determined by the Central or the State Government, as the case may be, in the case of each class of consumers and shall be payable by the consumer in advance.

(b) In the event of the failure of any consumer to pay the fees on or before the date specified in the fee-notice supply to the installation of such consumer shall be liable to be disconnected under the direction of the Inspector. Such disconnection, however, shall not be made by the supplier without giving to the consumer seven clear days notice in writing of his intention so to do.

(c) In the event of the failure of the owner of any installation to rectify the defects in his installation pointed out by the Inspector or by any officer appointed to assist him and authorized under sub-rule (2) of rule 4A in the form set out in Annexure IX and within the time indicated therein, such installation shall be liable to be disconnected under the directions of the Inspector after serving the owner of such installation with a notice:

Provided that the installation shall not be disconnected in case an appeal is made under rule 6 and the appellate authority has stayed the orders of disconnection:

Provided further that the time indicated in the notice shall not be less than forty-eight hours in any case:

Provided also that nothing contained in this clause shall have any effect on the application of rule 49.

3. Not withstanding the provisions of this rule, the consumer shall at all times be solely responsible for the maintenance of his installation in such conditions as to be free from danger.

CHAPTER-V

14.8 GENERAL CONDITIONS RELATING TO SUPPLY AND USE OF ENERGY

Rule 47. Testing a Consumer's Installation

1. Upon receipt of an application for a new or additional supply of energy and before connecting the supply or reconnecting the same after a period of six months, the supplier shall inspect and test the applicant's installation.
   The supplier shall maintain a record of test results, obtained at each supply point to a consumer, in a form to be approved by the inspector.
2. If as result of such inspection and test the supplier is satisfied that the installation is likely to constitute danger he shall serve on the applicant a notice in writing requiring him to make such modifications as are necessary to render the installation safe. The supplier may refuse to connect or reconnect the supply until the required modifications have been completed and he has been notified by the applicant.

Rule 47A. Installation and Testing of Generating Units

Where any consumer or occupier installs a generating plant, he shall give a thirty-day notice of his intension to commission the plant to the supplier as well as the Inspector:

Provided that no consumer or occupier shall commission his generating plant of a capacity exceeding 10 kW without the approval in writing of the Inspector.

Rule 48. Precautions Against Leakage before Connection

1. The supplier shall not connect with his works the installation or apparatus on the premises of any applicant for supply unless he is reasonably satisfied that the connection will not at the time of making the connection cause a leakage from that installation or the apparatus of a magnitude detrimental to safety. Compliance with the rule shall be checked by measuring the insulation resistance as provided below:
   (i) All the electrical equipments shall have the "IR" values as stipulated in the relevant Indian Standards.
   (ii) At a pressure of 500 V DC applied between each live conductor and earth for a period of one minute, the insulation resistance of low voltage and medium voltage equipments shall be at least 1 mega-ohm or as specified in the relevant Indian Standard.
   (iii) At a pressure of 2.5 kV DC applied between each live conductor and earth for a period of one minute, the insulation resistance of high voltage equipment shall be at least 5 mega-ohm or as specified in the relevant Indian Standard.

2. If the supplier declines to make connection under the provisions of sub-rule (1), he shall serve upon the applicant a notice in writing, stating his reason for so declining.

Rule 49. Leakage on Consumer's Premises

1. If the Inspector or any officer appointed to assist the Inspector and authorized under sub-rule (2) of rule 4A or the supplier has reason to believe that there is in the system of consumer leakage which is likely to affect injuriously the use of energy by the supplier or by other persons, or which is likely to cause danger he may give the consumer reasonable notice in writing that he desires to inspect and test the consumer's installation.
2. If, no such notice being given:
   (a) the consumer does not give all reasonable facilities for inspection and testing of his installation, or
   (b) when an insulation resistance at the consumer's installation is so low as to prevent safe use of energy. The supplier may, and if directed to do so by the Inspector, shall discontinue the supply of energy to the installation but only after giving to the consumer forty-eight hour's notice in writing of disconnection of supply and shall not commence the supply until he or the Inspector is satisfied that the cause of the leakage has been removed.

Rule 50. Supply and Use of Energy

1. The supplier shall not be supplied, transformed, converted or used or continued to be supplied, transformed, converted or used unless provisions as set out below are observed:
   (a) The following controls of requisite capacity to carry and break the current are placed after the point of commencement of supply as defined in rule 58, so as to be readily, accessible and capable of being easily operated to completely isolate the supply to the installation, such equipment being in addition to any equipment installed for controlling individual circuits or apparatus:
      (i) a linked switch with fuse(s) or a circuit breaker by low and medium voltage consumers;
      (ii) a linked switch with fuse(s) or a circuit breaker by HV consumers having aggregate installed transformer/apparatus capacity up to 1000 kVA to be supplied at voltage up to 11 kV and 2500 kVA at higher voltages (above 11 kV and not exceeding 33 kV;
      (iii) a circuit breaker by HV consumers having an aggregate installed transformer/apparatus capacity above 1000 kVA and supplied at 11 kV and above 2500 kVA supplied at higher voltages (above 11 kV and not exceeding 33 kV);
      (iv) a circuit breaker by EHV consumer:

      Provided that where the point of commencement of supply and the consumer apparatus are near each other one linked switch with fuse(s) or circuit breaker near the point of commencement of supply as required by this clause shall be considered sufficient for the purpose of this rule.

(b) In case of every transformer the following shall be provided:

(i) On the primary side of transformers a linked switch with fuse(s) or circuit breaker of adequate capacity:

Provided that the linked switch on the primary side of the transformer may be of such capacity as to carry the full load current and to break only the magnetizing current of the transformer.

Provided further that for all transformers:

(A) having a capacity of 5000 kVA and above and installed before the commencement of the Indian Electricity (1st Amendment) Rules, 2000, and

(B) having a capacity of 1000 kVA and above and installed on or after the commencement of the Indian Electricity (1st Amendment) Rules, 2000, a circuit breaker shall be provided.

(ii) In respect of all transformers installed on or after the commencement of the Indian Electricity (1st Amendment) Rules, 2000, on the secondary side of transformers HV to EHV, MV or LV a circuit breaker of adequate rating shall be installed:

Provided that for supplier's transformers of capacity up to 630 kVA, a linked switch with fuse or circuit breaker of adequate rating shall be installed on the secondary side.

(c) Except in the case of composite control gear designed as a unit every distinct circuit is protected against excess energy by means of suitable cut-out or a circuit breaker of adequate breaking capacity suitably located and so constructed are to prevent danger from overheating, arcing or scattering of hot metal when it comes into operation and to permit for ready renewal of the fusible metal of the cut-out without danger.

(d) The supply of energy to each motor or other apparatus meant for operating one particular machine is controlled by a suitable linked switch or a circuit breaker or an emergency tripping device with manual reset of requisite capacity placed in such a position as to be adjacent to the motor or group of motors or other apparatus readily accessible to and easily operated by the person in-charge and so connected in circuit that by its means all supply of energy can be cut off from the motor or group of motors or apparatus and from any regulating switch, resistance or other device associated therewith.

(e) All insulating materials are chosen with special regard to the circumstances of its proposed use and their the mechanical strength is sufficient for its purpose and so far as is practicable, of such a character or so protected as to maintain adequately its insulating property under all working conditions in respect of temperature and moisture; and

(f) adequate precautions shall be taken to ensure that no live parts are so exposed as to cause danger.

2. Where energy is being supplied, transformed, converted or used, the consumer, supplier or the owner of the concerned installation shall be responsible for the continuous observance of the provisions of sub-rule (1) in respect of his installations.

3. Every consumer shall use all reasonable means to ensure that where energy is supplied by a supplier, no person other than the supplier shall interfere with the service lines and apparatus placed by the supplier on the premises of the consumer.

Rule 50A. Additional Provisions for Supply and Use of Energy in Multi-storeyed Buildings (more than 15 metres in height)

1. Before making an application for commencement of supply or recommencement of supply after an installation has been disconnected for a period of six months or more the owner/occupier of a multi-storeyed building shall give not less than 30 days notice in writing to the Inspector together with particulars. The supply of energy shall not be commenced or recommenced within this period, without the approval or otherwise in writing of the Inspector.
2. The supplier/owner of the installation shall provide at the point of commencement of supply a suitable insulation device with cut-out or breaker to operate on all phases except neutral in the three-phase four-wire circuit and fixed in a conspicuous position at not more than 2.75 metres above the ground so as to completely isolate the supply to the building in case of emergency.
3. The owner/occupier of a multi-storeyed building shall ensure that electrical installations/works inside the building are carried out and maintained in such a manner as to prevent danger due to shock and fire hazards, and the installation is carried out in accordance with the relevant codes of practices.
4. No other service pipes shall be taken along the ducts provided for laying power cables. All ducts provided for power cables and other services shall be provided with fire-barrier at each floor crossing.

Rule 51. Provisions Applicable to Medium, High or Extra High Voltage Installations

The following provisions shall be observed where energy at medium, high or extra-high voltage is supplied, converted, transformed or used:

1. (a) All conductors (other than those of overhead lines) shall be completely enclosed in mechanically strong metal casing or metallic covering which is electrically and mechanically continuous and adequately protected against mechanical damage unless the said conductors are accessible only to an authorized person or are installed and protected to the satisfaction of the Inspector so as to prevent danger.

   Provided that non-metallic conduits confirming to the relevant Indian Standard Specifications may be used for medium voltage installations, subject to such conditions as the Inspector or Officer appointed to assist an Inspector may think fit to impose.

   (b) All metal work enclosing, supporting or associated with the installation other than that designed to serve as a conductor shall be connected with an earthing system as per standards laid down in the Indian Standards in this regard and also in accordance with rule 61(4).

(c) Every switchboard shall comply with the following provisions, namely:
   (i) a clear space of not less than 1 metre in width shall be provided in front of the switchboard;
   (ii) if there are any attachment or bare connections at the back of the switchboard, the space (if any) behind the switchboard shall be either less than 20 centimetres or more than 75 centimetres in width measured from the farthest outstanding part of any attachment or conductor;
   (iii) if the space behind the switchboard exceeds 75 centimetres in width, there shall be a passageway from either end of the switchboard clear to a height of 1.8 metres.

(d) In case of installation provided in premises where inflammable materials including gases and/or chemicals are produced, handled or stored, the electrical installations, equipment and apparatus shall comply with the requirements of flame proof, dust tight, totally enclosed or any other suitable type of electrical fittings depending upon the hazardous zones as per the relevant Indian Standard Specifications.

2. Where an application has been made to a supplier for supply of energy to any installation, he shall not commence, or where the supply has been discontinued for a period of one year and above, recommence the supply unless he is satisfied that the consumer has complied in all respects with the conditions of supply set out in sub-rule (1) of this rule, rules 50, 63 and 64.
3. Where a supplier proposes to supply or use energy at medium voltage or to recommence supply after it has been discontinued for a period of six months he shall, before connecting or reconnecting the supply, give notice in writing of such intention to the Inspector or any officer of specified rank and class appointed to assist the Inspector.
4. If at any time after connecting the supply, the supplier is satisfied that any provision of sub-rule (1) of the rule, or of rules 50 and 64 is not being observed, he shall give notice of the same in writing to the consumer and the Inspector specifying how the provisions has not been observed and to rectify such defects in a reasonable time and if the consumer fails to rectify such defects pointed out, he may discontinue the supply after giving the consumer a reasonable opportunity of being heard and recording reasons in writing, unless the Inspector directs otherwise. The supply shall be discontinued only on written orders of an officer duly notified by the supplier in this behalf. The supply shall be restored with all possible speed after such defects are rectified by the consumer to the satisfaction of the supplier.

Rule 52. Appeal to Inspector in Regard to Defects

1. If any applicant for supply or a consumer is dissatisfied with the action of the supplier in declining to commence, to continue or to recommence the supply of energy to his premises on the grounds that the installation is defective or is likely to constitute danger, he may appeal to the Inspector to test the installation and the supplier shall not, if the Inspector or, under his orders, any other officer appointed to assist the Inspector is satisfied that the installation is free from the defect or danger complained of, be entitled to refuse supply to the consumer on the grounds aforesaid, and shall within twenty-four

hours after the receipt of such intimation from the Inspector, commence, continue or recommence the supply of energy.

2. Any test for which application has been made under the provision of sub-rule (1) shall be carried out within seven days after the receipt of such application.
3. This rule shall be endorsed on every notice given under the provision of rules 47, 48 and 49.

Rule 53. Cost of Inspection and Test of Consumer's Installation

1. The cost of the first inspection and the test of consumer's installation carried out in pursuance of the provisions of 47 shall be borne by the supplier and the cost of every subsequent inspection and test shall be borne by the consumer unless in the appeal under rule 52, the Inspector directs otherwise.
2. The cost of any inspection and test made by the Inspector or any officer appointed to assist the Inspector, at the request of the consumer or other interested party, shall be borne by the consumer or other interested party, unless the Inspector directs otherwise.
3. The cost of each and every such inspection and test by whomsoever borne shall be calculated in accordance with the scale specified by the Central or the State Government as the case may be in this behalf.

Rule 54. Declared Voltage of Supply to Consumer

Except with the written consent of the consumer or with the previous sanction of the State Government a supplier shall not permit the voltage at the point of commencement of supply as defined under rule 58 to vary from the declared voltage:

(i) in case of low or medium voltage, by more than 6 per cent;
(ii) in the case of high voltage, by more than 6 per cent on the higher side or by more than 9 per cent on the lower side; or
(iii) in the case of extra high voltage, by more than 10 per cent on the higher side or by more than 12.5 per cent on the lower side.

Provided that in the case of high voltage, the voltage variation limit of 12.5 per cent may continue till the 31st March 1974.

Rule 55. Declared Frequency of Supply to Consumer

Except with the written consent of the consumer or with the previous sanction of the State Government a supplier shall not permit the frequency of an alternating current supply to vary from the declared frequency by more than 3 per cent.

Rule 56. Sealing of Metres and Cut-outs

1. A supplier may affix one or more seals to any cut-out and to any meter, maximum demand indicator, or other apparatus placed upon a consumer's premises in accordance with Section-26, and no person other than the supplier shall break any such seal.

2. The consumer shall use all reasonable means in his power to ensure that no such seal is broken otherwise than by the supplier.

Rule 57. Meters, Maximum Demand Indicators and other Apparatus on Consumer's Premises

1. Any meter, or maximum demand indicator or other apparatus placed upon a consumer's premises in accordance with Section-26 shall be of appropriate capacity and shall be deemed to be correct it its limits or error are within the limits specified in the relevant Indian Standard Specification and where no such specification exists, the limits of the error do not exceed 3 per cent, above or below absolute accuracy at all loads in excess of one-tenth of full loads and up to full load:

   Provided that for extra high voltage consumers, the limit of error shall be ± 1%.
2. No meter shall register at no load.
3. Every supplier shall provide and maintain in proper condition such suitable apparatus as may be prescribed or approved by the Inspector for the examination, testing and regulation of meters used or intended to be used in connection with the supply of energy.

   Provided that the supplier may with the approval of the Inspector and shall, if required by the Inspector enter into a joint arrangement with any other supplier for the purpose aforesaid.
4. Every supplier shall examine, test and regulate all meters, maximum demand indicators and other apparatus for ascertaining the amount of energy supplied before their first installation at the consumer's premises and at such other intervals as may be directed by the State Government in this behalf.
5. Every supplier shall maintain a register of meters, showing the date of the last test, the error recorded at the time of the test, the limit of accuracy after adjustment and final test, the date of installation, withdrawal, reinstallation, etc. for the examination of the Inspector or his authorized representative.
6. Where the supplier has failed to examine, test and regulate the meters and keep records thereof as aforesaid, the Inspector may cause such meters to be tested and sealed at the cost of the owner of the meters in case these are found defective.

Rule 58. Point of Commencement of Supply

The point of commencement of supply of energy to consumer shall be deemed to be the point at the incoming terminal of the cutouts installed by the consumer under rule 50.

Rule 59. Precautions Against Failure of Supply: Notice of Failures

1. The layout of the electric supply lines of the supplier for the supply of energy throughout his area of supply shall under normal working conditions be sectionalized and so arranged, and provided with cut-outs or circuit breakers so located as to restrict within reasonable limits the extent of the portion of the system affected by any failure of supply.

2. The supplier shall take all responsible precautions to avoid any accidental interruptions of supply, and also to avoid danger to the public or to any employee or authorized person when engaged or any operation during and in connection with the installation, extension and replacement, repair and maintenance of any works.
3. The supplier shall send to the Inspector or any officer of a specified rank and class appointed to assist the Inspector, notice of failure of supply of such kind as the Inspector or any officer of a specified rank and class appointed to assist the Inspector may from time to time require to be notified to him, and such notice shall be sent by the earliest practicable post after the failure occurs or after the failure becomes known to the supplier and shall be in such form and contain such particulars as Inspector may from time to time specify.
4. For the purposes of testing or for any other purpose connected with the efficient working of the undertaking the supply of energy may be discontinued by the supplier for such period as may be necessary, subject (except in cases of emergency) to not less than twenty-four hours notice being given by the supplier to all classes of consumers specified by the Inspector likely to be affected by such discontinuance:
   Provided that the supply of energy shall be discontinued during such hours as are likely to interfere the least with the use of energy by consumers and the energy shall not be discontinued if the inspector so directs.

CHAPTER-VI

14.9 ELECTRIC SUPPLY LINES, SYSTEMS AND APPARATUS FOR LOW AND MEDIUM VOLTAGES

Rule 60. Test for the Resistance of Insulation

1. Where any electrical supply line for use at low or medium voltage has been disconnected from a system for the purpose of addition or alternation or repair, such electric supply line shall not be reconnected to the system until the supplier or the owner has applied the test prescribed under rule 48.
2. The provisions of sub-rule (1) shall not apply to the overhead lines except overhead insulated cables unless the Inspector otherwise directs in any particular case.

Rule 61. Connection with Earth

1. The following provisions shall apply to the connection with earth of systems at low voltage in the case where the voltage normally exceeds 125 volts and of the systems at medium voltage:
   (a) Neutral conductor of a three-phase, four-wire system and the middle conductor of a 2-phase, three-wire system shall be earthed by not less than two separate and

distinct connections with a minimum of two different earth electrodes or such large number as may be necessary to bring the earth resistance to a satisfactory value both at the generating station and at the substation. The earth electrodes so provided, may be interconnected to reduce earth resistance. It may also be earthed at one or more points along the distribution system or service line in addition to any connection with earth, which may be at the consumer's premises.

(b) In the case of a system comprising electric supply lines having concentric cables, the external conductor of such cables shall be earthed by two separate and distinct connections with earth.

(c) The connection with earth may include a link by means of which the connection may be temporarily interrupted for the purpose of testing or for locating a fault.

(d) (i) In a direct current three-wire system the middle conductor shall be earthed at the generating station only, and the current from the middle conductor to earth shall be continuously recorded by means of a recording ammeter, and if at any time the current exceeds one-thousandth part of the maximum supply current, immediate steps shall be taken to improve the insulation of the system.

(ii) Where the middle conductor is earthed by means of a circuit breaker with a resistance connected in parallel the resistance shall not exceed 10 ohms and on the opening to the circuit breaker, immediate steps shall be taken to improve the insulation of the system, and the circuit breaker shall be reclosed as soon as possible.

(iii) The resistance shall be used only as a protection for the ammeter in case of earths on the system and until such earths are removed, immediate steps shall be taken to locate and remove earth.

(e) In the case of an alternating current system, there shall not be inserted in the connection with earth any impedance (other than that required solely for the operation of switchgear or instruments), cut-out or circuit breaker, and the result (if any) passing through the connection with earth is normal, shall be duly recorded by the supplier.

(f) No person shall make connection with earth by the aid of, nor shall he keep it in contact with, any water main not belonging to him except with the consent of the owner thereof and of the Inspector.

(g) Alternating current systems, which are connected with earth as aforesaid may be electrically interconnected:

Provided that each connection with the earth is bonded to the metal sheathing and metallic armouring (if any) of the electric supply lines concerned.

2. The frame of every generator, stationary motor, portable motor, and metallic parts (not intended as conductors) of all transformers and any other apparatus used for regulating or controlling energy and all medium voltage energy-consuming apparatus shall be earthed by the owner by two separate and distinct connections with earth.

3. All metal casings or metallic coverings containing or protecting any electric supply-line or apparatus shall be connected with earth and shall be so joined and connected across all junction boxes and other openings as to make good mechanical and electrical connection throughout their whole length.

Provided that where the supply of energy is at low voltage, this sub-rule shall not apply to isolated wall tubes or to brackets, electro-liners switches, ceiling fans, regulators covers or other fittings (other than portable hand lamps and portable and transportable apparatus) unless provided with an earth terminal and to Class II apparatuses/appliances.

Provided further that where the supply of energy is at low voltage and where the installations are either new or renovated all plug sockets shall be of the three-pin types, and the third pin shall be permanently and efficiently earthed.

Explanation: The words "Class II apparatus-appliance will have the same meaning as assigned to these words in the relevant ISS.

4. All earthing systems shall:
   (a) consist of equipotential bonding conductors capable of carrying the prospective earth fault current and a group of pipe/rod/plate electrodes for dissipating the current to the general mass of earth without exceeding the allowable temperature limits as per relevant Indian Standards in order to maintain all non-current carrying metal works reasonably at earth potential and to avoid dangerous contact potentials being developed on such metal works;
   (b) limit earthy resistance sufficiently low to permit adequate fault current for the operation of protective devices in time and to reduce neutral shifting;
   (c) be mechanically strong, withstand corrosion and retain electrical continuity during the life of the insulation. All earthing systems shall be tested to ensure efficient earthing, before the electric supply lines or apparatus are energized.
5. All earthing systems belonging to the supplier, shall, in addition, be tested for resistance on dry day during the dry season not less than once every two years.
6. A record of every earth-test made and the result thereof shall be kept by the supplier for a period of not less than two years after the day of testing and shall be available to the Inspector or any officer appointed to assist the Inspector and holding gazetted rank and authorized under sub-rule (2) of rule 4A when required.

Rule 61A. Earth Leakage Protective Device

The supply of energy to every electrical installation other than low voltage installation below 5 kW and those low voltage installations, which do not attract provisions of Section 30 of the Indian Electricity Act, 1910, shall be controlled by an earth leakage protective device so as to disconnect the supply instantly on the occurrence of earth fault or leakage of current:

Provided that the above shall not apply to overhead supply lines having protective devices which are effectively bonded to the neutral of supply transformers and confirming to rule 91 of the Indian Electricity Act, 1956.

Rule 62. System at Medium Voltage

Where a medium voltage supply system is employed the voltage between earth and any conductor forming part of the said system shall not, under normal conditions, exceed low voltage.

CHAPTER-VII

14.10 ELECTRIC SUPPLY LINES, SYSTEMS AND APPARATUS FOR HIGH AND EXTRA HIGH VOLTAGES

Rule 63. Approval by Inspector

1. Before making an application to the Inspector for permission to commence or recommence supply after an installation has been disconnected for one year or above at high or extra high voltage to any person, the supplier shall ensure that the high or extra high voltage electric supply lines or apparatus belonging to him are placed in position, properly joined and duly completed and examined. The supply of energy shall not be commenced by the supplier unless and until the Inspector is satisfied that the provisions of rules 65 to 69 (both inclusive) have been complied with and the approval in writing of the Inspector have been obtained by him.

   Provided that the supplier may energize the aforesaid electric supply lines or apparatus for the purpose of tests specified in rule 65.
2. The owner of any high or extra high voltage installation shall, before making application to the Inspector for approval of his installation or additions thereto, test every high or extra voltage circuit or additions thereto, other than an overhead line, and satisfy himself that they withstand the application of the testing voltage set out in sub-rule (1) of rule 65 and shall duly record the results of such tests and forward them to the Inspector.

   Provided that, an Inspector may direct such owner to carry out such tests as he deems necessary or if he thinks fit, accept the manufacturer's certified tests of any particular apparatus in place of the tests required by this sub-rule.
3. The owner of any high or extra high voltage installation who makes any additions or alternations to his installation shall not connect to the supply his apparatus or electric supply lines, comprising the said alternations of additions have been approved in writing by the Inspector.

Rule 64. Use of Energy at High and Extra High Voltage

1. The Inspector shall not authorize a supplier to commence supply or where the supply has been discontinued for a period of one year and above, to recommence the supply at high or extra high voltage to any consumer, unless.
   (a) All conductors and apparatus situated on the premises of the consumer are so placed as to be inaccessible except to an authorized person and all operations in connection with the said conductors and apparatus are carried out by an authorized person.
   (b) The consumer has provided and agrees to maintain a separate building or a locked weatherproof and fireproof enclosure of agreed design and location, to which the supplier at all times have access, for the purpose of housing his high or extra high

voltage apparatus and metering equipment, or where the provisions for a separate building or enclosure is impracticable, the consumer has segregated the aforesaid apparatus of the supplier from any other part of his own apparatus.

Provided that such segregation shall be by the provision of fireproof walls, if the Inspector considers it to be necessary:

Provided further that in the case of an outdoor installation, the consumer shall suitably segregate the aforesaid apparatus belonging to the supplier from his own to the satisfaction of the Inspector.

(c) All the pole type substations are constructed and maintained in accordance with rule 69.

2. The following provisions shall be observed where energy at high or extra high voltage is supplied, converted, transformed or used.

(a) (i) Clearances as per Indian Standard Code shall be provided for electrical apparatus so that sufficient space is available for easy operation and maintenance without any hazard to the operating and maintenance personnel working near the equipment and for ensuring adequate ventilation:

Provided that the owner of the transformer installation shall not allow any encroachment below such installations. The Electrical Inspector shall direct appropriate law enforcing authorities to remove such encroachments, if in his/her opinion; such encroachments pose a danger to the life of the operating personnel/ public person or property.

(ii) Following minimum safety working clearance shall be maintained for bare conductors or live parts of any apparatus in outdoor substations, excluding overhead lines of HV and EHV installations:

Highest system voltage	Safety working clearance (meters)
12 kV	2.6
36 kV	2.8
72.5 kV	3.1
145 kV	3.7
245 kV	4.3
420 kV	6.4
800 kV	10.3

Note:

(1) The above values are valued for altitude not exceeding 1000 metres. A correction factor of 1.25% per 100 m is to apply for increasing the clearance for altitudes more than 1000 m and up to 3000 m.

(2) The above safety working clearances are based on an insulation height of 2.44 m, which is the height of lowest point on the insulator (where it meets the earthed metal) from the ground.

(3) "Safety Working Clearance" is the minimum clearance to be maintained in air between the live part of the equipment on one hand and earth or another piece of equipment or conductor on which it is necessary to carry out the work, on the other.

(4) The "Highest System Voltage" is defined as the highest rms phase-to-phase voltage, which occurs under normal operating conditions at any time and at any point of the

system. It excludes voltage transients (such as those due to system switching) and temporary voltage variations due to abnormal system conditions (such as those due to fault conditions or the sudden disconnection of large loads).

(b) The windings of motors or other apparatus within each from any position in which a person may require to be suitably protected so as to prevent danger.

(c) Where transformer or transformers are used, suitable provision shall be made, either by connecting with earth a point of the circuit and lower voltage or otherwise, to guard against danger by reason of the said circuit becoming accidentally charged above its normal voltage by leakage from or contact with the circuit at the higher voltage.

(d) A substation or a switch-station with apparatus having more than 2000 litres of oil shall not be located in the basement where proper oil draining arrangement cannot be provided.

(e) Where a substation or a switch-station with apparatus having more than 2000 litres of oil is installed, whether indoor or outdoor, the following measures shall be taken, namely:

  (i) The baffle walls of four hours fire rating shall be provided between the apparatus in the following cases:

    (a) single-phase banks in the switch-yards of generating station and substations:

    (b) on the consumer premises;

    (c) where adequate clearance between the units is not available.

  (ii) Provisions shall be made for suitable oil soak pit where use of more than 9000 litres of oil in any one oil tank, receptacle or chamber is involved, provision shall be made for the draining away or removal of any oil which may leak or escape from the tanks receptacles or chambers containing the same, special precaution shall be taken to prevent the spread of any fire resulting from the ignition of the oil from any cause and adequate provision shall be made for extinguishing any fire which may occur. Spare oil shall not be stored in any such substation or switch-station.

  (iii) All the transformers and switchgears shall be maintained in accordance with the maintenance schedule given in the relevant code of practices of BIS and the authorized person shall keep a record thereof as required under rule 3(6).

  (iv) Notwithstanding anything contained in rules 64(2)(d) and 64(2)(f)(ii), only dry type of transformers shall be used for installations inside the residential/commercial buildings.

(f) (i) Without prejudice to the above measures, adequate fire protection arrangement shall be provided for quenching the fire in the apparatus;

(ii) Where it is necessary to locate the substation/switch-station in the basement, following measures shall be taken:

  (a) the room shall necessarily be in the first basement at the periphery of the basement;

  (b) the entrances to the room shall be provided with fire resisting doors of two hours fire rating. A curb (sill) of a suitable height shall be provided at the entrance in order to prevent the flow of oil from a ruptured transformer into other parts of the basement. Direct access to the transformer room shall be provided from outside;

  (c) the transformers shall be protected by an automatic high velocity water spray system or by carbon dioxide or BCF (bromochlorodifluromethane) or BTM Bromotri-fluromethane) fixed installation system or nitrogen injection and drain method.

(iii) Oil filled transformers installed indoors shall not be on any floor above the ground or below the first basement.

(g) Cable trenches inside substations and switch-stations containing cables shall be filled with sand, pebbles or similar noninflammable materials or completely covered with non-inflammable slabs.

(h) Unless the conditions are such that all the conductors and apparatus may be made dead at the same time for the purpose of cleaning or for other work, the said conductors and apparatus shall be so arranged that they may be made dead in sections, and that work on any section made dead may be carried on by an authorized person without danger;

(i) Only person authorized under sub-rule (1) of rule 3, shall carry out the work on live lines and apparatus.

3. All EHV apparatus shall be protected against lightning as well as against switching over voltage. The equipment used for protection and switching shall be adequately coordinated with the protected apparatus to ensure safe operation as well as to maintain the stability of the interconnected units of the power system.

Rule 64A. Additional Provisions for Use of Energy at High and Extra High Voltage

The following additional provisions shall be observed where energy at high or extra high voltage is supplied, converted, transferred or used, namely:

1. **Interlocks:** Suitable interlocks shall be provided in the following cases:
   (a) Isolators and the controlling circuit breakers shall be interlocked so that the isolators cannot be operated unless the corresponding breaker is in open position.
   (b) Isolators and the corresponding earthing switches shall be interlocked so that no earthing switch can be closed unless and until the corresponding isolator is in open position.
   (c) Where two or more supplies are not intended to be operated in parallel, the respective circuit breakers or linked switches controlling the supplies shall be interlocked to prevent possibility of any inadvertent paralleling or feedback.
   (d) When two or more transformers are operated in parallel, the system shall be so arranged as to trip the secondary breaker of a transformer in case the primary breaker of that transformer trips.
   (e) All gates or doors which give access to live parts of an installation shall be interlocked in such a way that these cannot be opened unless the live parts are made dead. Proper discharging and earthing of these parts should be ensured before any person comes in close proximity of such parts.
   (f) Where two or more generators operate in parallel and neutral switching is adopted, interlock shall be provided to ensure that generator breaker cannot be closed unless one of the neutrals is connected to the earthing system.
2. **Protection:** All systems and circuits shall be so protected as to automatically disconnect the supply under abnormal conditions:
   The following protection shall be provided, namely;

(a) Over current protection to disconnect the supply automatically if the rated current of the equipment, cable or supply line is exceeded for a time, which the equipment, cable or supply line is not designed to withstand.

(b) Earth fault/earth leakage protection to disconnect the supply automatically if the earth fault current exceeds the limit of current for keeping the contact potential within reasonable value.

(c) Gas pressure type and winding and oil temperature protection to give alarm and tripping shall be provided on all transformers of rating 1000 kVA and above.

(d) Transformers of capacity 10 MVA and above shall be protected against incipient faults by differential protection; and

(e) All generators with rating of 1000 kVA and above shall be protected against earth fault leakage. All generators of rating 1000 kVA and above shall be protected against faults within the generator winding using restricted earth fault protection or differential protection or by both.

(f) High-speed busbar differential protection along with local breaker back-up protection shall be commissioned and shall always be available at all 220 kV and above voltage substations and switchyards and generating stations connected with the grids:

Provided that in respect of existing 220 kV substations and switchyard having more than one incoming feeders, the high-speed busbar differential protection along with local breaker back up protection, shall be commissioned and shall always be available:

Provided further that the provisions contained in this sub-rule shall have effect in respect of the existing installations from the date to be specified by the appropriate government but such a date shall not be later than a period of three years from the date this rules comes into force.

Rule 65. Testing, Operation and Maintenance

1. Before approval is accorded by the Inspector under rule 63 the manufacturer's test certificates shall, if required, be produced for all the routine tests as required, under the relevant Indian Standard.
2. No new HV or EHV apparatus, cable or supply line shall be commissioned unless such apparatus, cable or supply line are subjected to site tests as per relevant Code of Practice of the Bureau of Indian Standards.
3. No HV or EHV apparatus, cable or supply line which has been kept disconnected, for a period of 6 months or more, from the system for alterations or repair shall be connected to the system until such apparatus, cable or supply line are subjected to the relevant tests as per Code of Practice of the Bureau of Indian Standards.
4. Notwithstanding the provisions of sub-rules (1) to (3) (both inclusive) the Inspector may require certain additional tests to be carried out before charging the installation or subsequently.
5. All apparatus, cables and supply lines shall be maintained in healthy conditions and tests shall be carried out periodically as per the relevant Code of Practice of the Bureau of Indian Standards.

6. Records of all tests, trappings, maintenance works and repairs of all equipment, cables and supply lines shall be duly kept in such a way that these records can be compared with earlier ones; and
7. It shall be the responsibility of the owner of all HV and EHV installations to maintain and operate the installations in a condition free from danger and as recommended by the manufacturer and/or by the relevant Code of Practice of the Bureau of Indian Standards and/or by the Inspector.
8. Failure of transformers and reactors of 20 MVA and higher capacity shall be reported by the consumers and the suppliers of Electricity within 48 hours of the occurrence of the failure, to the Electrical Inspector and Central Electricity Board. The reasons for failure and measures to be taken to avoid occurrence of failure shall be sent to the Electrical Inspector and Central Electricity Board within one month of the occurrence in the format given in Annexure XV.

Rule 66. Metal Sheathed Electric Supply Lines: Precautions Against Excess Leakage

The following provisions shall apply to electric supply lines (other than overhead lines) for use at high or extra high voltage:

1. (a) The conductors shall be enclosed in metal sheathing which shall be electrically continuous and connected with earth, and the conductivity of the metal sheathing shall be maintained and reasonable precautions taken where necessary to avoid corrosion of the sheathing.

   Provided that in the case of thermoplastic insulated and sheathed cables with metallic armour, the metallic wire or taps armour shall be considered as metal sheathing for the purpose of this rule.

   Provided further that this rule shall not apply to cables with thermoplastic insulation without any metallic screen or armour.

   (b) The resistance of the earth connection with metallic sheath shall be kept low enough to permit the controlling circuit breaker or cutout to operate in the event of any failure of insulation between the metallic sheath and the conductor.

   (c) Where an electric supply line as aforesaid has concentric cables and the external conductor is insulated from an outer metal sheathing and connected with earth, the external conductor may be regarded as the metal sheathing for the purpose of this rule, provided that the foregoing provisions as to conductivity are complied with.
2. Nothing in the provisions of sub-rule (1) shall preclude the employment in generating stations, substations and switch-stations (including outdoor substations and outdoor switch-stations) of conductors for use at high or extra high voltage, which are not enclosed in metal sheathing or preclude the use of electric supply lines laid before the prescribed date to which the provisions of these rules apply.

Rule 67. Connection with Earth

1. All non-current carrying metal parts associated with HV/EHV installation shall be effectively earthed to a grounding system or mat which will:

(a) limit the touch and step potential to tolerable values;
(b) limit to ground potential rise to tolerable values so as to prevent danger due to transfer of potential through ground, earth wires, cable sheath fences, pipelines, etc.
(c) maintain the resistance of the earth connection to such a value as to make operation of the protective device effective;

1A. In the case of star-connected systems with earthed neutrals or delta-connected system with earthed artificial neutral point:

(a) The neutral point of every generator and shall be earthed by connecting it to the earthing system as defined in rule 61(4) and hereinabove by not less than two separate and distinct connections:

Provided that the neutral point of a generator may be connected to the earthing system through an impedance to limit the fault current to the earth:

Provided further in the case of multi-machine system neutral switching may be resorted to, for limiting the injurious effect of harmonic current circulating in the system;

(b) In the event of an appreciable harmonic current flowing in the neutral connection so as to cause interference with communication circuits the generator or transformer neutral shall be earthed through suitable impedance.

(c) In case of the delta-connected system the neutral point shall be obtained by the insertion of grounding transformer and current limiting resistance or impedance wherever considered necessary at the commencement of such a system.

1B. In case of generating stations, extra high voltage substations and extra high voltage industrial installations, the system neutral earthing and protective frame earthing may be, if system design so warrants, integrated into common earthing grid provided the resistance to earth of combined mat does not cause to exceed the step and touch potential beyond its permissible values.

2. Single phase high or extra high voltage systems shall be earthed in a manner approved by the Inspector.

3. In the case of a system comprising electric supply lines having concentric cables, the external conductor shall be the one to be connected with earth.

4. Where a supplier proposes to connect with earth an existing system for use at high or extra high voltage which has not hitherto been so connected with earth, he shall give not less than fourteen days' notice in writing together with particulars to the telegraph authority of the proposed connection with earth.

5. (a) Where the earthing lead and earth connection are used only in connection with earthing guards erected under high or extra high voltage overhead lines where they cross a telecommunication line or a railway line, and where such lines are equipped with earth leakage relays of a type and setting approved by the Inspector, the resistance shall not exceed 25 ohms.

(b) Every earthing system belonging to either the supplier or the consumer shall be tested for its resistance to earth on a dry day during dry season not less than once a year. Records of such tests shall be maintained and shall be produced, if required before the Inspector or any officer appointed to assist him and authorized under sub-rule (2) of rule 4A.

6. Insofar as the provisions of rule 61 are consistent with the provisions of this rule, all connections with earth shall also comply with the provisions of that rule.

Rule 68. General Conditions as to Transformation and Control of Energy

1. Where energy at high or extra high voltage is transformed, converted, regulated or otherwise controlled in substations or switch-stations (including outdoor substations and outdoor switch-stations) or in street boxes constructed underground, the following provisions shall have effect:
   (a) Substations and switch-stations shall preferably be erected above ground, but where necessarily constructed underground, due provision for ventilation and drainage shall be made and any space housing switchgear shall not be used for storage of any materials especially inflammable and combustible materials or refuse.
   (b) Outdoor substations except pole type substations and outdoor switch-stations shall (unless the apparatus is completely enclosed in a metal covering connected with earth, the said apparatus also being connected with the system by armoured cables) be efficiently protected by fencing not less than 1.8 metres in height or other means so as to prevent access to the electric supply lines and apparatus therein by an unauthorized person.
   (c) Underground street boxes (other than substations), which contain transformers shall not contain switches or other apparatus, and switches, cutouts or other apparatus required for controlling or other purposes shall be fixed in separate receptacles above ground wherever practicable.
2. Where energy is transformed, suitable provisions shall be made either by connecting with earth a point of the system at the lower voltage or otherwise to guard against danger by reason of the said system becoming accidentally charged above its normal voltage by leakage from a contact with the system at the higher voltage.

Rule 69. Pole Type Substations

Where platform type construction is used for a pole type substation and sufficient space for a person to stand on the platform is provided a substantial handrail shall be built around the said platform and if the handrail is of metal, it shall be connected with earth.

Provided that in the case of pole type substation on wooden support and wooden platform the metal handrail shall not be connected with earth.

Rule 70. Condensers

Suitable provisions shall be made for immediate and automatic discharge of every static condenser on disconnection of supply.

Rule 71. Additional Provisions for Supply to High Voltage Luminous Tube Sign Installation

1. Any person who proposes to use or who is using energy for the purpose of operating a luminous tube sign installation, or who proposes to transform or who is transforming energy to a high voltage for any such purpose shall comply with the following conditions:
   (a) All live parts of the installation including all apparatus and live conductors in the secondary circuit, but excluding the tubes except in the neighbourhood of their terminals) shall be inaccessible to unauthorized persons and such parts shall be effectively screened.
   (b) Irrespective of the method of obtaining the voltage of circuit, which feeds the luminous discharge tube sign, no part of any conductor of such circuit shall be in metallic connection (except in respect of its connection with earth) with any conductor of the supply system or with the primary winding of the transformer.
   (c) All live parts of an exterior installation shall be disposed so as to protect them against the effects of the weather, and such installations shall be so arranged and separated from its surroundings as to limit, as far as possible, the spreading of fire.
   (d) The secondary circuit shall be permanently earthed at the transformer and the core of every transformer shall be earthed.
   (e) Where the conductors of the primary circuit are not in metallic connection with the supply conductors (e.g., where a motor-generator or a double-wound converter is used), one phase of such one phase of such primary circuit shall be permanently earthed at the motor-generator or converter, or at the transformer.
      (ee) An earth leakage circuit breaker of sufficient rating shall be provided on the low voltage side to detect the leakage in such luminous tube sign installations.
   (f) A final sub-circuit, which forms the primary circuit of a fixed luminous discharge-tube sign installation, shall be reserved solely for such purpose.
   (g) A separate primary final sub-circuit shall be provided for each transformer or each group of transformers having an aggregate input not exceeding 1000 volt-amperes, of a fixed luminous discharge tube sign installation.
   (h) An interior installation shall be provided with suitable adjacent means for disconnecting all phases of the supply except the "neutral" in a three-phase four-wire circuit.
   (i) For installations on the exterior of a building, a suitable emergency fireproof linked switch to operate on all phases except the neutral in a three-phase four-wire circuit shall be provided and fixed in a conspicuous position at not more than 2.75 metres above the ground.
   (j) A special "caution" notice shall be affixed in a conspicuous place on the door of every high voltage enclosure to the effect that low voltage supply must be cut off before the enclosure is opened.
   (k) Where static condensers are used, they shall be installed on the load side of the fuses and the primary side of the transformer.
   (l) Where static condensers are used on primary side, means shall be provided or automatically discharging the condensers when the supply is cut-off.

Provided that static condensers or any circuit interrupting devices on the high or extra high voltage side shall not be used without the approval in writing of the Inspector.

2. The owner or user of any luminous tube sign or similar high voltage installation shall not bring the same into use without giving to the Inspector not less than 14 days' notice in writing of the intention to do so.

Rule 72. Additional Provisions for Supply to High Voltage Electrode Boilers

1. Where a system having a pint connected with earth is used for supply of energy at high or extra high voltage to an electrode boiler, which is also connected with earth, the following conditions shall apply:
   (a) The metal work of the electrode boiler shall be efficiently connected to the metal sheathing and metallic armouring (if any), of the high voltage electric supply line where by energy is supplied to the electrode boiler.
   (b) The supply of energy at high or extra high voltage to the electrode boiler shall be controlled by a suitable circuit breaker so set as to operate in the event of the phase currents becoming unbalanced to the extent of 10 per cent of the rated current consumption of the electrode boiler under normal conditions of operation.
   Provided that if in any case a higher setting is essential to ensure stability of operation of the electrode boiler, the setting may be increased so as not to exceed 15 per cent of the rated current consumption of the electrode boiler under normal conditions of operation.
   (c) An inverse time element device may be used in conjunction with the aforesaid circuit breaker to prevent the operation thereof unnecessarily on the occurrence of unbalanced phase currents of momentary or short duration.
   (d) The supplier shall serve a notice in writing on the telegraph-authority at least seven days prior to the date on which such supply of energy is to be afforded specifying the location of every point (including earth connection of the electrode boiler) at which the system is connected with earth.
2. The owner or user of any high or extra high voltage electrode boiler shall not bring the same into use without giving the Inspector not less than fourteen days notice in writing of his intention so to do.

Rule 73. Supply to X-ray and High Frequency Installation

1. Any person who proposes to employ or who is employing energy for the purpose of operating an X-ray or similar high frequency installation shall comply with the following conditions:
   (a) Mechanical barriers shall be provided to prevent too close an approach to any high voltage parts of the X-ray apparatus, except the X-ray tube and its leads, unless such high voltage parts have been rendered shockproof by being shielded by earthed metal or adequate insulating material.

(b) Where extra high voltage generators operating at 300 peak kV or more are used, such generators shall be installed in rooms separate from those containing the other equipment and any step-up transformer employed shall be so installed and protected as to prevent danger.

(c) A suitable switch shall be provided to control the circuit supplying a generator, and shall be so arranged as to be open except while the door of the room housing the generator is locked from the outside.

(d) X-ray tubes used in therapy shall be mounted in an earthed metal enclosure.

(e) Every X-ray machine shall be provided with a milli-ammeter or other suitable measuring instrument, readily visible from the control position and connected, if practicable, in the earthed lead but guarded if connected in the high voltage lead.

(ee) Notwithstanding the provisions of clause (e), an earth leakage circuit breaker of sufficient rating shall be provided on the low voltage side to detect the leakage in such X-ray installations.

(f) This sub-rule shall not apply to shockproof portable units or shockproof self-contained and stationary units.

2. (a) In the case of non-shockproof equipment, overhead high voltage conductors, unless suitably guarded against personal contact, shall be adequately spaced and high voltage leads on tilting tables and fluoroscopes shall be adequately insulated or so surrounded by barriers as to prevent inadvertent contact.

(b) The low voltage circuit of the step up transformer shall contain a manually operated control device having overload protection, in addition to the overcurrent device for circuit protection, and these devices shall have no exposed live part and for diagnostic work, there shall be an additional switch in the said circuit which shall be one of the following types:

(i) A switch with a spring or other mechanism what will open automatically except while held closed by the operator, or

(ii) A time switch, which will open automatically after a definite period of time for which it has been set.

(c) If more than one piece of apparatus be operated from the same high or extra high voltage source, each shall be provided with a high or extra high voltage switch to give independent control.

(d) Low frequency current carrying parts of a machine of the quenched gap or open gap type shall be so insulated or guarded that they cannot be touched during operation, the high frequency circuit proper which delivers high frequency current normally for the therapeutic purposes, being exempted.

(e) All X-ray generators having capacitors shall have suitable means for discharging the capacitors manually.

(f) Except in the case of self-contained units, all 200 peak kV, or high X-ray generators shall have a sphere gap installed in high voltage system adjusted so that it will break down on over voltage surges.

3. (a) All non-current carrying metal parts of tube stands, fluoroscopes and other apparatus shall be properly earthed and insulating floors, mats or platforms shall be provided for operators in proximity to high or extra high voltage parts unless such parts have been rendered shockproof.

(b) Where short wave therapy machines are used, the treatment tables and examining chairs shall be wholly non-metallic.

4. The owner of any X-ray installation or similar high frequency apparatus shall not bring the same into use without giving to the Inspector not less than 14 days notice in writing of his intention to do so.

   Provided that the aforesaid notice shall not be necessary in the case of shockproof portable X-ray and high frequency equipment which have been inspected before the commencement of their use and periodically, thereafter.

CHAPTER-VIII

14.11 OVERHEAD LINES, UNDERGROUND CABLES AND GENERATING STATIONS

Rule 74. Material and Strength

1. All conductors of overhead lines other than those specified in sub-rule (1) of rule 86 shall have a breaking strength of not less than 350 kg.
2. Where the voltage is low and the span is of less than 15 metres and is on the owner's or consumer premises, a conductor having an actual breaking strength of not less than 150 kg may be used.

Rule 75. Joints

Joints between conductors of overhead lines shall be mechanically and electrically secured under the conditions of operation. The ultimate strength of joint shall not be less than 95 per cent of that of the conductor and the electrical conductivity not less than that of the conductor.

Provided that no conductor of an overhead line shall have more than two joints in a span.

Rule 76. Maximum Stresses, Factor of Safety

1. The load and permissible stresses on the structural members, conductors and ground wire of self-supporting steel lattice towers for overhead transmission lines shall be in accordance with the specifications laid down, from time to time, by the Bureau of Indian Standards.
2. Overhead lines not covered in sub-rule (1) shall have the following minimum factors of safety:

for metal supports	1.5
for hand-moulded concrete supports	3.0
for mechanically processed concrete supports	2.5
for wood supports	3.0

Explanation I

The minimum factors of safety shall be based on such load as may cause failure of the supports to perform its function (assuming that the foundation and other components of the structure are intact).

Explanation II

The load shall be equivalent to the yield point stress or the moulds of rupture, as the case may be, for supports subject to bending and vertical loads and the crippling load for support used as struts.

Explanation III

The strength of the supports of the overhead lines in the direction of the line shall not be less than one-fourth of the strength required in the direction transverse to the line.

Explanation IV

The minimum factor of safety for the stay wires, guard wires or breaker wires shall be 2.5 based on the ultimate tensile strength of the wire.

Explanation V

The minimum factor of safety for conductors shall be based on their ultimate tensile strength. In addition the conductor tension at 32°C without external load shall not exceed the following percentages of the ultimate strength of the conductor:

Initial unloaded tension	35 per cent
Final unloaded tension	25 per cent

Provided that the conductors having a cross-section of a generally triangular shape, such as conductors composed of 3 wires, the final unloaded tension at 32°C shall not exceed 30 per cent of the ultimate tensile strength of such conductor.

3. For the purpose of calculating the factors of safety prescribed under sub-rule (2).
   (a) The maximum wind pressure shall be as specified in the relevant Indian Standards.
   (b) For cylindrical bodies, the effective area shall be taken as full projected area exposed to wind pressure; and
   (c) The maximum and minimum temperature shall be as specified in the relevant Indian Standards.
4. Notwithstanding anything contained in sub-rules (2) and (3), in localities where overhead lines are liable to accumulations of ice or snow the appropriate Government may, by order in writing, specify the loading conditions for the purpose of calculating the factor of safety.

Rule 77. Clearances Above Ground of the Lowest Conductor

1. No conductor of an overhead line, including service lines erected across a street shall at any part thereof be at height less than:
   (a) for low or medium voltage lines — 5.8 metres
   (b) for high voltage lines — 6.1 metres
2. No conductor of an overhead line, including service lines, lines erected along any street shall at any part thereof be at a height less than:
   (a) for low and medium voltage lines — 5.5 metres
   (b) for high voltage lines — 5.8 metres
3. No conductor of an overhead line including service lines, lines erected elsewhere than along or across any street shall be at a height less than:
   (a) for low, medium and high voltage lines up to and including 11,000 volts, if bare — 4.6 metres
   (b) for low, medium and high voltage lines up to and including 11,000 volts if insulated — 4.0 metres
   (c) for high voltage lines above 11,000 volts — 5.2 metres
4. For extra high voltage lines the clearance above ground shall not be less than 5.2 metres plus 0.3 metres for every 33,000 volts.

   Provided that the minimum clearance along or across any street shall not be less than 6.1 metres.

Rule 78. Clearance between Conductors and Trolley-wires

1. No conductor of an overhead line crossing a tramway or trolley bus route using trolley-wires shall have less than the following clearances above any trolley-wire.
   (a) low and medium voltage lines — 1.2 metres

   Provided that where an insulated conductor suspended from a bearer wire crosses over a trolley-wire, the minimum clearance for such insulated conductor shall be 0.6 metre.

   (b) high voltage lines up to and including 11,000 volts — 1.8 metres
   (c) high voltage lines above 11,000 volts — 2.5 metres
   (d) extra high voltage lines — 3.0 metres
2. In any case of crossing referred to in sub-rule (1), whoever lays his line later in time, shall provide the clearance between his own lines and the line which will be crossed in accordance with the provisions of said sub-rule:

   Provided that if the later entrant is the owner of the lower line and is not able to provide adequate clearance, he should bear the cost for modification of the upper line so as to comply with this rule.

Rule 79. Clearances from Buildings of Low and Medium Voltage Lines and Service Lines

1. Where a low or medium voltage overhead line passes above or adjacent to or terminates on any building, the following minimum clearances from any accessible point, on the basis of maximum sag, shall be observed:

(a) For any flat roof, open balcony, verandah roof and lean-to-roof:
   (i) When the line passes above the building, a vertical clearance of 2.5 metres from the highest point; and
   (ii) When the line passes adjacent to the building, a horizontal clearance of 1.2 metres from the nearest point, and

(b) For pitched roof:
   (i) When the lines passes above the building, a vertical clearance of 2.5 metres immediately under the lines, and
   (ii) When the line passes adjacent to the building, a horizontal clearance of 1.2 metres.

2. Any conductor so situated as to have a clearance less than that specified in sub-rule (1) shall be adequately insulated and shall be attached at suitable intervals to a bare earthed bearer wire having a breaking strength of not less than 350 kg.
3. The horizontal clearance shall be measured when the line is at a maximum deflection from the vertical due to wind pressure.

Rule 80. Clearance from Buildings of High and Extra High Voltage Lines

1. Where a high or extra high voltage overhead line passes above or adjacent to any building or part of a building it shall have on the basis of maximum sag a vertical clearance above the highest part of the building immediately under such line, of not less than:

(a) for high voltage lines up to and including 33,000 volts	3.7 metres
(b) for extra high voltage line	3.7 metres plus 0.3 metre for every additional 33,000 volts or part thereof.

2. The horizontal clearance between the nearest conductor and any part of such building shall, on the basis of maximum deflection due to wind pressure, be not less than:

(a) for high voltage lines up to and including 11,000 volts	1.2 metres
(b) for big voltage lines above 11000 volts and up to and including 33,000 volts	2.0 metres
(c) for extra high voltage lines	2.0 metres plus 0.3 metre for every additional 33,000 volts, or part thereof

Rule 81. Conductors at Different Voltages on Same Supports

Where conductors forming parts of systems at different voltages are erected on the same supports, the owner shall make adequate provision to guard against danger to linesmen and others from the lower voltage system being charged above its normal working voltage by leakage from or contact with the higher voltage system; and the methods of construction and the clearances between the conductors of the two systems shall be subject to the prior approval of the Inspector.

Rule 82. Erection of or Alteration to Buildings, Structures, Flood Banks and Elevation of Roads

1. If at any time subsequent to the erection of an overhead line (whether covered with insulating material or bare), any person proposes to erect a new building or structure or flood bank or to raise any road level or to carry out any other type of work whether permanent or temporary or to make in or upon any building or structure or flood bank or road, any permanent or temporary addition or alteration, he and the contractor whom he employs to carry out the erection, addition or alteration, shall, if such building, structure, flood bank, road or additions and alterations, thereto would, during or after the construction result in contravention of any of the provisions of rules 77, 79 or 80 give notice in writing of his intention to the supplier, and to an Inspector and shall furnish therewith scale drawing showing the proposed building, structure, addition or alteration and scaffolding required during its construction.
2. (a) On receipt of the notice referred to in sub-rule (1) or otherwise, the supplier shall examine, whether the line under reference was lawfully laid and whether the person was liable to pay the cost of alteration and if so, send a notice without undue delay, to such person together with an estimate of the cost of the expenditure likely to be incurred to so alter the overhead line and require him to deposit, within thirty days of the receipt of the notice with the supplier, the amount of the estimated cost.

(b) If the person referred to in sub-rule (1) disputes the supplier's estimated cost of alteration of the overhead line or even the responsibility to pay such cost, the dispute may be referred to the Inspector by either of the parties whereupon the same shall be decided by the Inspector.

3. No work upon such building, structure, flood bank, road and addition or alteration thereto shall be commenced or continued until the Inspector has certified that the provisions of rules 77, 79 or 80 are not likely to be contravened either during or after the aforesaid construction:

   Provided that the Inspector may, if he is satisfied that the overhead line has been so guarded as to secure the protection of persons or property from injury, or risk of injury, permit the work to be executed prior to the alteration of the overhead line, or, in case of temporary addition or alteration, without alteration of overhead line.
4. On receipt of the deposit, the supplier shall alter the overhead line within one month of the date of deposit or within such longer period as the Inspector may allow and ensure that it shall not contravene the provisions of rules 77, 79 or 80 either during or after such construction.
5. In the absence of an agreement to the contrary between the parties concerned, the cost of such alteration of the overhead line laid down shall be estimated on the following basis, namely:
   (a) The cost of additional material used on the alteration giving due credit for the depreciated cost of the material which would be available from the existing line;
   (b) The wages of labour employed in effecting the alteration;
   (c) Supervision charges to the extent of 15 per cent of item (b); and

(d) Any change incurred by the supplier to complying with the provisions of Section 16 in respect of such alterations.

6. Where the estimated cost of the alteration of the overhead line is not deposited, the supplier shall be considered an aggrieved party for the purpose of this rule.

Rule 82A. Transporting and Storing of Material Near Overhead Lines

1. No rods, pipes or similar materials shall be taken below or in the vicinity of any bare overhead conductors or lines if they are likely to infringe the provisions for clearances under rules 79 and 80, unless such materials are transported under the direct supervision of a competent person authorized in this behalf by the owner of such overhead conductors or lines.
2. Under no circumstances rods, pipes or other similar materials shall be brought within the flashover distance of bare live conductors or lines; and
3. No material or earth work or agricultural produce shall be dumped or stored or trees grown below or in the vicinity of bare overhead conductors lines so as to reduce the requisite safety clearances specified under rules 79 and 80.

Rule 83. Clearance: General

1. For the purpose of computing the vertical clearance of an overhead line, the maximum sag of any conductor shall be calculated on the basis or the maximum sag in still air and maximum temperature as specified by the State Government under rule 76 (2) (d). Similarly, for the purpose of computing any horizontal clearance of an overhead line, the maximum deflection of any conductor shall be calculated on the basis of the wind pressure specified by the State Government under rule 72 (2) (a) or may be taken as 35°, whichever is greater.
2. No blasting for any purpose shall be done within 300 metres from the boundary of substation or from the High Voltage or Extra High Voltage electric supply lines or tower structure without the consultation of the owner of such substation or electric supply lines or tower structures and in case of mining leasehold area, without the written permission of the Chief Inspector of Mines or the Electrical Inspector of Mines.

Rule 84. Routes: Proximity to Aerodrome

Overhead lines shall not be erected in the vicinity of aerodrome until the aerodrome authorities have approved in writing the route of the proposed lines.

Rule 85. Maximum Intervals between Supports

All conductors shall be attached to support at intervals not exceeding the safe limits based on the ultimate tensile strength of the conductor and the factor of safety prescribed in rule 76.

Provided that in the case of overhead lines carrying low or medium voltage conductors, when erected in, over, along or across any street, the interval shall not, without the consent in writing of the Inspector, exceed 65 metres.

Rule 86. Conditions to Apply Where Telecommunication Lines and Power Lines are Carried on Same Supports

1. Every overhead telecommunication line erected on supports carrying power line shall consist of conductors each having a breaking strength of not less than 270 kg.
2. Every telephone used on a telecommunication line erected on supports carrying a power line shall be suitably guarded against lighting and shall be protected by cutouts.
3. Where a telecommunication line is erected on supports carrying a high or extra high voltage power line arrangement shall be made to safeguard any person using the telephone against injury resulting from contact, leakage or induction between such power and telecommunication lines.

Rule 87. Lines Crossing or Approaching Each Other

1. Where an overhead line crosses or is in proximity to any telecommunication line, either the owner of the overhead line or the communication line, whoever lays his line later, shall arrange to provide for protective devices or guarding arrangement, in a manner laid down in the Code of Practice or the guide line prepared by the Power and Telecommunication Coordination Committee and subject to the provisions of the following sub-rules.
2. When it is intended to erect a telecommunication line or an overhead line which will cross or be in proximity to an overhead line or a telecommunication line, as the case may be, the person proposing to erect such line shall give one month's notice of his intention so to do along with the relevant details of protection and drawings to the owner of the existing line.
3. Where an overhead line crosses or is in proximity to an overhead line, guarding arrangements shall be provided so as to guard against the possibility of their coming into contact with each other.

Where an overhead line crosses another overhead line, clearances shall be as under:

Minimum clearances in metres between lines when crossing each other

S/N	Nominal system voltage	11-66 kV	110-132 kV	220 kV	400 kV	800 kV
1.	Low & Medium	2.44	3.05	4.58	5.49	7.94
2.	11-66 kV	2.44	3.05	4.58	5.49	7.94
3.	110-132 kV	3.05	3.05	4.58	5.49	7.94
4.	220 kV	4.58	4.58	4.58	5.49	7.94
5.	400 kV	5.49	5.49	5.49	5.49	7.94
6.	800 kV	7.94	7.94	7.94	7.94	7.94

Provided that no guarding are required when an extra voltage line crosses over another extra high voltage, high voltage, medium or low voltage line or a road or a tram subject to the condition that adequate clearances are provided between the lowest conductor of the extra high voltage line and the topmost conductor of the overhead line crossing underneath the extra high voltage line and the clearances as stipulated in rule 77 from the topmost surface of the road is maintained.

4. A person erecting or proposing to erect a line which may cross or be in proximity with an existing line, may normally provide guarding arrangements on his own line or require the owner of the other overhead line to provide guarding arrangements as referred to in sub-rule (3).
5. In all cases referred to in the preceding sub-rules, the expenses of providing the guarding arrangements shall be borne by the person whose line was last erected.
6. Where two lines cross, the crossing shall be made as nearly at right angles as the nature of the case admits and as near the support of the line as practicable and the support of the lower line shall not be erected below the upper line:

   Provided that the angle of crossing of power lines shall not be less than 60°.
7. The guarding arrangement shall ordinarily be carried out by the owner of the supports on which it is made and he shall be responsible for its efficient maintenance.
8. All work required to be done by or under this rule shall be carried out to the satisfaction of the Inspector.

Rule 88. Guarding

1. Where guarding is required under these rules, the provisions of sub-rules (2) to (4) shall apply.
2. Every guard wire shall be connected with earth at each point at which its electrical continuity is broken.
3. Every guard wire shall be an actual breaking strength of not less than 635 kg and if made of iron or steel, shall be galvanized.
4. Every guard wire of cross-connected system of guard wires, shall have sufficient current carrying capacity to ensure the rendering dead, without risk of fusing of the guard wire or wires till the contact of any line wire has been removed.
5. *Lines crossing trolley wires*: In the case of crossing over a trolley wire the guarding shall fulfil the following conditions, namely:
   - (a) Where there is only one trolley-wire, two guard-wires shall be erected as in diagram A;
   - (b) Where there are two trolley-wires and the distance between them does not exceed 40 cm, two guard-wires shall be erected as in diagram B;
   - (c) Where there are two trolley-wires and the distance between them exceed 40 cm but does not exceed 1.219 metres, three guard-wires shall be erected as in diagram C;
   - (d) Where there are two trolley-wires and the distance between them exceeds 1.2 metres, each trolley-wire, shall be separately guarded as in diagram D;

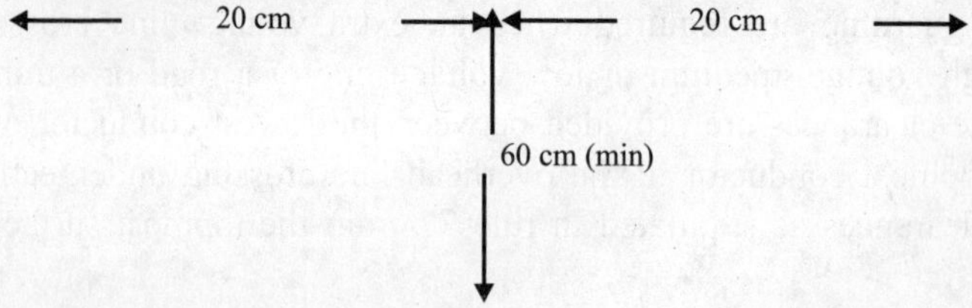

Diagram A

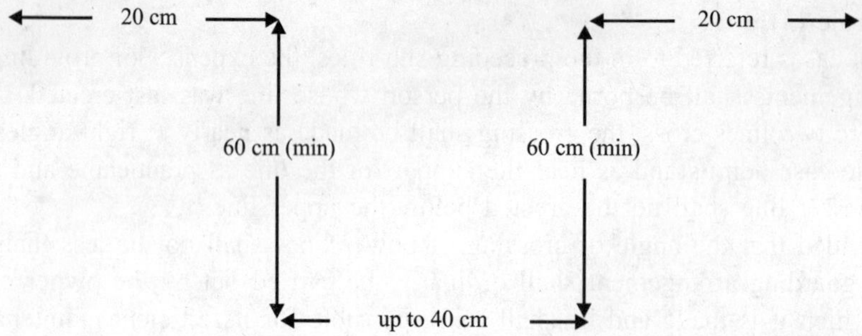

Diagram B

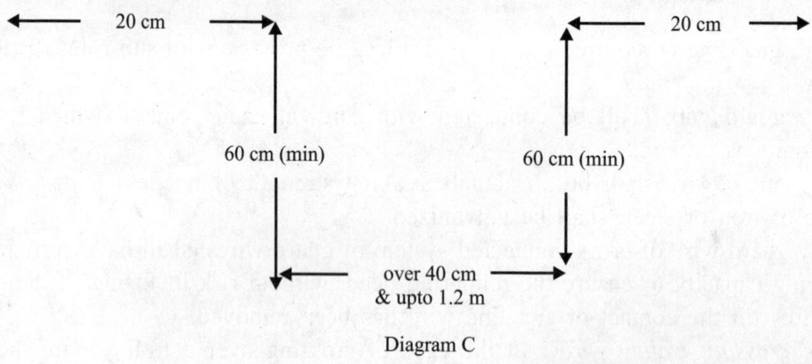

Diagram C

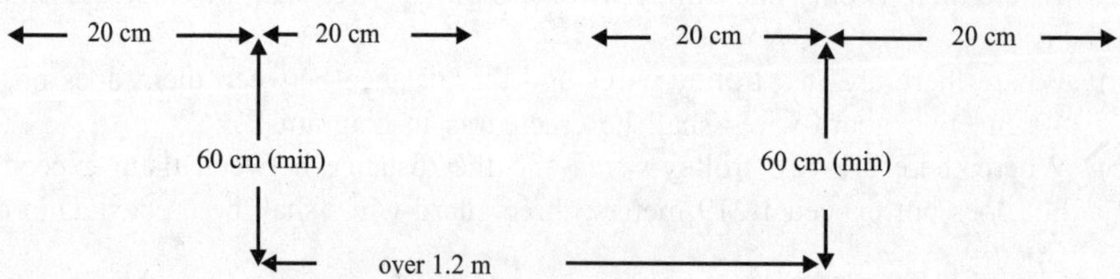

Diagram D

(e) The rise of trolley boom shall be so limited that if the trolley leaves the trolley wire, it shall not foul the guard, wire; and

(f) Where a telegraph line is liable to fall or be blown down upon an arm, stay wire or span wire and so slide down upon a trolley-wire, guard hooks shall be provided to prevent such sliding.

Rule 89. Service Lines from Overhead Lines

No service line or lapping shall be taken off from an overhead line except at a point of support.

Provided that the number of trappings per conductor shall not be more than four is case of low and medium voltage connections.

Rule 90. Earthing

1. All metal supports and all reinforced and pre-stressed cement concrete supports of overhead lines and metallic fittings attached thereto shall be permanently and efficiently earthed. For this purpose, a continuous earth wire shall be provided and securely fastened to each pole and connected with earth ordinarily at four points in every km, the spacing between the points being as nearly equidistant as possible. Alternatively, each support and the metallic fitting attached thereto shall be efficiently earthed.

1A. Metallic bearer wire used for supporting insulated wires of low and medium voltage overhead service lines shall be efficiently earthed or insulated.

2. Each stay wire shall be similarly earthed unless and insulator has been placed in it at a height not less than 3.0 metres from the ground.

Rule 91. Safety and Protective Devices

1. Every overhead line (not being suspended from a dead bearer wire and not being covered with insulating material and not being a trolley-wire) erected over any part of a street or other public place or in any factory or mine or on any consumer's premises shall be protected with a device approved by the Inspector for rendering the line electrically harmless in case it breaks.
2. An inspector may by notice in writing require the owner of any such overhead line wherever it may be erected to protect it in the manner specified in sub-rule (1).
3. The owner of every high and extra high voltage overhead line shall make adequate arrangements to the satisfaction of the Inspector to prevent unauthorized persons from ascending any of the supports of such overhead lines which can be easily climbed upon without the help of a ladder or special appliances. Rails, reinforced cement concrete poles and pre-stressed cement concrete poles without steps, tubular poles, wooden supports without steps, I-section and channels shall be deemed as supports which cannot be easily climbed upon for the purpose of this rule.

Rule 92. Protection Against Lightning

1. The owner of every overhead line (substation or generating station) which is so exposed as to be liable to injury from lightning shall adopt efficient means for diverting to earth any electrical surges due to lightning.
2. The earthing lead for any lightning arrestor shall not pass through any iron or steel pipe, but shall be taken as directly as possible from the lightning arrestor to a separate earth electrode and/or junction of the earth mat already provided for the high and extra high voltage substation subject to the avoidance of bends wherever practicable.
   Note: A vertical ground electrode shall be connected to this junction of the earth mat.

Rule 93. Unused Overhead Lines

1. Where an overhead line ceases to be used as an electric supply line, the owner shall maintain it in a safe mechanical condition in accordance with rule 76 or shall remove it.
2. Where any overhead line ceases to be used as an electric supply line, an Inspector may, by a notice in writing served on the owner, require him to maintain it in a safe mechanical condition or to remove it within fifteen days of the receipt of the notice.

CHAPTER-IX

14.12 ELECTRIC TRACTION

Rule 94. Additional Rules for Electric Traction

1. The rules in this chapter apply only where energy is used for purposes of traction:
   Provided that nothing in this chapter shall apply to energy used for the public carriage of passengers, animals or goods on, or for the lightning or ventilation of the rolling stock of any railway or tramway subject to the provisions of the Indian Railways Act 1890 (IX of 1890).
2. In this chapter, the conductor used for transmitting energy to a vehicle is referred to as the "line" and the other conductor as the "return".
3. The owner of the line, return, rails or trolley-wire, as the case may be, shall be responsible for the due observance of rules 95 to 108.

Rule 95. Voltage of Supply to Vehicle

No person shall supply energy at high or extra-high voltage to any trolley-wire or other conductor used in direct electrical and mechanical connection with any vehicle, except with the written approval of the Central or the State Government as the case may be, and subject to such conditions as the Central or the State Government may think reasonable and proper in the circumstances.

Rule 96. Insulation of Lines

Every line shall be insulated throughout.

Rule 97. Insulation of Returns

1. Where any rails on which cars run, or any conductors laid between or within 0.9 metre of such rails, form any part of a return, such part may be uninsulated. All other returns or parts of a return, shall be insulated, unless they are of such conductivity as to secure the conditions required by sub-rules (2) and (3) of rule 98.
2. Where any part of a return is uninsulated, it shall be connected with the negative or neutral of the system.

Rule 98. Proximity to Metallic Pipes, etc.

1. Where an uninsulated return is in proximity to any metallic pipe, structure or substance, not belonging to the owner of the return, he shall, if so required by the owner of such pipe, structure or substance, connect his return therewith at the latter's expense.
2. Where the return is partly or entirely uninsulated, the owner shall, in the construction and maintenance of his system, adopt such means for reducing the difference produced by the current between the potential of the uninsulated return at any one point and the potential of the uninsulated return at any other point as to ensure that the difference of potential between the uninsulated return and metallic pipe, structure or substance in the vicinity shall not exceed four volts where the return is relatively positive, or one and one-third volts where the return relatively negative.
3. The owner of any such pipe, structure or substance as is referred to in sub-rule (2) may, in respect of it, require the owner of the uninsulated return at reasonable times and intervals to ascertain by test in his presence, or in that of his representative, whether that condition specified in sub-rule (2) is fulfilled, and, if such condition is found to be fulfilled, all reasonable expenses of, and incidental to, the carrying out of the test shall be borne by the owner of the pipe, structure or substance.

Rule 99. Difference of Potential on Return

Where the return is partly of entirely uninsulated, the owner shall keep a continuous record of the difference of potential, during the working of his system, between every junction of an insulated return with an uninsulated return and the point on the route most distant from that junction, and the difference of potential shall not, under normal running conditions, exceed a mean value of seven volts between the highest momentary peak and the average for the hour of maximum load.

Rule 100. Leakage on Conduit System

Where both the line and the return are placed within a conduit, the following conditions shall be fulfilled in the construction and maintenance of the system:

(a) Where the rails are used to form any part of the return, they shall be electrically connected (at distances not exceeding 30 metres apart), with the conduit by means of copper strips having a cross-sectional area of at least 0.40 sq. cm, or by other means of equal conductivity. Where the return is wholly insulated and contained within the conduit, the latter shall be connected with earth at the generating station or substation through an instrument suitable for the indication of any contact or partial contact of either the line or the return with the conduit; and

(b) The leakage current shall be ascertained daily, before or after the hours of running, when the line is fully charged; and if at any time it is found to exceed 0.6 ampere per km of single tramway track, the transmission and use of energy shall be suspended unless the leakage is stopped within twenty-four hours.

Rule 101. Leakage on System other than Conduit System

Where both the line and the return are not placed within a conduit, the leakage-current shall be ascertained daily before or after the hours of running, when the line is fully charged; and if at any time it is found to exceed 0.3 ampere per km of single tramway track, the transmission and use of energy shall be suspended unless the leakage is stopped within twenty-four hours.

Rule 102. Passengers Not to have Access to Electric Circuit

Precautions to the satisfaction of an Inspector shall be taken by the owner of every vehicle to prevent:

(a) the access of passengers to any portion of the electric circuit where there is danger from electric shock;

(b) any mental, handrail or other metallic substance liable to be handled by passengers, becoming charged.

Rule 103. Current Density in Rails

Where rails on which cars run are used as a return, the current density in such rails, shall not under ordinary working conditions, exceed 1.4 amperes per sq. cm of cross-sectional area.

Rule 104. Isolation of Sections

Every trolley-wire shall be constructed in sections not exceeding 1.6 km in length, and means shall be provided for isolating each section.

Rule 105. Minimum Size and Strength of Trolley-wire

No trolley-wire shall be of less cross-sectional area than 0.5 sq. cm or shall have an actual breaking load of less than 2000 kg.

Rule 106. Height of Trolley-wire and Length of Span

A trolley-wire or a traction feeder on the same supports as a trolley-wire shall nowhere be at a height from the surface of the street of less than 5.2 metres except, where it passes under a bridge or other fixed structure, or through or along a tunnel or mine-shaft or the like in which case it shall be suspended to the satisfaction of an Inspector.

Rule 107. Earthing of Guard Wires

Every guard wire shall be connected with earth at each point at which its electrical continuity is broken and shall also be connected with the rails at intervals of not more than five spans.

Rule 107A. Proximity to Magnetic Observatories and Laboratories

Traction works shall not be carried out in the vicinity of geomagnetic observatories and laboratories without the concurrence of the Central Government or of any officer authorized by it in this behalf.

Rule 108. Records

1. The owner shall, so far as is consistent with his system of working, keep the following records, namely:
   - (a) daily records showing:
     - (i) the maximum working current from the source of supply;
     - (ii) the maximum working voltage at the source of supply;
     - (iii) the difference of potential, as required by rule 99; and
     - (iv) the leakage current (if any), as required by rule 100 and rule 101, and
   - (b) occasional records showing:
     - (i) every test made under sub-rules (2) and (3) of rule 98;
     - (ii) every stoppage of leakage, together with the time occupied; and
     - (iii) particulars of any abnormal occurrence affecting the electrical working of the system.
2. Such records shall be open to examination by an Inspector or by any person authorized in writing by an Inspector.

CHAPTER-X

14.13 ADDITIONAL PRECAUTIONS TO BE ADAPTED IN MINES AND OILFIELDS

Rule 109. Application of Chapter

1. The rules in this chapter shall supply only where energy is used in mines as defined in the Mines Act, 1952 (35 of 1952).

2. In mines and oilfields, the rules in this chapter shall not apply to apparatus above the ground level except where such apparatus may directly affect the safety of the persons employed in underground, opencast and oilfields.

Rule 110. Responsibility for Observance

1. It shall be the duty of the owner, agent, engineer or manager of a mine, or of the agent, engineer of any company operating in an oilfield, or of the owner, engineer of one or more drilled wells situated in an oilfield, to comply with and enforce the following rules and it shall be duty of all persons employed to conduct their work in accordance with such rules.
2. Adequate number of authorized supervisors and electricians shall be on duty in every mine or oilfield while energy is being used therein.

   Explanation: For the purpose of this rule, the word "engineer" shall:
   (a) in the case of a coal mine, have the same meaning as assigned to it in the Coal Mines Regulations, 1957;
   (b) in the case of a metalliferous mine, have the same meaning as assigned to it in the Metalliferous Mines regulations, 1961; and
   (c) in the case of an oil mine, mean the "Installation Manager" under the Oil Mines Regulations, 1984.

Rule 111. Notices

1. On or before the first day of February in every year, in respect of every mine or oil-field, returns giving the size and type of apparatus, together with such particulars in regard to circumstances of its use which may be required by the Inspector, shall be sent to the Inspector by the persons specified in rule 110 in the form set out in Annexure X or XI, whichever is applicable.
2. The persons specified in rule 110 shall also give to the Inspector not less than seven days notice in writing of the intention to bring into use any new installation in a mine or oilfield giving details of apparatus installed and its location.

   Provided that in case of any additions or alterations to an existing low and medium voltage installation, immediate notice in writing shall be sent to the Inspector before such additions or alterations are brought into use.
3. This rule shall not apply to telecommunication or signalling apparatus.

Rule 112. Plans

1. A correct plan, on the same scale as the plan kept at the mine in fulfilment of the requirements of the Mines Act, 1952 (XXXV of 1952), shall be available in the office at the mine showing the position of all fixed apparatus and conductors therein, other than lights, telecommunication or signalling apparatus, or cables for the same.

2. A similar plan, on the scale not less than 25 cm to a km (1 : 4000) shall keep by the manager or owner of one or more wells in any oilfields.
3. A similar plan, on such scale, as the Central Government may direct showing the position of all electricity supply lines, shall be kept in the office of a licensee or other person transmitting or distributing energy in a mine or oilfield.
4. The plans specified under the provisions of this rule shall be examined, and corrected as often as necessary to keep them reasonably up-to-date. The dates of such examinations shall be entered thereon by the manager or owner of the mine or wells and such plans shall be available to the Inspector, or an Inspector of Mines, at any time.

Rule 113. Lighting, Communications and Fire Precautions

1. In a mine illuminated by electricity, one or more flame safety lamps, other lights approved by the Inspector of Mines, shall be maintained in a state continuous illumination in all places where failure of the electric light at any time would be prejudicial to safety.
2. Efficient means of communication shall be provided in every mine between the point where the switchgear provided under sub-rule (1) of rule 121 erected and the shaft-bottom or other distributing centres in the mines.
3. Fire extinguishing appliances of adequate capacity and of an approval type shall be installed and properly maintained in every place in a mine containing apparatus, other than cables, telecommunication and signalling apparatus.
4. In case of mines, minimum clearance above ground of the low conductor of overhead lines or overhead cables where dumpers or trackless vehicle are being operated, shall not be less than twelve metres in height from the ground across the road where dumpers or trackless vehicles cross.

Rule 114. Isolation and Fixing of Transformer, Switchgear, etc.

1. Where necessary to prevent danger of mechanical damage, transformer and switchgear shall be placed in a separate room, compartment or box.
2. Unless the apparatus is so constructed, protected and worked as to obviate the risk of fire, no inflammable material shall be used in the construction of a room, compartment or box containing apparatus, or in the construction of any fittings therein. Each such room, compartment or box shall be substantially constructed and shall be kept dry and illuminated and efficient ventilation shall be provided for all apparatuses installed therein.
3. Adequate working space and means of access, clear of obstruction and free from danger, shall so far as circumstances permit, be provided for all apparatuses that have to be worked or attended to and all handles intended to be operated shall be conveniently placed for that purpose.

Rule 115. Method of Earthing

Where earthing is necessary in a mine it shall be carried out by connection to an earthing system at the surface of the mine, in a manner approved by the Inspector.

Rule 116. Protective Equipment

1. In the interest of safety, appropriate equipment shall be suitably placed in the mines for automatically disconnecting supply to any part of the system, where a fault, including an earth fault occurs. Fault current shall not be more than 750 milliamps in 550/1100 volt systems for underground/oilfields and 50 amps in 3.3 kV/6.6 kV systems in open cast mines. The magnitude of the earth fault current shall be limited to these specified values by employing suitably designed, restricted neutral system of power supply.
2. The operation of the switchgear and the relays shall be recorded daily at the generating station, substation or switch-station in a register kept for the purpose.
3. The effectiveness of the switchgear and the protective system shall always be kept and maintained in working order, shall be checked once every three months and the result thereof shall be recorded in a separate register kept for the purpose.

Rule 117. Earthing Metal, etc.

1. All metallic sheaths, coverings, handles, joint-boxes, switchgear frames, instrument covers, switch and fuse covers or boxes, all lamp-holders unless efficiently protected by an insulated covering made of fire-resisting material, and the frames and bed-plates of generators, transformers and motors (including portable motors), shall be earthed by connection to an earthing system in the manner prescribed in rule 115.
2. Where cables are provided with a metallic covering constructed and installed in accordance with clause (d) of rule 122, such metallic covering may be used as a means of connection to the earthing system.
3. All conductors of an earthing system shall have conductivity, at all parts and at all joints, at least equal to 50 per cent, of that of the largest conductor used solely to supply the apparatus, a part of which it is desired to earth:

   Provided that no conductor of an earthing system shall have a cross-sectional area less than 0.15 sq. cm, except in the case of the earth conductor of a flexible cable used with portable apparatus where the voltage does not exceed 125 volts, and the cross-sectional area and conductance of the earth-core in not less than that of the largest of the live conductors in the cable.
4. All joints in earth conductors and all joints in the metallic covering of cables shall be properly soldered or otherwise efficiently made.
5. No switch, fuse or circuit breaker shall be inserted in any earth conductor.
6. This rule shall not apply (except in the case of portable apparatus) to any system in a mine in which the voltage does not exceed 30 volts.

Rule 118. Voltage Limits

Energy shall not be transmitted into a mine at a voltage exceeding 11,000 volts and shall not be used therein at a voltage exceeding 6,600 volts:

Provided that:

(a) Where hand-held portable apparatus is used, the voltage shall not exceed 125 volts;

(b) Where electric lighting is used:

  (i) in underground mines, the lighting system shall have a mid or neutral point connected with earth and the voltage shall not exceed 125 volts between phases;

  (ii) on the surface of a mine or in an open cast mine, the voltage may be raised to 250 volts, if the neutral or the mid-point of the system is connected with earth and the voltage between the phases does not exceed 250 volts;

(c) Where portable hand-lamps are used in underground working of mine, the voltage shall not exceed 30 volts.

(d) Where any circuit is used for the remote control or electric interlocking of apparatus, the circuit voltage shall not exceed 30 volts.

Provided that in fixed plants, the said voltage may be permitted up to 650 volts, if the bolted type plug is used.

Rule 119. Transformers

Where energy is transformed, suitable provision shall be made to guard against danger by reason of the lower voltage apparatus becoming accidentally charged above its normal voltage by leakage from or contact with the higher voltage apparatus.

Rule 120. Switchgear and Terminals

Switchgear and all terminals, cable-ends, cable-joints and connections to apparatus shall be totally enclosed and shall be so constructed, installed and maintained, as to comply with the following requirements:

(i) all parts shall be of mechanical strength sufficient to resist rough usage;

(ii) all conductors and contact areas shall be of adequate current carrying capacity and all joints in conductors shall be properly soldered or otherwise efficiently made;

(iii) the lodgment of any matter likely to diminish the insulation or affect the working of any switchgear shall be prevented.

(iv) all live parts be so protected or enclosed as to prevent persons accidentally coming into contact with them and to prevent danger from arcs, short circuits, fire, water, gas or oil;

(v) where there may be risk of igniting gas, coal-dust, oil or other inflammable material, all parts shall be so protected as to prevent open sparking and

(vi) every switch or circuit breaker shall be so constructed as to be capable of opening the circuit it controls, and dealing with any short circuit without danger.

Rule 121. Disconnection of Supply

1. Properly constructed switchgear for disconnecting the supply of energy to a mine or oil-field shall be provided at a point approved by the Inspector. During the time any cable supplying energy to the mine room the aforesaid switchgear is live, a person authorized to operate the said switchgear shall be available within easy reach thereof:

Provided that in the case of gassy coal seam of degree-II and degree-III, the main mechanical ventilator operated by electricity shall be interlocked with the switchgear so as to automatically disconnect the power supply in the event of stoppage of main mechanical ventilator.

2. When necessary in the interest of safety, appropriate apparatus suitably placed, shall be provided for disconnecting the supply from every part of a system.
3. Where considered necessary by the Inspector in the interests of safety, the apparatus specified in sub-rule (2) shall be so arranged as to disconnect automatically from the supply any section of the system subjected to a fault.
4. Every motor shall be controlled by switchgear, which shall be so arranged as to disconnect the supply from the motor, and from all apparatuses connected thereto. Such switchgear shall be so placed as to be easily operated by the person authorized to operate the motor.
5. Wherever required by the Inspector the motor shall be controlled by switchgear to disconnect automatically the supply in the event of conditions of over-current, over-voltage and single phasing.

Rule 122. Cables

All cables, other than flexible cables for portable or transportable apparatus, shall comply with the following requirements:

(a) All such cables (other than the outer conductor of a concentric cable) shall be covered with insulating material and shall be efficiently protected from mechanical damage and supported at sufficiently frequent intervals, and in such a manner as to prevent damage to such cables;

(b) (i) except as provided in clause (c), no cables other than concentric cables or two-core or multi-core cables protected by a metallic covering, or single-core cables protected by a metallic covering and which contain all the conductors of a circuit shall be used.

1. Where the voltage exceeds 125 volts; or
2. When an Inspector considers that there is risk of igniting gas or coal-dust or other inflammable material, and so directs;

(ii) the sheath of metal-sheathed cables and the metallic armouring of armoured cables shall be of a thickness not less than that recommended from time to time in the appropriate standards of the Bureaus of Indian Standards;

(c) where a medium voltage direct current system is used, two single-core cables may be used for any circuit provided that their metallic coverings are bonded together by earth conductors so placed that the distance between any two consecutive bonds is not greater than 30 metres measured along either cable;

(d) the metallic covering of every cable shall be:

(i) electrically and mechanically continuous throughout;

(ii) earthed, if it is required by sub-rule (1) of rule 117 to be earthed by a connection to the earthing system of conductivity not less than that of the same length of the said metallic covering;

(iii) efficiently protected against corrosion where necessary;

(iv) of a conductivity at all parts and at all joints at least equal to 50 per cent of the conductivity of the largest conductor enclosed by the said metallic covering; and

(v) where there may be risk of igniting gas, coal-dust, or other inflammable material, so constructed as to prevent, as far as is practicable, the occurrence of open sparking as the result of any fault or leakage from live conductors;

(e) cables and conductors where connected to motors, transformers, switchgear, and other apparatus, shall be installed so that:

(i) they are mechanically protected by securely attaching the metallic covering to the apparatus; and

(ii) the insulating material at each cable and is efficiently sealed so as to prevent the diminution of its insulating properties;

(f) where necessary to prevent abrasion or to secure gas-tightness, properly constructed glands or bushes shall be provided;

(g) unarmoured cables or conductors shall be conveyed either in metallic pipes or metal casings or suspended from efficient insulators by means of non-conducting materials which will not cut the covering and which will prevent contact with any timbering or metal work. If separate insulated conductors are used, they shall be installed at least 3.75 cm apart and shall not be brought together except at lamps, switches and fittings.

Rule 123. Flexible Cables

1. Flexible cables for portable or transportable apparatus shall be two-core or multi-core (unless required for electric welding), and shall be covered with insulating material, which shall be efficiently protected from mechanical injury. If a flexible metallic covering is used either as the outer conductor of a concentric cable or as a means of protection from mechanical injury, it shall not be used by itself to form an earth conductor for such apparatus, but it may be used for that purpose in conjunction with an earthing core.
2. Every flexible cable intended for use with portable or transportable apparatus shall be connected to the system and to such apparatus by properly constructed connectors;

   Provided that for high voltage machines a bolted type connector shall be used and the trailing cable shall be suitably anchored at the machine end;

   Provided further that where there are space limitation for multiple on-board motors and equipment for transportable or portable machines, direct entry flexible cable with elastometric ceiling rings, compression gland, packing gland or sealing box which does not alter the flameproof property may be permitted if a cable entry can accept any sealing right with same outside diameter but different internal dimension, the ring shall have a minimum uncompressed axial height of 20 mm for circular cables of diameter not greater than 20 mm and 25 mm for circular cables of diameter greater than 20 mm.
3. At every point where flexible cables are joined to main cable, a circuit breaker shall be provided which is capable of automatically disconnecting the supply from such flexible cables.
4. Every flexible cable attached to a portable or transportable machine shall be examined periodically by the person authorized to operate the machine, and, if such cable is used

underground, it shall be examined at least once in each shift by such person. If such cable is found to be damaged or defective, it shall forthwith be replaced by a cable in good condition.

5. If the voltage of the circuit exceeds low voltage, all flexible cables attached to any transportable apparatus shall be provided with flexible metallic screening or pliable armouring and cables of portable apparatus shall be provided with flexible metallic screening on all the power and pilot cores:

   Provided that the provision of this sub-rule shall not apply to flexible cables attached to any transportable apparatus used in open cast mines where reeling and unreeling of such cables is necessary as per design features of the equipment.

6. All flexible metallic screening or armouring specified in sub-rule (5) shall comply with the provisions of rule 122 (d).

   Provided that in the case of separately screened flexible cables the conductance of each such screen shall not be less than 25 per cent of that of the power conductor and the combined conductance of all such screens shall in no case be less than that of 0.15 sq cm copper conductors.

7. Flexible cable exceeding ninety metres in length shall not be used with any portable or transportable apparatus:

   Provided that such flexible cable when used with coal-cutting machines or cutter or loader or armoured face conveyor for long wall operation, or with shuttle cars or load haul dumper or cutter loader or all alike equipment, for development and de-pillaring operation, shall not exceed two hundred and fifty metres in length:

   Provided further that the aforesaid cable in case of an open cast mine when used with electrically operated heavy earth moving machinery shall not exceed three hundred metres in length and for bucked wheel excavator at 11 kV shall not exceed one thousand metres in length.

8. Flexible cable, when installed in a mine, shall be efficiently supported and protected from mechanical injury.

9. Flexible cables shall not be used with apparatus other than portable or transportable apparatus.

10. When flexible cables are used they shall be detached or otherwise insulated from the source of supply when not in use, and arrangements shall be made to prevent the energizing of such cables by unauthorized persons.

Rule 124. Portable and Transportable Machines

The person authorized to operate an electrically driven coal-cutter, or other portable or transportable machine, shall not leave the machine while it is in operation and shall, before leaving the area in which such machine is operating, ensure that the supply is disconnected from the flexible cable, which supplies the machine. When any such machine is in operation, steps shall be taken to ensure that the flexible cable is not dragged along by the machine:

Provided that all portable and transportable machines used in underground mines shall operate on remote control from the concerned switchgear with Pilot Core Protection.

Rule 125. Sundry Precautions

1. All apparatuses shall be maintained reasonably free from dust, dirt and moisture, and shall be kept clear of obstruction.
2. All apparatuses other than portable and transportable apparatus shall be housed in a room, compartment or box so constructed as to protect the contents from damage occasioned by falling material or passing traffic.
3. Inflammable or explosive material shall not be stored in any room, compartment, or box containing apparatus, or in the vicinity of any apparatus.
4. Should there be a fault in any circuit, the part affected shall be made dead without delay, and shall remain so until the fault has been remedied.
5. While lamps are being changed the supply shall be disconnected.
6. No lamp-holder shall be in metallic connection with the guard or other metal work of a portable hand-lamp.
7. The following notices, in Hindi and local language of the district, so designed and protected as to be easily legible at all times, shall be exhibited:
   (a) at all places where electrical apparatus is in use, a notice forbidding unauthorized persons to operate or otherwise interfere with such apparatus;
   (b) at those places in the interior or at the surface of the mine where a telephone or other means of communication is provided, a notice giving full instructions to persons authorized to effect the disconnection, at the surface of the mine, of the supply of energy to the mine.
8. All apparatuses, including portable and transportable apparatus, shall be operated only by those persons who are authorized for the purpose.
9. Where a plug-and-socket-coupling other than of bolted type is used with flexible cables, an electrical interlock or other approved device shall be provided to prevent the opening of the coupling while the conductors are live.

Rule 126. Precautions Where Gas Exists

1. In any part of a coal seam of the first-degree gassiness:
   (a) all cables shall be constructed, installed, protected, operated and maintained in such a manner as to prevent risk of open sparking;
   (b) at any place, which lies in-bye of the last ventilation connection, all signalling, telecommunication and remote control circuits shall be so constructed, installed, protected, operated and maintained as to be intrinsically safe;
   (c) all apparatuses including portable and transportable apparatus including lighting fittings used at any place, which lies in-bye of the last ventilation connection, shall be flameproof:

   Provided that electrically operated or battery operated portable or transportable apparatus such as shuttle car, men or material transporting equipment of increased safety type 'e' shall be permitted at any place with suitable monitoring devices for detection of gases, if any.

2. At any place which lies in any part of a coal seam of second and third degree gassiness:
   (a) all signalling, telecommunication and remote control circuits shall be so constructed, installed, protected, operated and maintained as to be intrinsically safe;
   (b) all cables shall be constructed, installed, protected, operated and maintained in such a manner as to prevent risk or open sparking;
   (c) all apparatus, including portable and transportable apparatus used at any place within 90 metres of any working face or goaf in case of a second degree gassy mine and within 270 metres of any working face or goaf in case of third degree gassy mine or at any place which lies in bye of the last ventilation connection or in any return airways shall be flameproof;
   (d) all electric lamps shall be enclosed in flameproof enclosures.
3. In any oil-mine or oilfield, at any place within the Danger Areas:
   (a) all signalling, telecommunication and remote control circuits shall be so constructed, installed, protected, operated and maintained as to be intrinsically safe;
   (b) all cables shall be constructed, installed, protected, operated and maintained in such a manner as to prevent risk or open sparking;
   (c) all apparatus including portable and transportable apparatus shall be flameproof;
   (d) all electric lamps shall be enclosed in flameproof enclosures.
4. In any coal seam of degree second and third degree gassiness or the danger zone of oil-mine the supply shall be disconnected:
   (a) immediately, if open sparking occurs;
   (b) during the period required for examination or adjustment of the apparatus, which would necessitate the exposing of any part liable to open sparking;
   (c) the supply shall not be reconnected until the apparatus has been examined by the electrical supervisor or one of his duly appointed assistants until the defect, if any, has been remedied or the necessary adjustment made;
   (d) a flame safety lamp shall be provided and maintained in a state of continuous illumination near an apparatus (including portable or transportable apparatus) which remains energized and where the appearance of the flame of such safety lamps indicates the presence of inflammable gas, the supply to all apparatus in the vicinity shall be immediately disconnected and the incident reported forthwith to an official of the mine, and such apparatus shall be interlocked with the controlling switch in such a manner as to disconnect power supply automatically in the event of percentage of inflammable gas exceeding one and one quarter in that particular district.

       Provided that where apparatus for automatic detection of the percentage of inflammable gas or vapour are employed in addition to the flame safety lamps, such apparatus shall be approved by the inspector of mines and maintained in perfect order.
5. (i) In any part of a coal-sea of any degree of gassiness or in any hazardous area of an oil-mine, if the presence of inflammable gas in the general body of air is found any time to exceed one and one quarter per cent, the supply of energy shall be immediately disconnected from all cables and apparatus in the area and the supply shall not be reconnected so long as the percentage of inflammable gas remains in excess of one and one quarter per cent.

In an oil-mine where concentration of inflammable gas exceeds 20% of its lowest explosive limit, the supply of electric energy shall be cut-off immediately from all cables and apparatus lying within 30 metres of the installation and all sources of ignition shall also be removed from the said area and normal work shall not be resumed unless the area is made gas-free.

Provided that such disconnection shall not apply to intrinsically safe environmental monitoring scientific instruments. .

(ii) Any such disconnection or reconnection of the supply shall be noted in the log sheet which shall be maintained in the form set out in Annexure XII and shall be reported to the Inspector.

6. The provisions of this rule shall apply to any metalliferous mine, which may be notified by the Inspector of mines if inflammable gas occurs or if the Inspector of mines is of the opinion that inflammable gas is likely to occur in such mine.

Explanation: For the purpose of this rule:

1. The expressions "coal-seam of first degree gassiness", "coal-seam of second degree gassiness", "coal seam of third degree gassiness" and "flameproof apparatus", shall have the meanings respectively assigned to them in the Coal Mines Regulations, 1957.
2. The following areas in an oil-mine or oilfield shall be known as hazardous areas namely:
   (a) an area of not less than 90 metres, around an oil-well where a blow out has occurred or is likely to occur, as may be designated by the engineer-in-charge or the senior-most official present at the site;
   (b) an area within 90 metres of an oil-well, which is being tested by open flow;
   (c) an area within 15 metres of:
      (i) a producing well-head or any point of open discharge of the crude therefrom or other point where emission of hazardous atmosphere is normally likely to arise, or
      (ii) any wildcat or exploration well-head being drilled in an area where abnormal pressure conditions are known to exist or
      (iii) any exploration or interspersed well-head being drilled in the area where abnormal pressure condition are known to exist, or
   (d) any area within 4.5 metres of:
      (i) any producing well-head where a closed system of production is employed such as to prevent the emission or accumulation in the area in normal circumstances of a hazardous atmosphere; or
      (ii) exploration or interspersed well-head being drilled in an area where the pressure conditions are normal and where the system of drilling employed includes adequate measures, for the prevention in normal circumstances of emission or accumulation within the area of a hazardous atmosphere; or
      (iii) an oil well, which is being, tested other than by open flow.

      Explanation: For the purpose of clause (d) "hazardous atmosphere" means an atmosphere containing any inflammable gases or vapours in a concentration capable of ignition.

(e) all appliances, equipment and machinery that are or may be used in zone 0, zone 1 and zone 2 hazardous areas shall be of such type, standard and make as approved by the inspector by a general or special order in writing.

Explanation:

(i) "zone-0 hazardous area" means "an area in which hazardous atmosphere is continuously present";

(ii) "zone-1 hazardous area" means an "area in which hazardous atmosphere is likely to occur under normal operating conditions";

(iii) "zone-2 hazardous area" means "an area in which hazardous atmosphere is likely to occur only under abnormal operating conditions".

Rule 127. Shot-firing

1. When shot-firing is in progress adequate precautions shall be taken to protect apparatus and conductors other than those used for shot-firing from injury.
2. Current from lighting or power circuits shall not be used for firing shots.
3. The provisions of rule 123 shall apply in regard to the covering and protection of shot-firing cables, and adequate precautions shall be taken to prevent such cables touching other cables and apparatus.

Rule 128. Signalling

Where electrical signalling is used:

(a) adequate precautions shall be taken to prevent signal and telephone wires coming into contact with other cables and apparatus;

(b) the voltage used in any one circuit shall not exceed 30 volts; and

(c) contact-makers shall be so constructed as to prevent the accidental closing of the circuit;

(d) bare conductors, where used shall be installed in suitable insulators.

Rule 129. Haulage

Haulage by electric locomotives on the overhead trolley-wire system, at medium or low voltage, and haulage by storage battery locomotives, may be used with the prior consent in writing of the Inspector, and subject to such conditions as he may impose in the interests of safety.

Rule 130. Earthing of Neutral Points

Where the voltage of an alternating current system exceeds 30 volts, the neutral or mid-point shall be earthed by connection to an earthing system in the manner prescribed in rule 115:

Provided that when the system concerned is required for blasting and signalling purposes, the provisions of this rule shall not apply:

Provided further that in case of unearthed neutral system adequate protection shall be provided with the approval of the Inspector.

Rule 131. Supervision

1. (i) One or more electrical supervisors as directed by the Inspector shall be appointed in writing by the owner, agent or manager of a mine or by the agent or the owner of one or more wells in an oilfield to supervise the installation.
   (ii) The electrical supervisor so appointed shall be the person holding a valid Electrical Supervisor's Certificate of Competency, covering mining installation issued under sub-rule (1) of rule 45.
   (iii) If the inspector considers necessary for the compliance with the duties specified in this rule, he may direct the owner or agent of the mine to appoint one or more electricians who shall be persons holding licence under sub-rule (1) of rule 45.
2. Every person appointed to operate, supervise, examine or adjust any apparatus shall be competent to undertake the work, which he is required to carry out as directed by the engineer.
3. The electrical supervisor shall be responsible for the proper performance of the following duties, by himself or by electricians appointed under sub-rule (1):
   (a) through examination of all apparatuses (including the testing of earth conductors and metallic coverings for continuity) as often as may be necessary to prevent danger;
   (b) examination and testing of all new apparatuses, and of all apparatus, re-erected in the mine before it is put into service in a new position.
4. In the absence of any electrical supervisor for more than three days, the owner, agent or manager of the mine or the agent or owner of one or more oil-wells in an oilfields, shall appoint in writing as substitute electrical supervisor.
5. (i) The electrical supervisor or the substitute electrical supervisor appointed under sub-rule (4) to replace him shall be personally responsible for the maintenance at the mine or oilfield, of a logbook made up of the daily log-sheets prepared in the form set out in Annexure XII.
   (ii) The results of all tests carried out in accordance with the provisions of sub-rule (3) shall be recorded in the log-sheets prepared in the form set out in Annexure XII.

Rule 132. Exemptions

The provisions of rules 110 to 128 (both inclusive) and rule 131 shall not apply in any case where, on grounds of emergency or special circumstances, exemption is obtained from the Inspector. In granting any such exception, the Inspector may prescribe such conditions as he thinks fit.

CHAPTER-XI

14.14 MISCELLANEOUS

Rule 133. Relaxation by Government

(a) The State Government, or where mines, oilfields or railways or works executed for or an behalf of the Central Government, are affected, the Central Government may, by order in writing, direct that any of the provisions of rules in Chapter IV other than rules 44A and 46, and all rules in Chapters V, VI, VIII, and IX, shall be relaxed generally or in particular case to such extent and subject to such conditions as it may think fit.

(b) The Central Government may, by order in writing, direct that any of the provisions of Chapter X of these rules shall be relaxed in any particular case to such an extent and subject to such conditions as it may think fit.

Rule 134. Relaxation by Inspector

1. The Inspector may, by order in writing, direct that any of the provisions of rules 44, 50(1) (a), (b), and (d), 50 A (2), 51(1), 61(2), 63, 64 (2), 65, 71 to 73 (inclusive), 76 to 80 (inclusive), 90, 118,119(1) (a), 123 (5), 123(7), 123(9) and 130, shall be relaxed in any case to such extent and subject to such conditions as he may think fit.
2. Where the voltage of any system does not exceed 125 volts the Inspector may, by order in writing, direct that any of the provisions of rules 29 to 34 (inclusive), 36 to 39 (inclusive), 83, 92, 94 to 107 (inclusive), shall, in addition to the rules specified in sub-rule (1), be relaxed as regards such systems, to such extent and subject to such conditions as he may think fit.
3. Every relaxation so directed shall be reported forthwith to, and shall be subject to disallowance or revision by, the State Government, or where the relaxation affects mines, oilfields or railways, by the Central Government.

Rule 135. Supply and Use of Energy by Non-licensees and Others

Where any person other than a non-licensee is supplied with energy by a non-licensee or other person or has his premises for the time being connected to the conductors or plant of a non-licensee or other person, or himself generates energy and uses such energy or part thereof, such person shall be deemed to be a consumer for the purposes of rules 9, 10, 29 to 33 (inclusive), 45 to 70 (inclusive), 87 and 142 and the non-licensee or other person shall be subject to all the liabilities imposed on a licensee by these rules.

Rule 136. Responsibility of Agents and Managers

Where any person is responsible for the observance of any of these rules, every agent and manager of such person shall also be responsible for such observance in respect of matters under their respective controls.

Rule 137. Mode of Entry

All persons entering in pursuance of the Act or these rules, any building which is used as a human dwelling or place of worship shall, in making such entry, have due regard, so far as may be compatible with the exigencies of the purpose for which such entry is made, to the social and religious usages of the occupant of the building entered.

Rule 138. Penalty for Breaking Seal

Where, in contravention of rule 56 any seal referred to in that rule is broken:

(a) the person breaking the seal shall be punishable with fine which may extend to two hundred rupees, and

(b) the consumer when he has not himself broken the seal shall be punishable with fine, which may extend to fifty rupees unless he proves that he used all reasonable means in his power to ensure that the seal should not be broken.

Rule 138A. Penalty for Breach of Rule 44A

Where, in contravention of rule 44A, any person responsible for the generation, transformation, transmission, conversion, distribution, supply or use of energy fails to report to the Inspector and other authorities concerned the occurrence of accidents, such person shall be punishable with fine which may extend to three hundred rupees.

Rule 139. Penalty for Breach of Rule 45

Where any electrical installation work of the nature specified in sub-rule (1) of rule 45 has been carried out otherwise than:

(a) under the direct supervision of a person holding a certificate of competency issued by the State Government under that rule, and

(b) in the absence of any applicable exemption under the provision to sub-rule of that rule, by an electrical contractor licensed by the State Government in this behalf.

The consumer, owner or occupier, the contractor if any or the person through whom the work is being or was carried out and the person under shoes immediate supervision the work is being or was carried our, shall each be punishable with fine which may extend to three hundred rupees.

140. Penalty for Breach of Rule 82

(a) Where no notice is given under rule 82(1) or the amount of estimate as demanded under rule 82(2) is not deposited, both the persons proposing and the contractor engaged for erecting a new building or structure whether permanent or temporary or for making in or upon any building or structure, any permanent or temporary additions or alterations, shall be deemed to have committed a breach of rule 82 (1) and shall be punishable with a fine which may extend to three hundred rupees;

(b) If any person commences or continues any work in contravention of rule 82(3), in or upon any such building, structure, flood bank, road or carries out addition or alteration thereto, the person contravening the same shall be punishable with a fine which may extend to three hundred rupees.

In addition to this the suppliers shall, after obtaining the concurrence of the Inspector, discontinue the supply, if any, to such building, structure, flood bank or road, etc. but only after giving forty-eight hours notice to the person concerned in writing of disconnection of supply and shall not commence the supply until he and the Inspector are satisfied that the cause has been removed.

Rule 140A. Penalty for Breach of Rules 77, 79 or 80

Where a person is responsible for any construction which is or which results in contravention of the provisions of rules 77, 79 or 80, he and the contractor whom he employs shall be punishable with a fine which may extend to three hundred rupees, and in the case of a continuing breach, with a further delay fine which may extend to fifty rupees.

Rule 141. Penalty for Breach of Rules

Any person other than an Inspector or any officer appointed to assist the Inspector who, being responsible for the observance of any of these rules commits a breach thereof, shall be punishable for every such breach with fine which may extend to three hundred rupees, and in the case of a continuing breach with a further fine which may extend to fifty rupees, for everyday after the first during which the breach has continued.

Rule 142. Application of Rules

Subject to the provisions of subsection (2) of Section 58, these rules shall be binding on all persons, companies and undertakings to whom licences have been granted or with whom agreements have been made by or with the sanction of Government for the supply or use of electricity before the commencement of the Act.

Rule 143. Repeal

The Indian Electricity Rules, 1937 are herby repealed:

Provided that any order made, notification issued or anything done or any action taken under any of the said rules shall be deemed to have been made, issued, done or taken the corresponding provisions of the rules.

EXERCISES

1. Why is IE rules necessary to follow?
2. Explain the various tests recommended for lighting installations as per the IE rules.
3. Explain the periodical tests to be carried out on exiting electrical installations.
4. State the rules and regulations to be followed in providing service connections.
5. How would you select a conductor for low-tension line extension for a particular load?
6. Explain the necessity and requirement of a domestic earthing as per IE rule.
7. Explain the main factors to be considered for a substation earthing.
8. Describe IE rules pertaining to the clearances for distribution line crossing the road.
9. State different provisions applicable to medium, high and extra high voltage installations.
10. Write a brief note on sealing the energy meters and penalty for breaking the seal.
11. Describe salient features of IE rules applicable for Electric Traction.
12. Write a note on 'Executive Powers rendered to the Inspectors' in IE rules.
13. What are the additional precautions to be adopted in mines and oilfields?
14. Write the IE rules to be followed while testing, operation and maintenance of HV or EHV apparatus.
15. Write a note on 'Danger Notices' on the electrical installations.

SECTION-B

Experiments

Experiments

MOTIVATION

Engineering field is totally practical in nature. This expects a thorough understanding of theoretical concepts taught through practical work. Laboratory courses are also important because conducting the experiments means being familiar with the real performance of the machines. It gives idea to the students about what have they been taught in theory classes. Experimentation gives confidence to analyze and discuss the topics in depth.

GUIDELINES FOR THE TEACHERS

(1) Whether, it is Industrial Training Institute, Polytechnic or Degree level engineering college, it is very important to plan and design practical and the related theory in an integrated manner to correlate them.

(2) A well-organized laboratory manual should be provided to each student in advance for preparing the experiment methodology and procedure in advance. The manual should consist of instructions in step-by-step approach.

(3) The teacher should explain and assign the experiments to each student in advance so that they come prepared for the practical.

(4) Before handing over the apparatus to the students and allow them to handle instruments and machine, selected questions should be asked to adjudge the students preparedness.

(5) The teacher must ensure that each student is well aware of the precautions to be taken in electrical engineering laboratories. After all, they are handling electric current, which may be fatal, if not undertaken carefully.

(6) The teacher in association with the instructor should monitor the students carefully during circuit tracing/making and recording the data.

(7) Student should be assessed on regular basis for judicial evaluation of their performance in the laboratory experimentation. It should be based on his entire performance and not only on the final report submissions only.

Electrical Workshop

(8) The teacher should keep record of the performance of each student for every experiment through a well-formulated evaluation pro forma. A specimen copy of pro forma for evaluating the student's performance is given as Table B.1 for the teacher's reference.

Assessment Form of the Student's Performance

Student name	____________________
Roll no.	____________________
Class & group	____________________
Experiment name	____________________
Experimentation date	____________________
Date of submission of report	____________________

Table B.1: Student's performance assessment

S. No.	*Activity*	*Rating*						*Score*
		A 10/10	B 8/10	C 6/10	D 4/10	E 2/10	*Weight factor*	
1.	Preparedness in advance						4	
2.	Circuit tracing and connection						3	
3.	Observation and Data collection, etc.						3	
4.	Initiatives/group leadership						2	
5.	Accuracy in performing the experiment						5	
6.	Presentation of the laboratory report						6	
	Total score ___________							

Signature of the teacher

Date: ....

GUIDELINES FOR THE STUDENTS

(1) Students should get allocated the experiments to be performed in advance for prior preparation and confidence building.

(2) The aims and objectives of the experiment should be read and rewritten in specific terms.

(3) The list of apparatus/equipment required should be listed and get issued. Students are to choose the suitable range and type of the instruments.

(4) After selecting the instruments, the students should record the detail specification of the machines, apparatus and instruments to be used.

(5) Arrange the meters and other portable instruments on the experimental table in such a manner so that connecting leads, etc. be in close proximity of the related socket points.

(6) Connect the circuit systematically and correctly. Get it checked by the instructor before switching 'ON' the power supply.

(7) Complete the calculations and ensure that results are within acceptable limits before living the laboratory.

(8) Questions related to the experiments given in the instruction sheets should be answered in the report book.

(9) After completion of the practical works, disconnect all the instruments and accessories and replace them in their appropriate places.

Remember, your work does not finish till you submit the report and get it checked.

Experiment 1

AIM

Familiarization with electrical symbols.

INTRODUCTION

Electrical symbols are the heart of an electrical drawing. In electrical engineering drawings, also known, as language of engineers is the graphical representation, which is prepared by the use of symbols. Drawings are used to represent various components, equipment, accessories and types of connections in electrical planning, preparing estimates and layout. The symbols must convey the same meaning and understanding to everyone who reads the drawing. So, the standardization of the symbols has been done by the BIS (Bureau of Indian Standards). They have tried to adopt (as far as possible) symbols according to internationally agreed conventions as per IEC (International Electrotechnical Commission). The symbols should be self-explanatory and easy to draw.

SYMBOLS

Symbols may be categorized as given below:

1. General symbols
2. Variability symbols
3. Symbols for systems
4. Fundamental components
5. Transformers
6. Rotating machines
7. Meters
8. Switchgear and control gear
9. Fuses and fuse switches, etc.

Frequently needed symbols including types of currents, supply, conductors, lines, circuit elements, resistors, machines, transformers, contacts, relays, sockets, fuse boards, switch outlets, lighting apparatus, fans, household apparatus, etc. are listed hereunder.

LIST OF SYMBOLS

Current

Types of current	Symbol
DC	——
AC	∿
Power frequency	∿
Apparatus and machines suitable for both DC or AC (universal)	⏦

Power Distribution

Type of supply	Symbol
AC, single-phase, 50 Hz	1 ∿ 50 Hz
AC, three-phase, 50Hz	3 ∿ 50 Hz
AC, three-phase, 50 Hz, 415	3 ∿ 50 Hz, 415 V
AC, three-phase, with neutral, 50 Hz	3N ∿ 50 Hz
AC, three-phase, with neutral, 50 Hz, 415 V	3N ∿ 50 Hz, 415 V
DC 2-conductors 110 V	2 — 110 V
Positive polarity	+
Negative polarity	—
Terminal	o or •

Connections of Conductors

Junction of conductors	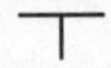
Double junction of conductors	
Crossing without electrical connection	
Crossing and connecting conductors	
(i) Single-line representation	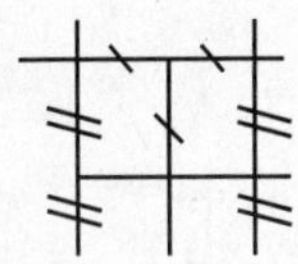
(ii) Multi-line representation	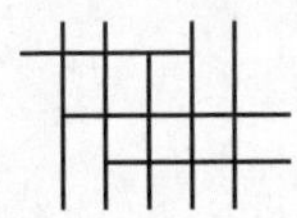

Symbols of Transmission/Distribution Lines

Lines	Symbol
Existing Line or cable	
Planned Line or cable	
Underground cable	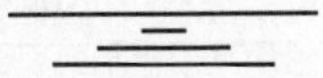
Overhead line	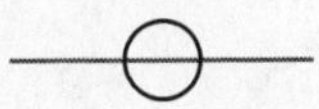

Conductors

Conductors	Symbol
Flexible conductor	
'n' No. of conductors (Single-line presentation)	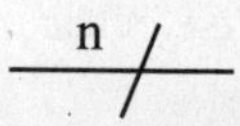 [e.g., for 4-conductors] 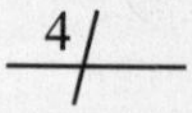
'n' No. of conductors (Multi-line presentation)	['n' No. of parallel lines] e.g., for 4-conductors 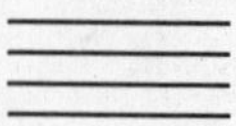
Three-phase circuit, 50 Hz, 6 kV, three conductors of 50 mm²	 50 Hz, 6 kV 3×50 mm²
Changing over from a single-line representation	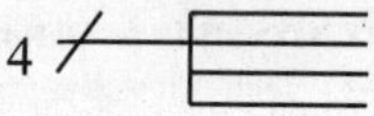
DC circuit, 110 V, two conductors of 125 mm² of aluminium	–110 V 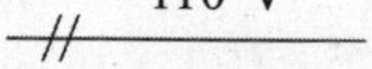 2×125 mm² Al
Group of 'n' conductors which follow the same way on the diagram	n 

Circuit Element

Element	Symbol
Non-reactive resistance	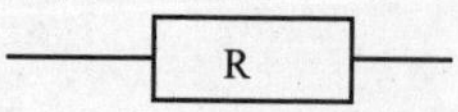
Impedance	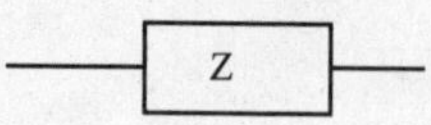
Inductance	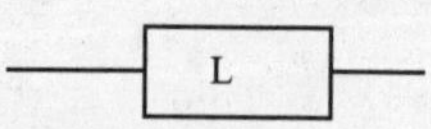

Capacitance	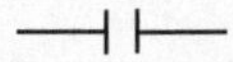
Winding	
Earth point	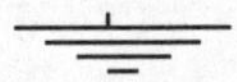
Fault	
Screened conductor or screened cable	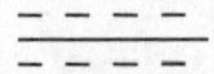
Magnetic core	
Laminated core	

Variable Resistors

General symbol	
Continuously variable	
Variable in steps	
With moving contact (general symbol)	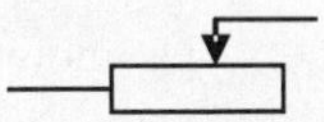
With moving contact continuously variable	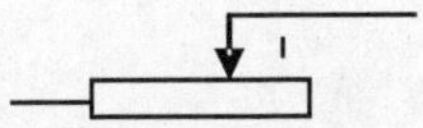
With moving contact, variable in steps	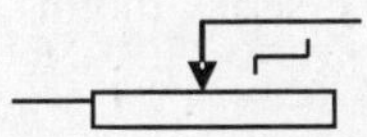
Voltage divider with moving contact	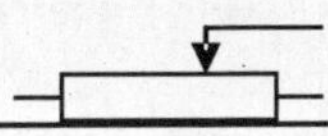

Rotating Machines

Machines	Symbol
Mechanically coupled machines	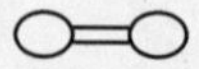
DC two-wire series motor	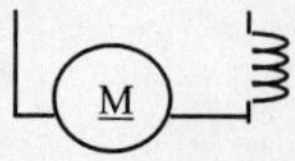
DC two-wire separately excited generator	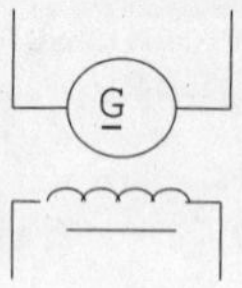
DC two-wire shunt motor	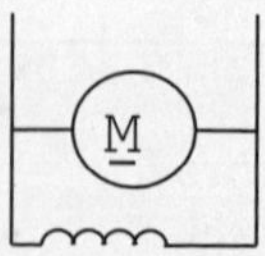
DC two-wire compound excited, short shunt generator, 240 V, 30 kW	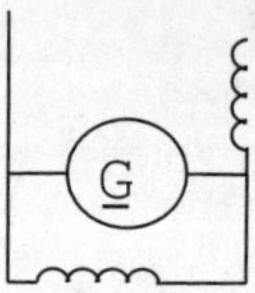 240 V 30 kW
AC single-phase series motor	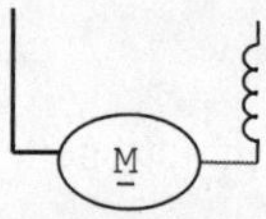
AC three-phase series motor	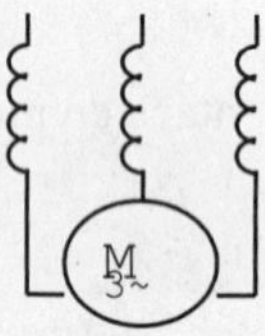
Single-phase synchronous generator	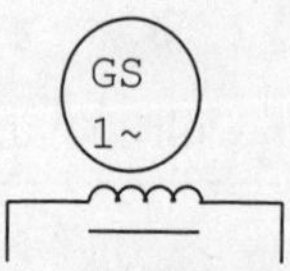
Three-phase, star connected, neutral not brought out synchronous generator	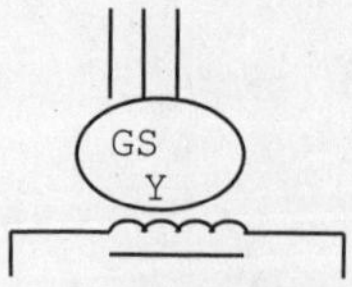

Three-phase, star-connected, neutral brought out, synchronous generator	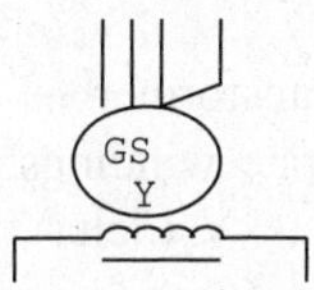
Squirrel cage, single-phase, leads of split phase brought out induction motor	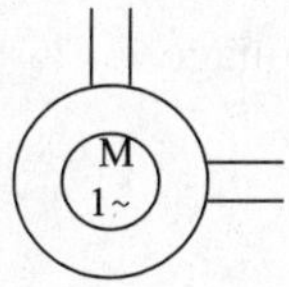
Squirrel cage, three-phase delta-connected induction motor	
Three-phase, squirrel cage, both leads of each phase brought out induction motor	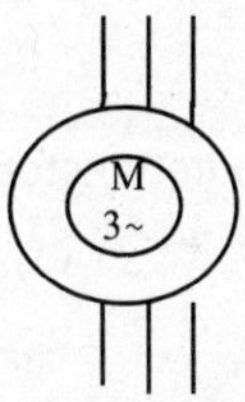

Transformers

Transformer specification	Single-line representation	Multi-line representation
Single-phase transformer with two windings (11 kV/415 V, 250 kVA, 50 Hz); Short-circuit voltage 4%	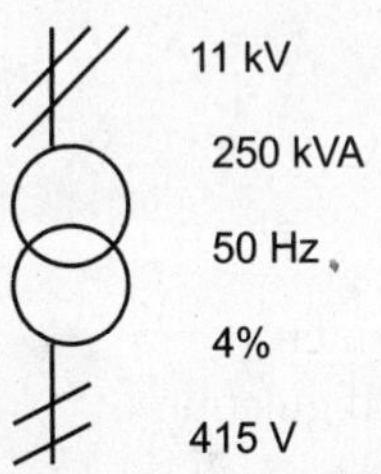	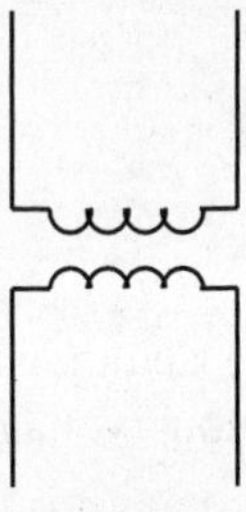

Three-phase transformer with two separate windings: (Star) 66 kV/11 kV (delta), 4 MVA, 50 Hz. Connection Y d 11
Short-circuit voltage 7.5%

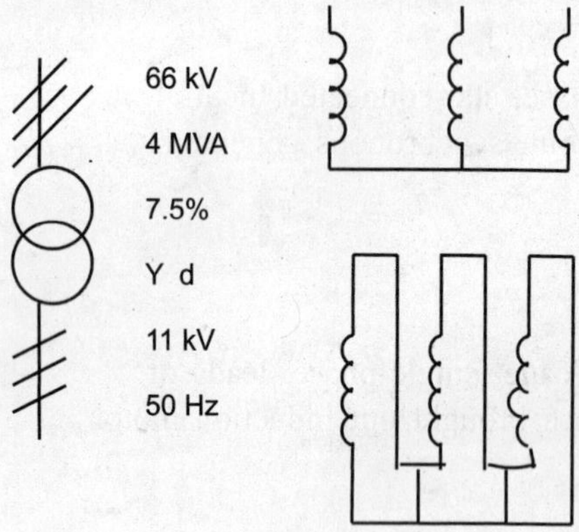

Single-phase autotransformer

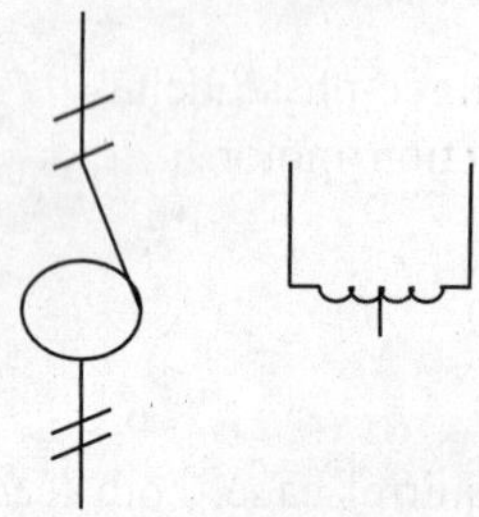

Three-phase star-connected autotransformer

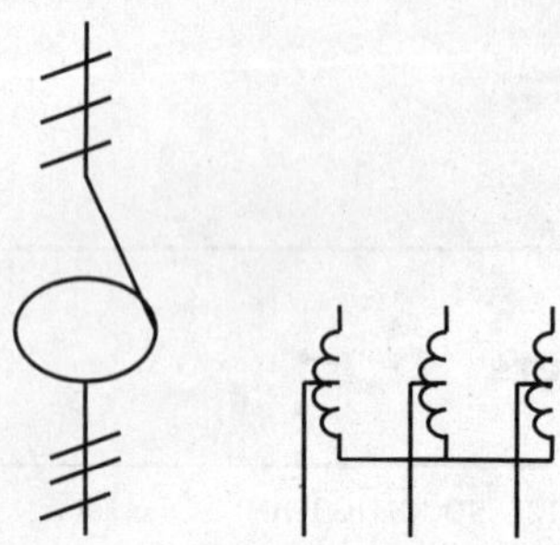

Single-phase autotransformer with continuous voltage regulation

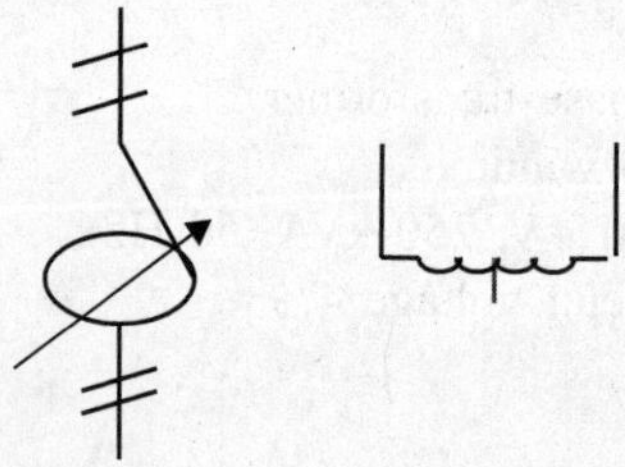

Switches, Circuit Breaker and Links

Switch

(i) General

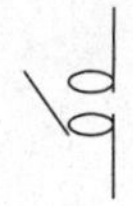

(ii) Two-way

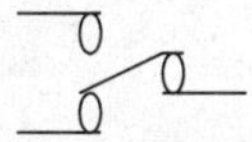

(iii) Intermediate

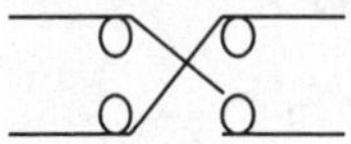

Three pole switch, multi-line representation

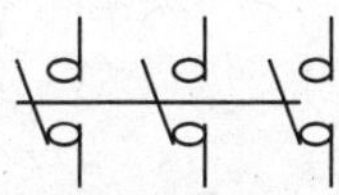

Three pole switch, single-line representation

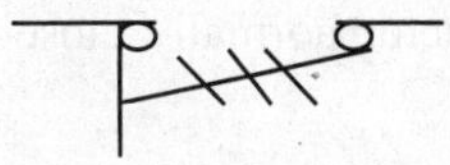

Circuit breaker

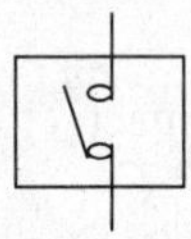

Isolator

Terminal strip

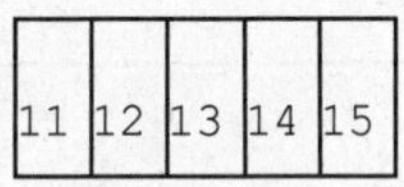

Link

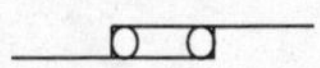

Open link

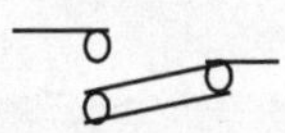

Distribution board cubical box

Contacts

Socket 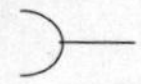

Plug 

Plug and socket 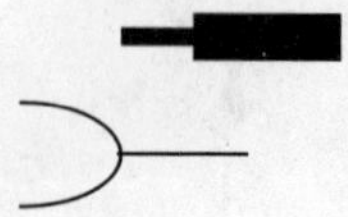

Relay or contactor contact, normally open 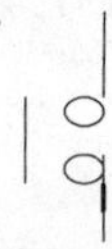

Relay or contactor normally closed 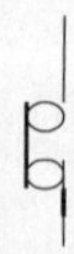

Thermal overload relay contact 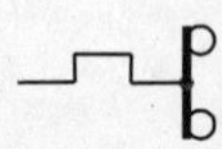

Socket Outlets

Socket outlet, 5 A 

Socket outlet, 15 A 

Combined switch and socket outlet, 5 A 

Relays and Contactors

Coil of electromagnetic relay or contactor

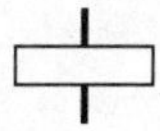

Thermal overload relay

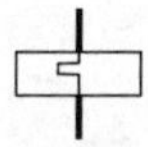

Electrically operated three-pole contactor

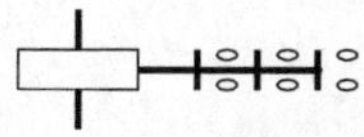

Electrically operated three-pole contactor with thermal overload device in all three poles

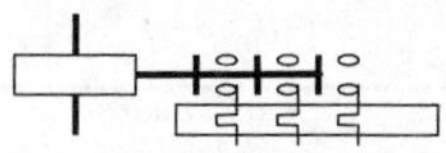

Time delay relay

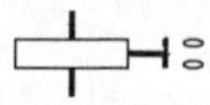

Fuse Boards

Main fuse board without switches

Main fuse board with switches

Distribution fuse board without switches

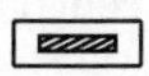

Distribution fuse board with switches

Switch Outlets

One way switch

(a) Single-pole

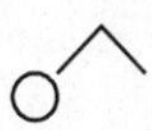

(b) Two-pole

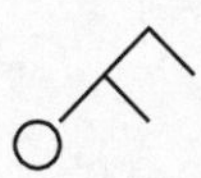

(c) Three-pole	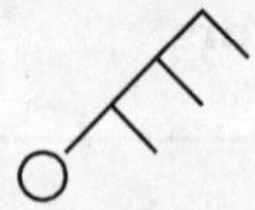
Two-way switch	
Intermediate switch	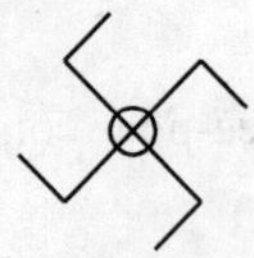
Push-button or bell-push	

Lighting Apparatus

Lamp or outlet for lamp	×
Group of three 40 W lamps	3 × 40W
Lamp mounted on wall or light bracket	
Lamp mounted on ceiling	
Group of three 40 W fluorescent lamps	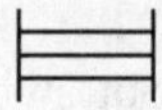
Fluorescent lamp	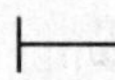

Fans

Ceiling fan	
Bracket fan	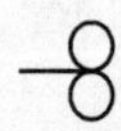
Exhaust fan	
Fan regulator	

Household Apparatus

Fuse	 or 
Signal lamp	
Indicator	
Horn	
Bell	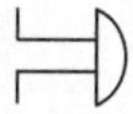
Buzzer	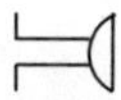
Siren	
Heater	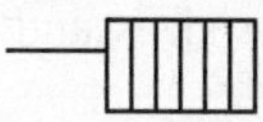
Storage type electric water heaters	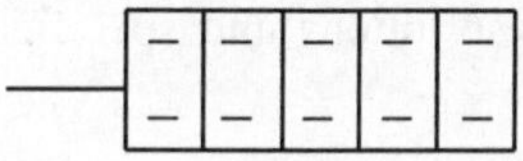

Conclusion: (Describe in brief, whatever you have learned.)

QUESTIONS

(To be answered in Practical Record Book)

1. What is significance of symbols in engineering?
2. How, drawing is said as the language of engineers?
3. How, multi-line representation is simplified?
4. Draw, the graphical symbols of:
   (i) Variable capacitor
   (ii) Ohmmeter
   (iii) Electrical fault
   (iv) Energy meter
   (v) Two-way switch
5. Describe the differences of symbols for siren, bell, buzzer and horn.

Experiment 2

AIM

Familiarization with tools used in electrical works.

INTRODUCTION

Selection of proper hand tools is very necessary for getting perfection in jobs. There are number of hand tools of different sizes and specifications, which are used in electrical works.

Note: Descriptions and specifications of hand tools are available in the Chapter 3.

Experiment 3

AIM

Familiarization with electrical materials.

INTRODUCTION

In electrical engineering, a large variety of materials are required. Different types of materials are employed in the manufacturing of electrical machines, apparatus, home appliances, relays, control equipment, wires, cables, transmission and distribution equipment, etc. The most common conducting material in use is aluminium and copper as conductors. To confine the flow of current in the conductor, the cable must be properly insulated, and the insulating material should possess a very high resistance so that leakage current is negligibly small or absent.

The choice of conductor depends upon:

(i) Current carrying capacity
(ii) Resistance and Impedance
(iii) Mechanical strength
(iv) Termination
(v) Flexibility
(vi) Conductor configuration.

The choice of insulating material depends upon:

(i) High resistance
(ii) High dielectric strength
(iii) High withstand capacity for temperature
(iv) High mechanical strength
(v) Should be non-inflammable

In general, nearly, almost, all the substance, are influenced by a magnet to varying degrees. However, a few substances such as iron, steel, nickel, cobalt, and some of their alloys are strongly attracted by a magnet and these exhibit pronounced magnetic properties.

Tungsten is a very hard metal whose resistivity is about twice that of aluminium. Its melting point (3300°C) is almost highest of all metals. Its property to be drawn into very thin wires, possess high tensile strength and high resistivity makes it suitable for making filament of bulbs.

Materials used in electrical engineering applications are categorized as:

S/N	Types of materials	Characteristics/Definition	Example
1.	Conducting	Good conductor of electricity having high electrical conductivity and very low electrical resistance	Copper, Silver, Aluminium, Tungsten, Steel, Molybdenum, Platinum, Nickel, Tantalum, Niobium, Mercury
2.	Semiconducting	Electrical conductivity is similar as conductor materials but resistivity of semiconductor materials varies with several factors like electrical field, voltage, temperature, impurity, illumination, etc.	Germanium, Silicon, Indium, Antimonide, Selenium, Tellurium, Boron, Silicon-carbide
3.	Superconducting	Superconductors possesses very low electrical resistance exceedingly small approaching to zero	Mercury at 4.2°K, Zinc at 0.79°K, Aluminium at 1.14°K, Cadmium at 0.54°K, Zirconium at 0.7°K, Thorium at 1.32°K, Lead at 7.26°K
4.	Magnetic	These are materials having high permeability and can be easily magnetized	Copper, Gold, Silicon, Diamond, Graphite, Germanium
5.	Ferromagnetic	These are materials having very high permeability and spontaneous magnetization	Iron, Cobalt, Nickel, Alnico, Carbon steel, Platinum
6.	Ferrimagnetic	These are ferromagnetic materials with very high permeability and spontaneous magnetization below the Curie temperature. These also have semiconducting property and are used in high frequency apparatus manufacturing	Permaalloys, Hypernick, Silicon, etc.
7.	Insulating	Materials having very high electrical resistance and negligibly small conductivity	Impregnated paper, Mica, Porcelain, Poly-vinyl chloride (PVC), Silica, Rubber, Glass, Ebonite Bakelite, Steatite, Vulcanized Bitumen, Cotton and Silk insulation, Empire tape (Varnished cambric)
8.	Ferroeletric (Ceramic-materials)	These are the insulating (dielectric) materials which performance is close to the ferromagnetic materials	Barium titanate, Ceramic, Dihydrogen Phosphate, Rochelle salt
9.	Gaseous insulators	The gases which have good insulating property at high pressures are called gaseous insulators	Air, Hydrogen, Carbon dioxide, Hydrogen, Ammonia, Florine, Sulphurhexafluoride, Hydrogen chloride
10.	Gaseous conductors	The gases which can be used as conducting medium for electrical applications are called gaseous insulators	Helium, Neon, Argon, Krypton, Xenon

Another very common material is Nichrome an alloy of Manganese (1.5%), Nickel (75%), Chromium (23%) and a very little percentage of iron. It is of very high resistivity and can be drawn into thin wires. These are used for making heating elements of electric heaters, electric furnaces, electric ovens, electric irons, etc.

Insulating materials play vital role in electrical engineering. These are classified on the basis of their thermal stability as given in the following table:

Classification of insulating materials on the basis of thermal stability in the service:

Class	Maximum working temperature	Materials
Y	90°C	Paper, Rubber, Cotton, Silk, Wood, Press board
A	105°C	Impregnated paper, Impregnated cotton and silk
E	120°C	'A' class materials with special treatment which possess higher degree of thermal stability
B	130°C	Mica, Glass fibre, Asbestos with bonding substances
F	155°C	Ebonite, Bakelite, Mica, Asbestos with bonding substances and combination of other materials capable for 155°C
H	180°C	Silicon elastomer, Mica, Impregnated Glass fibre, Asbestos, etc. bonding with silicon resins
C	Above 180°C	Porcelain, Mica, Quartz, Glass, etc. with inorganic binder

Conclusion: (Describe in brief, whatever you have learned.)

QUESTIONS

(To be answered in Practical Record Book)

1. Describe the resistivity. How is it significant in electrical engineering applications?
2. Name the materials being used in making of the following.
   (a) Filament of the incandescent lamp
   (b) Loading rheostat resistors
   (c) Electric heater element
   (d) Handle of the electric iron
3. Name the materials being used in making of the following:
   (a) Elements of fan regulator
   (b) Plug and socket
   (c) Thermocouple
   (d) DC motors starter elements
4. What do you understand by superconductors? Describe in brief.
5. List out the merits and demerits of aluminium as compared to copper as a conductor of electricity.
6. Describe the effects of impurities presence in the materials useful for electrical applications.
7. Explain the use of semiconducting materials.
8. Give the reason of the necessity of insulating materials in electrical machineries and appliances.

9. Mention few (minimum two) applications of the following materials:
   (a) PVC
   (b) Bakelite
   (c) Silicon
   (d) Mica
   (e) Ebonite
   (f) Porcelain
10. List out the applications of the following materials
    (a) Impregnated paper
    (b) Varnish
    (c) Bitumen
    (d) SF_6 gas
    (e) Ceramic

Experiment 4

AIM

Familiarization with abbreviations commonly used in electrical engineering.

INTRODUCTION

Many words or group of words appear very frequently in the description of the system and or in daily use. To avoid occupation of the space and time, it is wise to use group of words in a common short form. These are known as abbreviations. Some are given below.

ABBREVIATIONS

AAC	All Aluminium Conductor
ACB	Air Circuit Breaker
ACSR	All Conductor Steel Reinforced
AI	Artificial Intelligence
ALU	Arithmetic Logic Unit
BEL	Bharat Electronics Limited
BHEL	Bharat Heavy Electrical Limited
CEA	Central Electricity Authority
CT	Current Transformer
CVT	Capacitive Voltage Transformer
CCVT	Capacitor Coupled Voltage Transformer
DP	Double Pole
DPIC	Double Pole Iron Clad
DOL	Direct On Line Starter
DSP	Digital Signal Processing
EDMA	Extended Direct Memory Access
EPI	External Peripheral Interface
ELCB	Earth Leakage Circuit Breaker

EHV	Extra High Voltage
EMF	Electromotive Force
E-Mail	Electronic Mail
FFT	Fast Fourier Transform
ICDB	Iron Clad Distribution Board
HEC	Heavy Engineering Company
HPMVL	High Pressure Mercury Vapour Lamp
HRCF	High Rupturing Capacitive Fuse
HV	High Voltage
IE (I)	Institution of Engineers (India)
IEC	International Electrotechnical Committee
IEE	Institution of Electrical Engineers
IEEE	Institution of Electrical and Electronics Engineers
LA	Lightning Arrester
LDR	Light Dependent Resister
LED	Light Emitting Diode
LPMVL	Low Pressure Mercury Vapour Lamp
LV	Low Voltage
MCB	Miniature Circuit Breaker
MDB	Main Distribution Board
MNES	Ministry of Non-Conventional Energy Sources
MW	Mega Watt
NL	Neutral Link
NTPC	National Thermal Power Corporation
NHPC	National Hydropower Corporation
NPC	Nuclear Power Corporation
OLTC	Overload Trip Coil
OTEC	Ocean or Tidal Thermal Energy Conversion
PCI	Peripheral Component Interconnect
PGCL	Power Grid Corporation Limited
PT	Potential Transformer
PVC	Poly-vinyl chloride
RAM	Random Access Memory
RMS	Root Mean Square
RCCB	Residual Current Circuit Breaker
SCADA	Supervisory Control and Data Acquisition System
SMS	Sub Main Switch
SP	Single Pole
SPDT	Single Pole Double Through
SPST	Single Pole Single Through
SPV	Solar Photovoltaic
SVL	Sodium Vapour Lamp
SWG	Standard Wire Gauge
TPS	Triple Pole Switch

TPN	Triple Pole with Neutral
TPIC	Triple Pole Iron Clad
TPDT	Triple Pole Double Through
TPST	Triple Pole Single Through
TR	Thermal Relay
TRS	Tough Rubber Sheathed
VA	Volt Ampere
VAR	Volt Ampere Reactive
VIR	Vulcanized Indian Rubber
WT	Water Tight
WPC	Weather Proof Cable

Conclusion: (Describe in brief, whatever you have learned)

QUESTIONS

(To be answered in Practical Record Book)

1. Why abbreviations are preferred?
2. Name five organizations working for electricity.
3. Write extended forms of:
   (i) BIS
   (ii) IEC
   (iii) ELCB
   (iv) TPIC
   (v) OLTC
   (vi) PVC
4. Name four institutions looking after developments of electricity related departments.
5. Write extended forms of:
   (i) VIR
   (ii) MCB
   (iii) HRCF
   (iv) WPC
   (v) VA
   (vi) VAR

Experiment 5

AIM

To make (a) Straight joint and (b) T-joint of (7/20) PVC wire.

TOOLS AND APPARATUS REQUIRED

Combination pliers, Electrician knife, Side cutting pliers, Foot rule.

MATERIALS REQUIRED

PVC (7/20) wire–1 m; Cotton waste–20 gm; Sandpaper–one piece.

(A) STRAIGHT JOINT MAKING

Method

Flexible wires/cables are required for making the joints in electric wires or cables. Straight wire joining requires when either bundle length finishes or there is break or burn in the wire or cable.

Steps

1. Clean the wire/cable properly.
2. Cut the wire into two pieces.
3. Bend one of the edges of each piece.
4. Remove insulation about 9.5 cm length from one end of each piece.
5. Remove cotton pruding from both the ends of the wire/cable. It should be about 1.5 cm from the ends of both the wires/cables.
6. Separate the strands and clean with sandpaper. Re-twist the strands for about 2.5 cm and keep the remaining portion open.

7. Repeat the above process for the other piece of the wire/cable.
8. Take the help of the side cutting pliers and cut-off the central strand from both the wires/cables.
9. Bringing both the pieces together, mix them by intersecting each other.
10. Round off the ends after cutting off the remaining conductors.
11. Repeat the process on the other side with the second piece and complete the joint.

Figure B.5A demonstrates the making of straight joint.

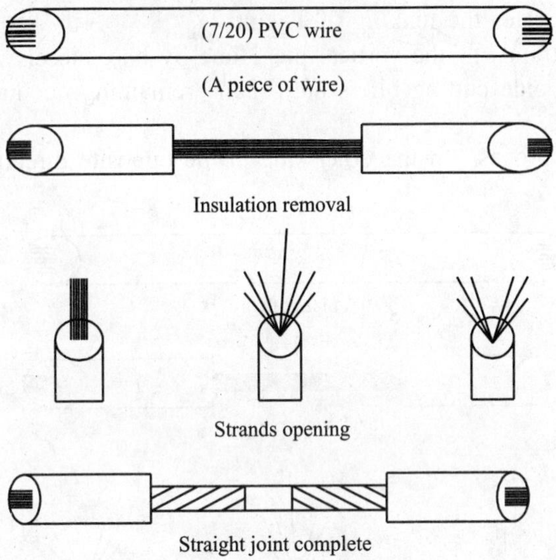

Fig. B.5A. Straight joint making.

(B) T-JOINT MAKING

Method

At termination in wiring or cabling system, T-joints are required. These are used in connecting the circuit in parallel.

Steps

1. Clean the wire/cable properly.
2. Cut wire/cable into two pieces.
3. Remove insulation up to a length of 6 cm at the centre of one wire. Use electrician knife.
4. Place the other piece vertically and remove insulation of about 10 cm from one end.
5. Open the strands and clean all the strands using sandpaper.

6. Hold the first piece of wire in both hands and twist it.
7. Segregate strands of the other piece of wire leaving the bare wire about 2.5 cm from the end of insulation and divide strands in two groups, generally 4-strands in one group and 3 strands in the other.
8. Bend the central strand of the vertical piece downward and twist it 3 to 4 times and cut the remaining conductors.
9. Hold the both pieces like the shape of a "T" in such a way that the second piece lies exactly in the centre of the horizontal piece with three strands on one side and the other three on the other.

Figure B.5B demonstrates the making of T-joint.

1. Wrap the three strands of the vertical piece one by one, closely and tightly.
2. With the help of side cutting pliers cut-off the remaining conductor and round-off the ends.
3. Repeat the same process on the other side in the opposite direction.

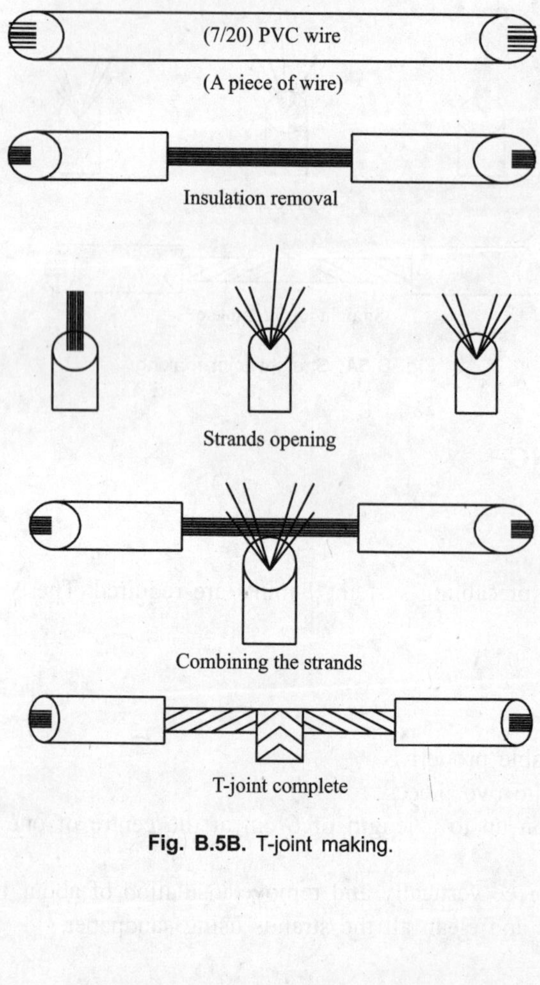

Fig. B.5B. T-joint making.

PRECAUTIONS

1. There should not be any mark of cut made by either pliers or punch on the conductors.
2. Each strand should be cleaned with the help of sandpaper.
3. There should not be any visible gap between the turns.
4. Joint should be made perfect, electrically as well as mechanically.
5. It is recommended to solder the joint.

Uses

1. T-joint is used to tap electrical energy from the horizontal transmission/distribution line.
2. Straight joint is used to increase the length of the wire.

Conclusion: (Write the result and discussion)

QUESTIONS

(To be answered in the Practical Record Book).

1. What is the purpose of making T-joint?
2. What is the purpose of making Straight joint?
3. Name the wire which is used for making joints.
4. List the instruments with proper specifications required for making joints.
5. Describe why cleaning of strands is essential before making joints.
6. Describe why would you recommend soldering the joints.
7. What will happen if there will remain gap between the turns at the time of making joint?

Experiment 6

AIM

To study fluorescent lamp.

APPARATUS REQUIRED

Fluorescent lamp (40 W, 220 V, 50 Hz, lumens-2450); Choke, (220 V, 50 Hz, 40 W, Cosϕ = 0.5), Starter (40 W, 220 V, 50 Hz); Multi-meter.

TOOLS REQUIRED

Line tester, screwdriver, combination pliers, and connecting wires

THEORY

The planning of lighting in the houses, farmhouses, street light, hotels, offices, multi-storey buildings, hospitals, etc. is the most important aspect of the illumination. Tubelight (fluorescent lamp) is mostly used everywhere. The comparison of fluorescent lamp over incandescent lamp (bulbs) is given in the Table B.6.1.

WORKING PRINCIPLE

When Ultraviolet rays (UV) fall on Phosphor (alloy of calcium tungsten, zinc silicate, calcium-borate, calcium silicate, etc.) re-radiate a substantial portion of it in the form of visible light. UV radiation can be produced by an electric discharge through a gas at low pressure (about 1/10,000 of atmospheric pressure). Loss of UV radiation can be minimized completely by surrounding the fluorescent materials or phosphor. Daylight-like light can be obtained by mixing up different powders in appropriate proportions.

CONSTRUCTION

Two tungsten electrodes, coated with an emissive material (like barium oxide or calcium oxide) are placed at both the ends of the fluorescent glass tube. Glass tube ends are closed with insulating materials like bakelite, fitted with two terminals of filament. Terminals are used for electrical connections. The glass tube is filled with argon gas at low pressure along with few drops of mercury. The filaments are connected to a starter, and a choke in series with a switch as shown in Figure B.6.1.

Table B.6.1: Comparison of fluorescent lamp and incandescent lamp

S/N	Fluorescent lamp	Incandescent lamp
1.	Overall running and maintenance cost is less	Overall running and maintenance cost is more
2.	Rapid reduction in output with age	Slow reduction in output with age
3.	Less electricity consumption	More electricity consumption
4.	High initial cost	Low initial cost
5.	Very low heat radiation	High heat radiation
6.	More brightness	Less brightness
7.	Light output is close to natural light	Colour rendering is defective
8.	Output affected by voltage fluctuation	Choke helps in maintaining constant light output

OPERATION

In the beginning, when electrodes are open there is no flow of current, hence electrodes remain cold. With the closing of switch, circuit completes and current starts flowing, which heats up the electrodes. Emission of the electrons starts up from the electrodes. After about 2-3 seconds, the bimetallic switch of the starter opens up automatically, and causes the choke to induce high voltage (1000 V) across the two electrodes. The high voltage across the electrodes starts ionization of argon gas. As temperature increases, mercury evaporates, and generates mercury vapour arc. This conducts the gas between the electrodes and causes the lamp glow.

STARTER

Starter completes the circuit initially for heating the electrodes, and after few moments opens the circuit to induce high voltage across the choke for starting ionization process.

CHOKE

Choke is like an inductor, which opposes the rate of rise of current. It serves mainly three purposes while connected with fluorescent lamp:

1. Produces high voltage about 1000 V to start ionization process of the gases in the main tube.
2. Limits the current produced through the circuit, when the tube is operating.
3. Provides stabilized impedance in series with the circuit.

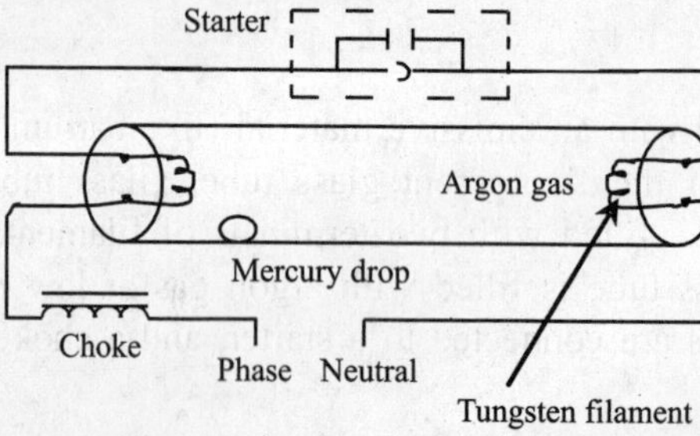

Fig. B.6.1. Fluorescent lamp connection diagram.

PROCEDURE

1. Understand the circuit connections carefully by observing the circuit diagram shown in Figure B.6.1.
2. Connect all the components as per the circuit diagram.

Observation Table

S/N	Name of the part to be tested	Testing method remarks	Observation
1.			
2.			
3.			
4.			
5.			

PRECAUTIONS

1. Check all the components before connecting it.
2. Do not touch live conductor.
3. Connections should be tight and correct.

Conclusion: (Write the result and discussion)

QUESTIONS

(To be answered in the Practical Record Book).

1. What is the function of a tube light starter?
2. What is the necessity of a choke for tube light?
3. How UV rays are produced inside the tube light?
4. What is the material used for making electrodes in the electric lamps?
5. Give the comparison between performance of an incandescent lamp and fluorescent lamp.

Experiment 7

AIM

To study High Pressure Mercury Vapour (HPMV) lamp.

APPARATUS REQUIRED

Mercury vapour lamp (220 V, 50 Hz, 1.15 A, 125 W), Choke, (220 V, 50 Hz, 1.15 A, $\cos\phi = 0.55$), Ammeter (0-15) Amps, Voltmeter (0-300 V), Wattmeter.

TOOLS REQUIRED

Line tester, screwdriver, combination pliers, connecting wires.

THEORY

This is an electric discharge lamp. In this lamp, light is produced by gaseous conduction. Two tubes are used in which inner tube is made up of quartz and outer tube made up of pearl glass. The space between the two tubes is completely evacuated to prevent heat loss. The inner tube contains argon gas and a small quantity of mercury. Two main electrodes along with one auxiliary electrode are placed in the inner tube. The main electrodes are made up of tungsten in coil shape, which is coated with barium oxide. The auxiliary electrode is connected through a 50 kΩ resistance. Ballast is like a choke, which is used in series to limit the current. See the Figure B.7.1.

The lumen efficiency of HPMV lamp is 30 to 40 lumens per watt. The average life of HPMV lamp is about 5000 hours.

OPERATION

An initial discharge for few seconds is established in the argon gas between one main electrode and auxiliary electrode by switching ON, the supply and after that in the region between both main electrodes. The heat produced due to discharge current causes heating up of the inside the inner tube and mercury gets vaporized. The mercury vapour pressure inside the inner tube increases, and potential difference across main electrodes grows from about 20 V to 150 V. After 4 to 5 minutes full brilliancy is achieved. Capacitor in parallel improves the power factor by compensating inductive effect of the ballast.

Uses

These are used for lighting the high ceiling bays, stage lighting, street lighting playgrounds, park, petrol pumps, etc.

PROCEDURE

Connect the lamp with meters and apparatus as per circuit diagram shown in Figure B.7.1 and take observations carefully.

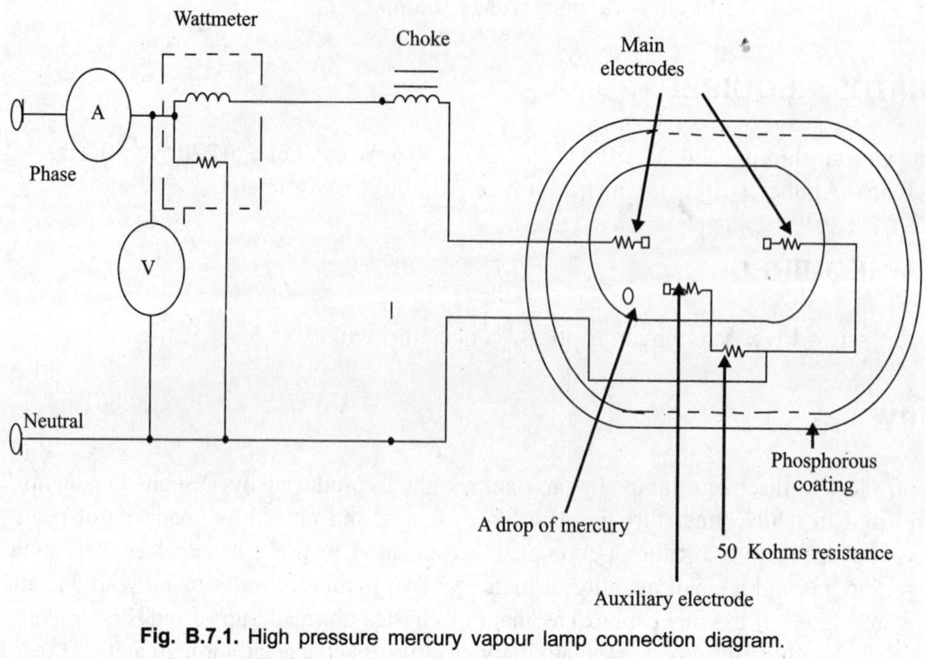

Fig. B.7.1. High pressure mercury vapour lamp connection diagram.

Observation Table

S/N	Time in seconds	Current in Amps	Power in watts
1.			
2.			
3.			
4.			
5.			

(***Note***: The current variation with respect to time is to be plotted on the graph sheet).

PRECAUTIONS

1. Check all the apparatus before connecting it.
2. Connections should be tight and correct.
3. Do not touch live conductor.
4. Take observations carefully.
5. Do not switch 'OFF' the lamp during taking the observations.

Conclusion: (Write result and discussion)

QUESTIONS

(To be answered in Practical Record Book).

1. How many electrodes are there in HPMV lamp and what are the functions of the electrodes?
2. What is the voltage required to the HPMV lamp during the starting and glowing?
3. What is the time required to glow the HPMV lamp fully?
4. Where is this lamp used?
5. Which gases are used in this lamp?

Experiment 8

AIM

To study Sodium Vapour lamp.

APPARATUS REQUIRED

Sodium vapour lamp (220 V, 50 Hz, 70 W), Choke (220 V, 50 Hz, 1.8 A, $\cos\phi = 0.55$)

TOOLS REQUIRED

Line tester, screwdriver, combination pliers, connecting wires.

THEORY

Sodium vapour lamps are of high lumen output with comparatively more or less same energy consumption and power input of HPMV lamp. This lamp is also made up with two tubes. Inner tube is made up of special chemical resistant glass capable to withstand chemical reaction of hot sodium.

Inner U-shape tube houses two coated tungsten electrodes, which are connected across a choke. The inner U-shape tube contains neon gas at a pressure of about 10 mm mercury and a small amount of sodium. This tube is enclosed within a double-walled vacuum flask type envelope and is well insulated in order to conserve the heat required to vaporize sodium.

The lumen efficiency of sodium vapour lamp is 40 to 50 lumens per watt. Its average life is about 8000 hours. These lamps remain unaffected by voltage variations.

OPERATION

Starting of sodium vapour lamp is affected by a high leakage reactance autotransformer by delivering an open circuit voltage of about 350 V for 45 W lamps and 410 V for 100 W lamps.

This voltage is sufficient to initiate the discharge through the neon gas. It takes few seconds to glow the lamp.

FUNCTION OF CHOKE/AUTOTRANSFORMER

Sharp rise of electric current damages the parts of appliances, lamps, etc. and choke is used for limiting the sharp rise in starting current. The use of autotransformer is:

1. To provide a high voltage of about 350-480 V at the time of starting.
2. To stabilize the current.

Uses

For street lighting, park lighting, railway yard lighting, etc.

PROCEDURE

Connect apparatus as per circuit diagram shown in Figure B.8.1 and take observations carefully.

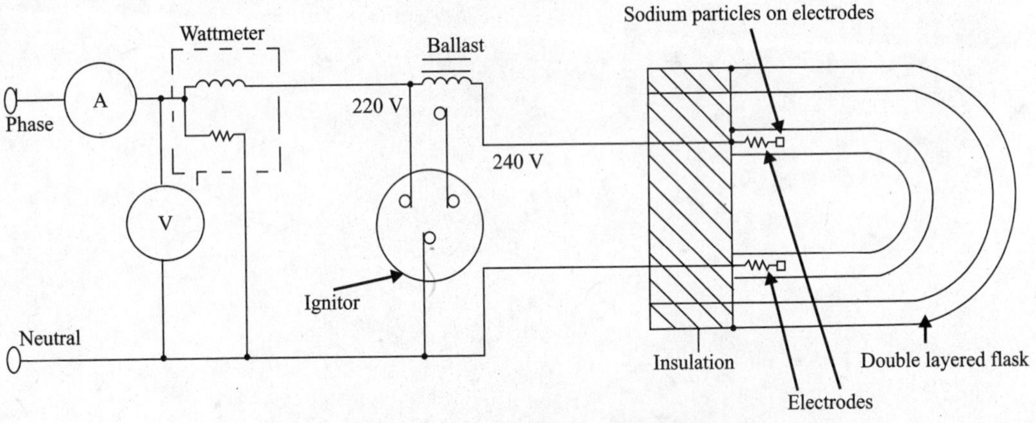

Fig. B.8.1. Sodium vapour lamp connection diagram.

Observation Table

S/N	Time in seconds	Current in Amps	Power in watts
1.			
2.			
3.			
4.			
5.			

(***Note***: The current variation with respect to time is to be plotted on the graph sheet)

PRECAUTIONS

1. Check all the apparatus properly before making connections.
2. Connections should be tight and correct.
3. Do not touch live conductor.
4. Take observations carefully.
5. Do not switch 'OFF' the lamp during taking the observations.

Conclusion: (Write result and discussion)

QUESTIONS

(To be answered in the Practical Record Book.)

1. What is voltage required to the sodium vapour lamp during the starting and glowing?
2. What is the time required to glow the sodium vapour lamp fully?
3. Where is sodium vapour lamp used?
4. Which gas is used in sodium vapour lamp?
5. What is function of the ballast in sodium vapour lamp?

Experiment 9

AIM

To do staircase wiring.

APPARATUS REQUIRED

Screwdriver, combination pliers, connecting wires, two-way switches (2 no.), bulb-holder, bulb, junction box, pipe and saddles with screws.

THEORY

In order to control one light point independently from two switches or more placed at different places on the staircases or in room corridors, this type of wiring is required. Wiring circuit diagram using two switches both of two-way type is shown in the Figure B.9.2 whose schematic diagram is shown in the Figure B.9.1. The phase wire is required to be connected to the common terminal C_1 of the first two-way switch. According to the requirement, moving the dolly of the switches, the connections can be made either to the terminal B_1 or A_1 and B_2 or A_2. Similarly, common terminal of the second two-way switch C_2 is connected with the neutral point (N) through a lamp. Two wires connect A_1 & A_2 and B_1 & B_2. One two-way switch is suitably installed near the first step of the stair and the other two-way switch is installed at the end of the stairs at first floor.

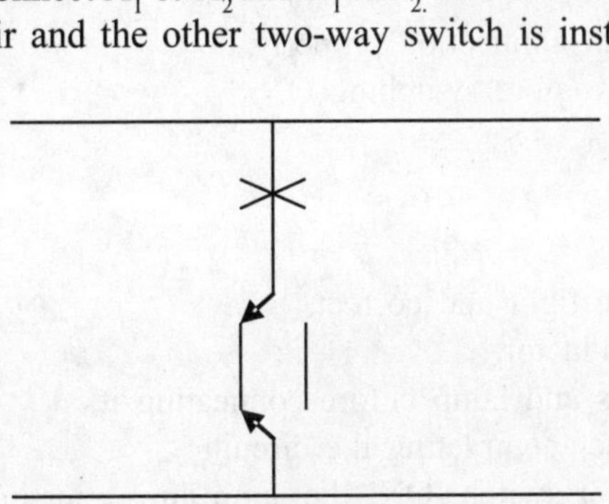

Fig. B.9.1. Schematic diagram of staircase lighting.

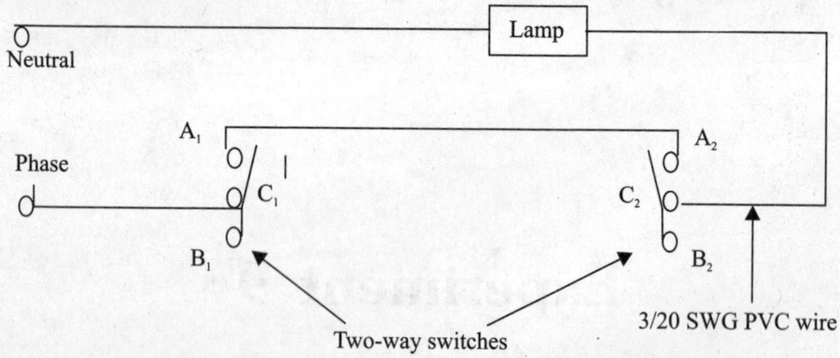

Fig. B.9.2. Staircase wiring connection diatram.

By making movement of the dolly of the two-way switches makes the circuit complete, and the lamp becomes 'ON'. Thus, either in position C_1-A_1-A_2-C_2 or C_1-B_1-B_2-C_2 of the switches, the lamp gets phase current, and hence, becomes 'ON'. On the other hand, in other alternate positions of the switches, the lamp does not get phase current, and hence, remains 'OFF'. So it is clear that the lamp can be made 'ON' or 'OFF' from any one of the two switches.

Suppose a person reaches near stair, where first switch is present and moves its dolly so that position C_1-A_1-A_2-C_2 or C_1-B_1-B_2-C_2 completes the circuit, and the lamp becomes 'ON' and hence glows. After reaching the upper floor, where second switch is situated, the person moves dolly of the second switch so as circuit gets open, and then the lamp become 'OFF'. Now on returning, he moves dolly of the second switch so that the circuit is again complete, and lamp becomes 'ON'. After reaching downstairs, he moves the dolly of the first switch and the lamp gets 'OFF'. So it is clear that when light is required on stairs during movement from down floor to the upper floor, the person switch 'ON' the lamp from entering the down floor and after reaching the upper floor he does not want light on stairs, he switches 'OFF' the lamp. The same can be repeated during movement in the reverse direction.

PROCEDURE

1. Connect the circuit and complete the wiring as per circuit diagram shown in Figure B.9.2.
2. Get checked the circuit connection properly.
3. Do the continuity test before switching ON.

PRECAUTIONS

1. Connections should be tight and correct.
2. Do not touch live conductor.
3. Check all the switches and lamp before connecting it.
4. Test the continuity after completing the circuit.
5. Do not switch 'OFF' or switch 'ON' the lamp during testing the circuit.

Conclusion: (Write result and discussion)

QUESTIONS

(To be answered in Practical Record Book).

1. Write two uses of the staircase wiring.
2. Which type of switch is used in staircase wiring and why?
3. What is the difference between schematic diagram and wiring diagram?
4. How do you perform continuity test?

Experiment 10

AIM

To do house wiring.

Note: Detail is given in Chapter 13.

Conclusion: (Write result and discussion)

QUESTIONS

(To be answered in the Practical Record Book)

1. Why is conduit pipe wiring preferred?
2. What do you mean by saddle, bend and tee?
3. Draw wiring diagram of a circuit consisting of a light point, plug point and fan point.
4. What are the wiring systems? Define them.
5. List out the disadvantages of the batten wiring.
6. List out the accessories required for conduit wiring.
7. What precautions should be taken for conduit wiring?
8. Compare the different types of wirings.
9. Distinguish between surface and concealed conduit wiring.
10. Explain the following types of wiring systems:
    (i) Cleat wiring
    (ii) Lead sheathed wiring

Experiment 11

AIM

To study single-phase induction motor.

APPARATUS REQUIRED

Single-phase induction motor (0.75 hp; 5 A; 220 V; single phase), connecting wires, etc.

THEORY

Fractional horsepower single-phase AC motors work on single-phase supply. These are required to do useful works in houses, offices, hospitals and business houses where low power appliances are to be run. Low power three-phase motors can also be built up but in the houses, offices, hospitals, schools, etc. single-phase supply is generally available. This is the reason that annual production of single-phase AC motors is about 80% of the total motors produced in number.

CONSTRUCTION

The rotor and the stator are main components of an electric motor. The rotor of a single-phase induction motor is exactly same as the squirrel cage rotor of a three-phase induction motor. The stator of a single-phase induction motor carries two windings, namely, main (or running) winding and auxiliary (or starting) winding. Single-phase supply cannot produce rotational field to produce torque, this is the reason that one more winding named as starting winding is required to make the motor able for starting. Both windings are of the single layer concentric type. The main winding is distributed over about two-thirds of the stator periphery. A single-phase induction motor has lower power output than three-phase induction motor of the same size because stator periphery of the single-phase induction motor is not fully utilized. It means in three-phase induction motor, all the stator slots are wound and useful but in single-phase induction motor only a fraction (say two-thirds) of the stator slots are useful.

WORKING

A varying sinusoidal magnetic field is produced when alternating current is fed to the stator winding. This varying sinusoidal magnetic field produces pulsating current in the rotor. This rotor current cannot produce rotating torque in the stationary rotor. This is the reason that a single-phase induction motor is not "self-starting motor". To make the motor capable for the starting a rotating field in the stator is necessary which can be achieved by providing the stator with two windings, generally, the main winding, and a starting winding on four poles (instead of two poles). The required phase difference of 90° between two windings is obtained by splitting the phase. That is the reason that this type of motor is called split phase induction motor. The combined effect of the field set up by the two windings causes rotating field that provides a starting torque. After starting the motor, current set up in the main winding carries out rotation of the field, and the starting (or auxiliary) winding is, generally, switched off from the stator circuit.

Let us describe how the rotating field is produced in the air gap around both pairs of windings. Let both the windings XX and YY axes carry the same alternating current of same maximum value (I_m), and frequency (f), but differ in phase by 90°. Let I_1 & I_2 be the values of the currents in XX and YY winding respectively. Then, $I_1 = I_m \sin(\omega t)$, $I_2 = I_m \sin(\omega t + 90°)$, and $\omega = 2\pi f$, here f = frequency. These currents, while flowing through the coils produce two crossed magnetic field, in phase with their respective currents, but differs from each other by 90°. Now, let the respective magnetic fields H_1 and H_2 are produced by I_1 in XX and I_2 in YY. H_m is the corresponding magnetic field produced by the maximum current I_m then, $H_1 = H_m \sin \omega t$ and $H_2 = H_m \sin(\omega t + 90°)$. Since the fields H_1 and H_2 are at right angles to each other, their resultant becomes:

$$H_m = \sqrt{[(H_1)^2 + (H_2)^2]}$$

and

$$\tan\phi = H_1/H_2$$

So, at $\omega t = 0$, H_m is along X-axis; and at $\omega t = 90°$, H_m is along Y-axis. Hence, the resultant magnetic field has a constant value of H_m, which rotates with a constant angular velocity (ω) in the anticlockwise direction. In other words, four poles and two winding single-phase stator produces a rotating magnetic field.

CAPACITOR-START SINGLE-PHASE INDUCTION MOTOR

In this type of motor phase splitting is done by the means of a capacitor connected in series with the auxiliary winding as shown in the Figure B.11.1. Here current drawn by the starting winding say (I_s) leads the supply voltage (V), while the current drawn by the main winding (I_m) lags the voltage (V) means currents are out of phase with each other. In the case angle between (starting current) I_s and (main winding current) I_m is increased up to 90° by applying capacitance, then starting torque increases because the resultant current (I) is small, and almost in phase with V. Consequently, an increase in the angle between I_s and I_m alone causes an increase in the starting torque.

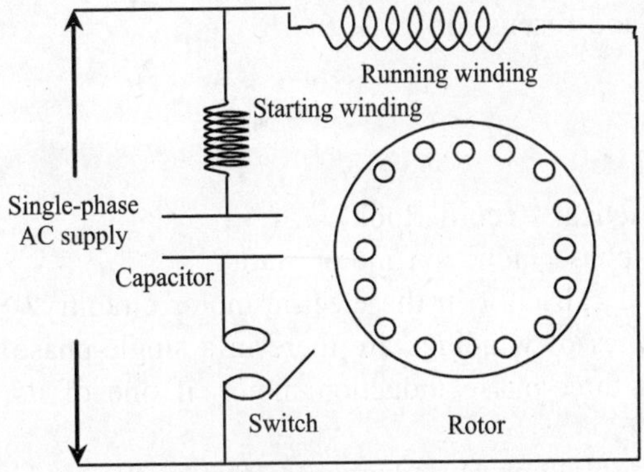

Fig. B.11.1. Capacitor start single-phase induction motor.

PROCEDURE

Steps

1. Open the bottom end cover of the body of the motor by opening the screws with the help of screwdriver.
2. Unscrew the screws of the top end cover of the motor.
3. After removing the bottom cover, winding coils will become visible.
4. Observe carefully the internal parts, i.e., stator, rotor, slots, windings, capacitor, bearings, etc.
5. Identify the running winding and starting winding terminals and resistance of these windings and trace the electric circuit for the motor.
6. Reassemble the motor.
7. Make a circuit diagram of the motor and make the connections as per circuit diagram.
8. Give the supply to the motor through the ammeter and note the current drawn by the motor.
9. Reverse the connections of the starting winding or running winding for changing its direction of rotation.
10. Change the capacitor connection and note the effect.
11. Test the motor by series testing board. Ensure that there is no leakage current through metallic body of the motor.

PRECAUTIONS

1. Before opening cover of the motor, supply must be disconnected.
2. Capacitor and the connections of the motor should be checked carefully.
3. Before connecting the supply voltage there should not be any discontinuity and weak point in the insulation.
4. Connections should be kept neat and tight.

Conclusion: (Write result and discussion)

QUESTIONS

(To be answered in the Practical Record Book)

1. How does a single phase induction motor starts?
2. What is the use of a capacitor in the electric motor circuitry?
3. State, how many types of windings are there in a single-phase motor?
4. What happens to a three-phase induction motor, if one of its three supply leads gets disconnected?
5. Discuss the uses of single-phase induction motors in the public life.
6. Describe how number of coils in the motor windings is decided.

Experiment 12

AIM

To study Megger.

APPARATUS REQUIRED

Megger (1000 V, 0-∞ mega-ohms, 160 rpm), Transformer (1 kW, 110-220 V), Fan (220-240 V, 50 Hz), single-phase motor (50 Hz, 220-240 V).

THEORY

Megger is a trade name of a hand driven portable ohmmeter, extensively used in electrical installations for:

(i) Measuring insulation resistance of the order of mega-ohm and
(ii) Checking continuity of the supply

The quality and the condition of insulation in industrial establishments are ensured by scheduled tests of the insulation. This is done with the help of Megger and this is part of preventive maintenance. The two terminals of the Megger are connected to the terminals under test, and then handle of the generator is rotated. The pointer then deflects over the calibrated scale. The scale is marked from 0 to ∞ in terms of ohms, kilo-ohms, and mega-ohms.

CONSTRUCTION

1. It consists of a DC generator, which is run by rotating the handle by the operator. By rotating at about 160 rpm, it induces about 1000 V.
2. The moving system basically an ohmmeter consists of two coils:
   (i) control coil
   (ii) deflecting coil

Both the coils are rigidly mounted on a common shaft and are pivoted so as to move freely in the magnetic field of a permanent magnet. Both the coils are connected parallel to a DC generator with polarities such that the torques produced by them is in opposition. The pointer connected to the coils directly gives the value of the resistance to be measured. The control coil is in series with a fixed control circuit. The deflecting coil is connected in series with a fixed deflecting circuit resistance and the resistance under test. The control çoil is basically in two parts, in series, the outer part a compensating coil. Both the parts are arranged with number of turns and radii of action such that, for external field of uniform intensity, their torques cancel one another. So in this condition an astatic combination is achieved.

CONNECTIONS

The generator connection is done across the potential coil through a fixed resistance and across the deflecting current coil through another resistance called the current limiting resistance. The external resistance 'R' to be measured is connected across the testing terminals. Circuit connection is shown in the Figure B.12.1.

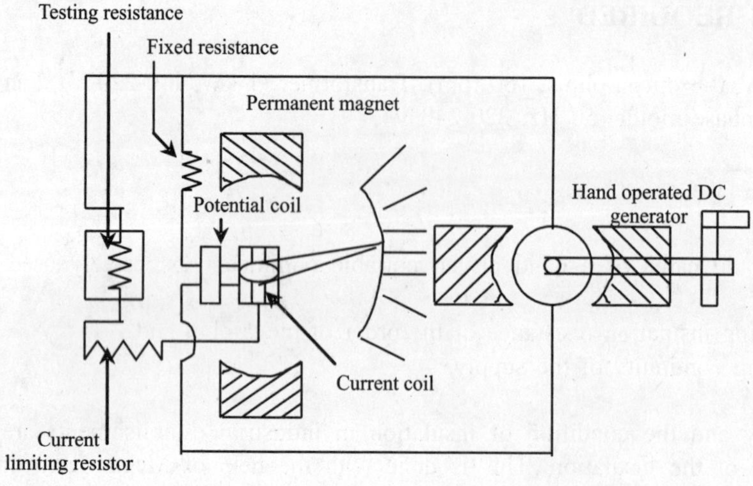

Fig. B.12.1. Internal connection of a Megger.

OPERATION

Firstly, let the testing terminals are open circuited means having very high resistance. Generator gets started by hand driving the handle, the voltage produced is applied across the potential coil and current flows through it, but at this point of time no current passes through current coil. The deflecting torque produced in the moving element rotates until the pointer points to infinity on the scale. And the external resistance of the circuit is too large to measure.

Now, let the testing terminals are connected to a low resistance or are short-circuited. Then comparatively higher amount of current passes through the deflecting current coil. The deflecting current produces deflecting torque. The difference of the deflecting torque produced in the current coil and potential coil moves the pointer till the pointer points to zero and in this case the external resistance is too small to measure.

PRECAUTIONS

1. It is a very sensitive type of instrument.
2. Wires should be free from leakage so better to use lead wires.
3. The two connecting leads must have electrical continuity.

Observation Table

S/N	Electrical instrument	Test	Resistance in ohms
1.	Fan	(i) Insulation test	
		(ii) Earth leakage test	
		(iii) Open circuit test	
		(iv) Short-circuit test	
		(v) Continuity test	
2.	Auto transformer	(i) Insulation test	
		(ii) Earth leakage test	
		(iii) Open circuit test	
		(iv) Short-circuit test	
		(v) Continuity test	
3.	Motor	(i) Insulation test	
		(ii) Earth leakage test	
		(iii) Open circuit test	
		(iv) Short-circuit test	
		(v) Continuity test	

USES OF MEGGER

(i) Continuity test
(ii) Insulation test which gives resistance value between conductors.

Conclusion: (Write result and discussion)

QUESTIONS

(To be answered in Practical Record Book)

1. Where is Megger used?
2. What are the functions of two coils of the Megger?
3. What is the need of DC generator in the Megger?
4. Justify a Megger is basically ohmmeter.

Experiment 13

AIM

To design and fabricate a single-phase small transformer.

APPARATUS REQUIRED

E&I shaped transformer core, bobbin, insulating paper, pliers, wire-cutter, coil-winding machine, soldering material, multimeter, insulation-tape and Megger.

THEORY

Transformer is a static device. It transfers AC electrical energy from one circuit to the other at the same frequency by magnetic coupling of the primary and secondary windings of the transformer. But the voltage level is changed at the secondary side. If the voltage on the secondary side of the transformer (V_2) is greater than the voltage at primary side (V_1), the transformer is called a step up transformer and if otherwise, it is called a step down transformer. In our country, usually, electrical power is generated at 11 kV of 50 Hz frequency. For economic reason, AC power is transmitted at very high voltages (220 kV, 400 kV or even more) over long distances; therefore, a step up transformer is applied at the generating station. Voltage is stepped down in different areas to different levels by means of the transformers at various substations. Ultimately for the utilization of electrical power at distribution sub-stations, the voltage is stepped down to 400/230 V and then supplied to the houses, offices and small industrial premises.

CONSTRUCTION

A transformer consists of two coils of insulated wire wound on a core. These coils are known as primary and secondary windings. Primary side of the transformer is where the main electrical power to be changed is connected. Secondary side of the transformer is where the electrical power is sent to the load. The two coils are insulated from each other as well as from the steel core. Paper or cloths are used as insulating material. Windings are usually wound on to a form and then wrapped with an insulating material, i.e., cloth or paper. The core is made of laminations of steel or silicon steel assembled to provide a continuous magnetic path. As usual, due to flux densities the silicon steel material has hysteresis and eddy current losses.

The core in the form of laminations is used because laminating the core minimizes eddy current loss. The laminations are insulated from each other by a light coating of varnish or by an oxide layer. The thickness of laminations varies from 0.35 mm to 0.5 mm for frequency of 50 Hz. On the basis of core, transformer is of two types:

(i) Core Type Transformer

A single-phase core type transformer consists of a magnetic frame with two cores. Upper and bottom yoke form the magnetic circuit. The low voltage and high voltage windings are placed on primary and secondary sides directly in very small transformers. But in large transformers, low voltage and high voltage windings are split into two parts and half of the turns of low voltage windings and half of the turns of high voltage winding are placed on each core. Symbolic representation of the single-phase core type transformer is shown in the Figure B.13.1.

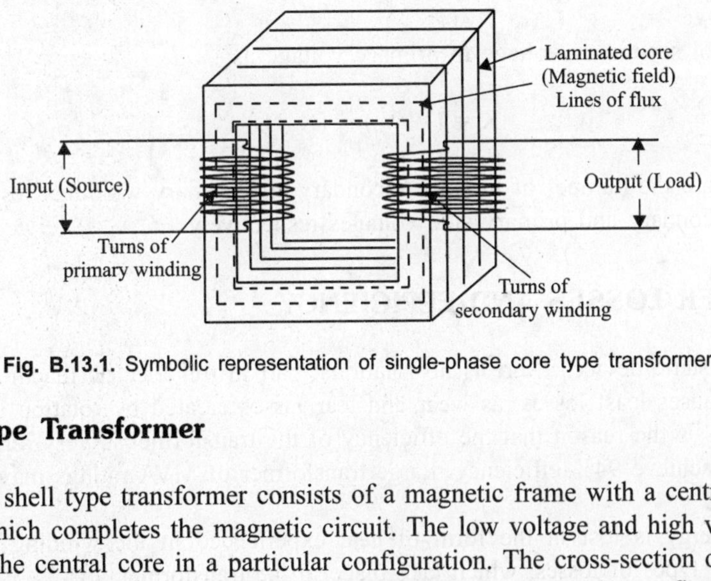

Fig. B.13.1. Symbolic representation of single-phase core type transformer.

(ii) Shell Type Transformer

A single-phase shell type transformer consists of a magnetic frame with a central core and two side's cores, which completes the magnetic circuit. The low voltage and high voltage windings are placed on the central core in a particular configuration. The cross-section of the yokes and side cores are only half the cross-section of the central core. Central core flux, φ gets divided at the yoke section equally, i.e., φ/2 towards each side cores. Symbolic representation of the single-phase shell type transformer is shown in the Figure B.13.2.

PRINCIPLE OF OPERATION

A symbolic representation of a transformer is shown in the Figure B.13.1. Supply is connected to the primary winding and load to the secondary winding. When supply voltage is applied to

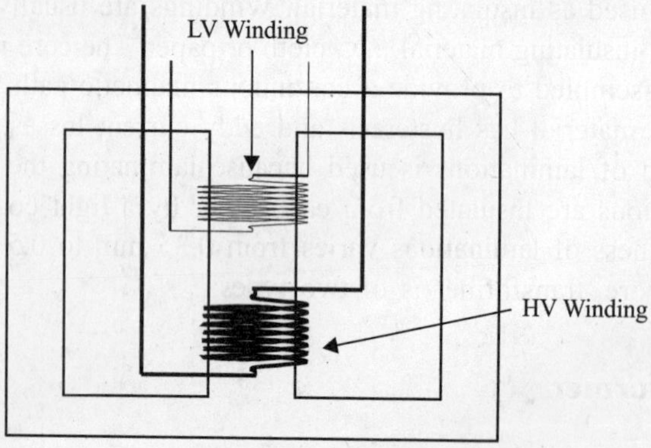

Fig. B.13.2. Symbolic representation of the single-phase shell type transformer.

the primary winding, flux is induced in the core due to which an e.m.f. is induced in the secondary winding. The direction of induced e.m.f. is opposite to the direction of flux.

TRANSFORMATION RATIO

This is the ratio of secondary voltage to primary voltage as:

$$K = (V_2/V_1) = (N_2/N_1)$$

Here, N_2 and N_1 are the number of turns in secondary and primary windings respectively. And V_2 and V_1 are secondary and primary side voltages respectively.

TRANSFORMER LOSSES AND EFFICIENCY

Transformer is a static device. There is no rotational part in this electric machine. Absence of rotational parts causes least losses, as wear and tear losses created by rotation is zero for the transformers. This is the reason that the efficiency of the transformer is very high, i.e., 5 kVA transformer may achieve 94% efficiency. Large transformer of MVA ratings may even achieve 99% efficiency.

But some energy is lost in the form of heat experienced in the winding and the core. Following are the types of losses, which take place in the transformer.

(i) Copper Loss

Windings of the transformers have certain resistive value; the flow of current in the windings causes i^2R-heating loss. This is known as copper loss.

(ii) Core Loss

The loss, which takes place in the magnetic core and depends on the voltage applied to the primary of the transformer, is known as core loss.

(iii) Eddy Current Loss

Eddy currents are the induced currents which circulate in the core. This circulating current causes loss due to resistive heating of the core. This loss is known as eddy current loss.

(iv) Hysteresis Loss

Reversal of the magnetic field needs small amount of energy and hence each time the magnetic field is reversed a small amount of energy is lost to hysteresis in the magnetic core. That is known as hysteresis loss.

(v) Stray Loss

Not all the magnetic fluxes produced by the primary winding are intercepted by the secondary winding. The remaining fluxes are absorbed by the other nearby objects and converted to heat. Any magnetic field not coupled to the secondary circuit contributes to leakage inductance and corresponding loss is known as stray loss.

(vi) Mechanical Losses

The alternating magnetic field causes fluctuating electromagnetic force between the coils of wire, core and other nearby metal works. This fluctuating electromagnetic forces cause vibrations of the parts of the transformer and create noise. In the process it consumes power and that is known as mechanical losses.

Specification of a Transformer

Transformers are specified in the following terms combined:

1. Input voltage (with taps if provided)
2. Input frequency
3. Input voltage fluctuating range
4. Output voltage at full rated load
5. Output current
6. Voltage regulation
7. Efficiency
8. Termination

9. Open type construction or enclosed type
10. Temperature allowed

Fabrication of a Single-phase Transformer

For designing and fabrication of a transformer, following four quantities are required to be known:

(1) Primary voltage
(2) Secondary voltage
(3) Output current and
(4) Efficiency

The following designed procedure is generally adapted:

Design Procedure

1. Understand the terminology
2. Understand the transformer equivalent circuit
3. Prepare the list of requirement
4. Select the core size
5. Compute number of primary turns
6. Compute number of secondary turns
7. Stack the laminations
8. Wind the coils
9. Test to check the continuity in the windings
10. Test to ensure for proper insulation

Example

Suppose, we want to fabricate a transformer of rating 110/45 V at an output current of 0.15 A and of efficiency 94%.

Here, the known values are:

$V_p = 110$ V, $V_s = 45$ V, $I_s = 0.15$ A, $\eta = 94\%$

Area of the core = 3.8 cm^2 (selected from the data sheet)

K = 0.077 (selected from the data sheet)

So,

O/P power = $V_s I_s = 45 \times 0.15 = 6.75$ W

Therefore,

I/P power = (O/P power/η) $\times$ 100 = 6.75 $\times$ 100/94 = 7.18 W

and

e.m.f. generated per turn;

$E = 4.44\ B\ A\ f \times 10^{-4}$ volts $turn^{-1}$

Where

E = volts turn^{-1}

B = flux density = 0.8 wb/m^2

A= area of core = 3.8 m^2

f = frequency = 50 Hz

Now

$E = 4.44 \times 0.8 \times 3.8 \times 50 \times 10^{-4}$ V turn^{-1}

= 0.0675 V turn^{-1}.

N_p = 110/0.0675 = 1629.6 ≡ 1630 turns

So

N_s = 45/0.0675 = 666.67 ≡ 667 turns

Due to voltage drop factor, add 10% to the number of turns

N_p = 1630 + 163 = 1793 turns

N_s = 667 + 67 = 734 turns

Size of primary winding wire = 32 SWG (select from the data sheet)

Size of secondary winding wire = 36 SWG (select from the data sheet)

Observations

Record the values of the insulation resistance of the transformer using Megger.

Primary to primary
Primary to secondary
Secondary to secondary
Primary to body
Secondary to body

Testing

Resistance of primary =

Resistance of secondary =

$V_{input} = V_p$ = 110 V

V_s = 45 V

Conclusion: (Write result and discussion)

PRECAUTIONS

1. Coil winding machine should be used properly.
2. SWG of the wire should be properly selected and verified before making the coils.
3. Wires should not be overlapped.
4. Transformer should be properly insulated.

QUESTIONS

(To be answered in Practical Record Book)

1. What are the various types of losses, which occur in a transformer?
2. What is all-day efficiency?
3. Distinguish between the step-up and step-down transformers.
4. Why core of a transformer is laminated?
5. Which type of material is used for making transformer core?
6. What is difference between core type and shell type transformers?
7. List out the design steps for a transformer.
8. How do you specify a transformer?

Experiment 14

AIM

To repair home appliances such as heater, electric iron and ceiling fan.

1. ELECTRIC HEATER

Electric heater is used for the purpose cooking and room heating. These are of two types.

(i) Ordinary plate heater

(ii) Room heater

The room heater having small fan for blowing the hot air is known as heat convector. This heats up the room faster than the ordinary room heater.

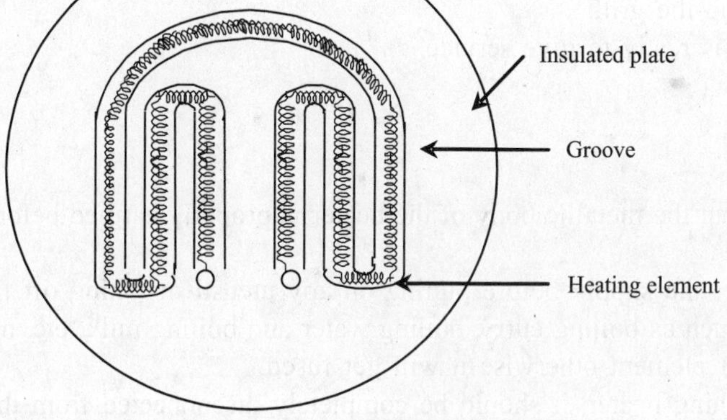

Fig. B14.1. An electric plate heater.

All the heaters have nichrome make heating coil. Fine nichrome wire is made in coil shape to accommodate larger length and in return to give more amount of heat. The coil is placed in the grooves made in the porcelain or china clay plates as shown in the Figure B.14.1. Both the ends of the continuous run (without break) nichrome coil is fitted with the terminals covered with the ebonite sheets. Plate with the coil is placed on a mild steel or iron made frame. In the room heater, nichrome heating element is wrapped on porcelain made roller shaped plate. This plate is safely surrounded by iron or mild steel made covering. Terminals for feeding the supply is fixed at covers and connected with the heating element.

METHOD OF REPLACING THE FUSED HEATER ELEMENT

Steps

1. Open the nuts and bolts of the outer body to which the heating element is fitted. Use screwdriver. Also, remove the steel grill, if any.
2. Take out the fused heating element from the heater grooves.
3. Clean the grooves of the heater plate carefully. Take care throughout the removing and placing the heating element otherwise groove may get damaged.
4. Measure the total length of the heater grooves from one nut bolt position to other with the help of thread and measuring tape.
5. Stretch the measured thread on the ground.
6. Take the new heater element and stretch it to the length of thread.
7. Connect two wires with the ends of heating element of the heater.
8. Give supply to the ends of element through extra-connected wires.
9. Check the continuity and ensure that coil is working.
10. Put off supply.
11. Remove the lead wires which were connected for supply.
12. Now place the element in the clean grooves of the heater.
13. Connect both ends of the heater element with nuts and bolts provided in heater plates.
14. Finally, fix up the grill.
15. Now, heater is ready to give service.

PRECAUTIONS

1. Make sure that the metallic body of the heater is properly earthed before switching 'ON' the heater.
2. Switch 'OFF' the supply before putting on any utensil or taking off from the heater.
3. Any liquid such as boiling curry, boiling water and boiling milk, etc. must not spill over on the heater element otherwise it will get fused.
4. While refreshing heater, it should be completely disconnected from the supply.

2. ELECTRIC IRON

It is used for ironing the uniforms, dresses, etc. Pressing the papers or clothes by hot objects makes the fibres of the papers or clothes straight and tight. In the electric iron, bottom of sole plate gets heated by the nichrome-made heating element. Heating element is fixed between the sole plate and a little bit heavy pressure plate. Asbestos sheet covers the heating element otherwise it may come in contact with the metallic parts. Handle is made up of ebonite an insulating material.

For cleaning, refreshing and sometimes for replacing the heating element, dismantling of the electric iron is required. Step by step method for dismantling and then replacement of heating element is given. In order to make the electric iron automatic, thermostat is provided. One can set the electric iron to be heated up to a certain temperature. After achieving preset temperature and heating up the sole plate up to a preset level, thermostat automatically cut 'OFF' the supply automatically. This way thermostat maintains temperature of the iron within a desired range.

METHOD OF REFRESHING ELECTRIC IRON

Steps

1. First, open the screw fitted in the outer knob with connector with the help of screwdriver. Now remove the knob by pulling it upward.
2. Now look into the indication lamp side. Unscrew the screw fitted in cover of indicating lamp. Remove the cover of the indicating lamp.
3. Now the terminals will be visible. Open the screws fitted with terminals to remove the wires. Pull the wire outside. Hold the handle and pull it upward. So, the cover will be removed.
4. Thermostat assembly will be visible now. Near the thermostat, nuts are used for fixing the pressure plate with sole plate. Open these nuts and remove pressure plate with thermostat. Do not try to disturb the thermostat assembly.
5. Check for heating element. Replace it, if element is fused as shown in the circuit Figure B.14.2. Also replace asbestos sheet.
6. Check the continuity of the heating element. If it is o.k. fix up the pressure plate back.
7. Fix up covers with handle by tightening the related screws.
8. Now join the terminals point and get your heater ready for service.

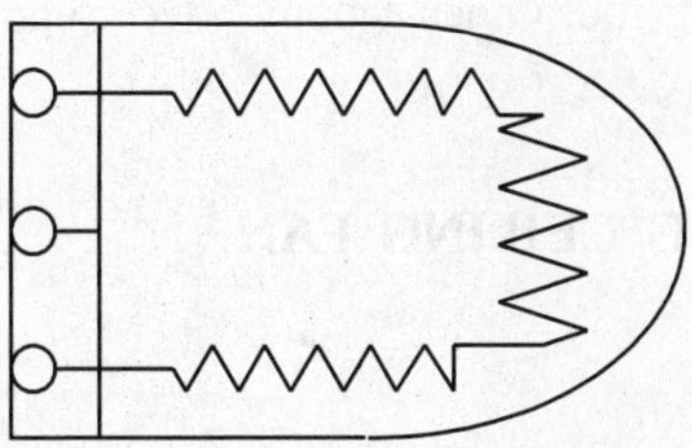

Fig. B.14.2. Electric circuit diagram for electric iron.

PRECAUTIONS

1. Always use three-core cable in which green wire is used for earthing purpose.
2. Make sure that wire should not come under hot iron.

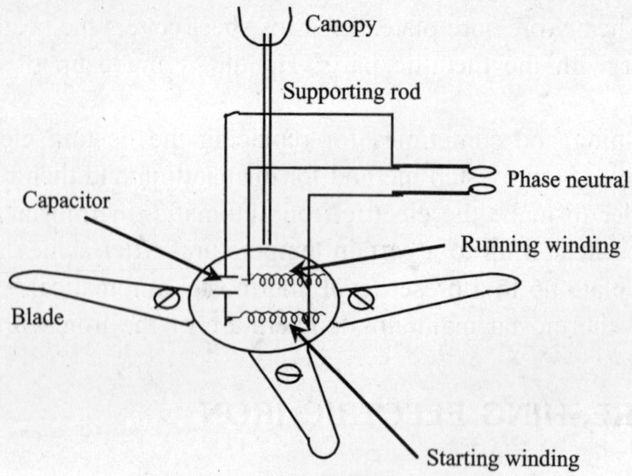

Fig. B.14.3. Circuit diagram for ceiling fan.

3. Do not overheat the iron. It may remove the nickel coating of sole plate and bottom becomes rusty.
4. Overheating of the sole plate sticks with the clothes.

3. CEILING FAN

Single-phase induction motor is commonly used in ceiling fans as well as in the table fans. DC motors are used in the fans installed in the trains and some industries where direct current supply is available. As single-phase induction motors are not self-starting, capacitors are used to make them able to start by applying the electric supply. Fan blades are made slightly curved to push the air to be circulated when blades rotate. Blades are made up of either aluminium or tin coated with paint. Nowadays plastic is also being used in making the blades for fans. Canopy is made up of plastic material and is nothing but cover, which covers the upper part of the motor at the bottom side of the hanging rod. Other canopy covers upper part of the rod connected at the roof near ceiling rose.

METHOD FOR REPAIRING CEILING FAN

Steps

1. First unscrew the screws fitted with body of the fan and blades. Now remove the blades.
2. Remove the nut fitted with the bolt used for hanging the fan with ceiling hook covered by the canopy.
3. Open the bottom end cover by opening the screws and with the help of screwdriver, batten and mallet.
4. Unscrew the screws of the top end covers.

5. As the bottom cover is removed, the winding coils will be visible.
6. Apply grease and oil wherever is required.
7. Do the needful repair and overhauling of the fan.
8. Reassemble the different parts.
9. Check the continuity before finally fixing the fan. Circuit diagram for a fan connection in given in Figure B.14.3.

Conclusion: (Write result and discussion)

PRECAUTIONS

1. Fan must be earthed by an earth wire coming from the wall socket by the third wire (green) in the lead.
2. While doing any type of repair, fan must be disconnected from wall socket.
3. Connections should be neat and tight.

QUESTIONS

(To be answered in Practical Record Book)

1. Why heating element of the appliances is made up of coil shape?
2. What will happen if boiling cooking material spills over the heating element of the heater?
3. How pressure plate is placed in an electric iron?
4. What is the frequency of AC supply available in our houses and what is the voltage?
5. How improper voltage affects the home appliances?
6. Which type of motors are used in fans installed in trains and why?
7. What is the role of capacitor placed in an electric fan?
8. What is the utility of canopy in the ceiling fan?
9. Which materials are suitable for making blades of electric fans?
10. Observe the difference between a ceiling fan and a table fan.

Experiment 15

AIM

To study moving iron, moving coil, electrodynamic and induction type meters.

INTRODUCTION

There are following three categories of meters.

(i) Indicating instruments, i.e., laboratory type ammeters, voltmeters
(ii) Electrostatic instruments, i.e., high voltage measuring instruments
(iii) Integrating instruments, i.e., energy meters and wattmeters.
(iv) Digital meters have become popular for accuracy

In the laboratories, mostly indicating type instruments are used. These instruments make use of a dial and a pointer to indicate the magnitude of the quantity to be measured.

All the voltmeters and ammeters are basically current measuring devices. In the ammeters, the current produces deflecting force and the main acting torque. In the voltmeter, current also produces the deflecting torque. But this current is proportional to the voltage to be measured. Ammeters are low resistance meters. Low resistance allows flow of maximum current creating negligibly small voltage drop. In the voltmeters, voltage is proportional to the current flow in the coil so the voltage drops in the coil. High resistance is generally connected in the voltmeters that creates proportionate voltage drop. That is why voltmeters are called high resistance meters.

The following three types of forces are required for satisfactory operation of indicating instruments.

(i) Deflecting force
(ii) Controlling force
(iii) Damping force

The deflecting system of an instrument converts the electric current or potential into a mechanical force. The *deflecting force* is produced on any one principle of electromagnetic, magnetic, electrostatic, heating or Hall effect. According to the flow of current, force is produced which deflects the pointer. The moving system due to deflecting force should be light in weight and its movement should be nearly frictionless.

The *controlling force* acts in opposition to the deflecting force. In the absence of deflecting force, it brings back the pointer on zero position on the scale. This force is generated with the help of springs. Phosphor bronze has been the most suitable material for the springs. To eliminate the effect of temperature variation upon the length of the spring, two spring coils in the opposite directions are used.

The *damping force* is essentially required to control and mitigate the oscillations of the pointers. This force calm down oscillations of the pointer, before coming it to the steady position under the influence of the deflecting force. The damping force acts in opposition to the deflecting force. The damping force is generated by friction. The frictional force is generated by:

(i) Air friction
(ii) Fluid friction
(iii) Eddy current

To obtain friction, a lightweight vane is attached to the moving system. The vane is made free to move in a tight chamber. If chamber is filled with air, it is said air friction damping and if filled with fluid (generally oil), it is said fluid friction damping. In the eddy current damping, a metallic disc experiences an electromagnetic damping force in opposition to the direction of movement.

Following are the main types of meters:

(i) Moving iron instrument
(ii) Permanent magnet moving coil instrument
(iii) Electrodynamic instrument
(iv) Induction type instrument

1. MOVING IRON INSTRUMENTS

Moving iron instruments are most common as far as ammeters and voltmeters are concerned. These are cheaper than other types of meters with equal accuracy and ruggedness. Moving iron is an oval-shaped disc of soft iron eccentrically mounted on a spindle. The stationary coil is semi-cylindrical in shape with narrow slot-like opening for the movement of iron. The controlling torque is usually provided by the springs, but gravity is provided only in case of vertically mounted instruments. Damping is usually of air friction type. Its constructional view is shown in Figure B.15.1.

These are classified as under:

(i) Attraction type
(ii) Repulsion type
(iii) Combination of the above two

In the attraction type meters, a vane or plate of soft iron of high permeability steel attached with the moving system is drawn into the field of a solenoid. Ampere-turn remains same for the both the ammeters and voltmeters. Excitation of the electromagnet causes movement in the iron vane in such a way as to increase the flux through it. This increases the inductance of the combination of the electromagnet and the iron vane.

There are two vanes inside the coil in the repulsion type instruments, one fixed vane and other movable. Both the vanes are magnetized identically and repulsive force, generally, moves the spindle attached with pointer.

In the combination type moving iron instruments, 240° movement and better uniform scale are possible. In this arrangement, as the deflection increases, the outer stationary vanes exert attractive force on the moving vane and add to the torque.

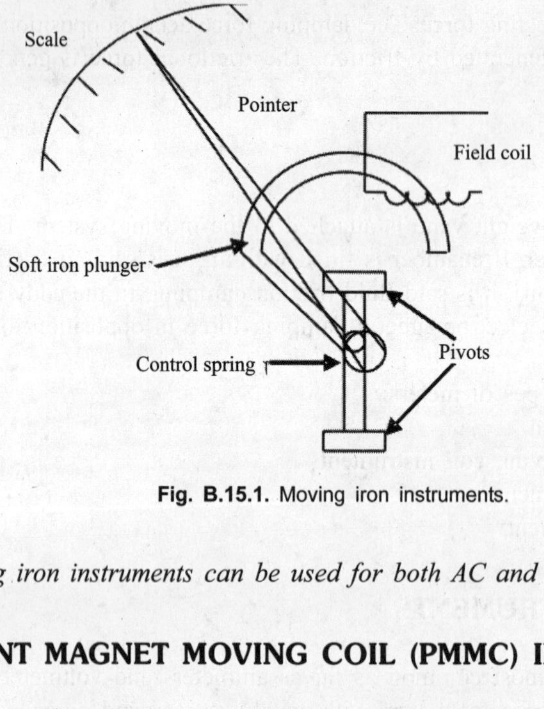

Fig. B.15.1. Moving iron instruments.

Note: *Moving iron instruments can be used for both AC and DC measurements.*

2. PERMANENT MAGNET MOVING COIL (PMMC) INSTRUMENTS

These instruments are known for measuring DC quantities accurately. The essential constructional features of the instruments are shown in the Figure B.15.2.

U-shape permanent magnet is fitted with two soft iron pole shoes. A soft iron cylinder is inserted between the poles, which provide a radial magnetic field of uniform density in the annular air gap. This arrangement is done to reduce reluctance of the magnetic circuit.

The moving coil is prepared in rectangular shape with many turns of fine copper wire. This coil of many copper turns, wound on a light aluminium. 'Former' is pivoted on jewelled bearings, which make it free to rotate in the air gap between the soft iron cylinder, and the pole shoes.

Two hairsprings provide the controlling torque. These hairsprings are made up of phosphor bronze. These springs also serve as lead-into, and lead-out for the current.

It is essential to make the pointer and scale statically and dynamically balanced. The steady deflection of pointers on the calibrated scale indicates the quantity under measurement. Since the deflection is directly proportional to the current passing through the meter, a linear scale is

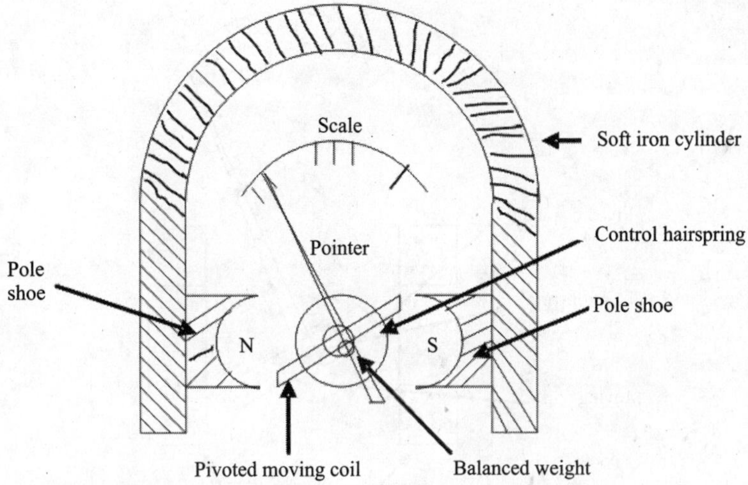

Fig. B.15.2. Permanent magnet moving coil instrument.

obtained. In low range ammeters (up to 20 mA), the entire current to be measured may be sent through the moving coil. DC ammeters design normally permits voltage drop of approximately 50 mV to 100 mV for full-scale deflection. Most DC voltmeters are designed to produce full-scale deflection with a current of 20, 10, or 1 mA.

When the moving coil moves in the magnetic field, eddy current get induced in the 'former', which produces damping torque.

Note: Since there is the presence of permanent magnet so such instruments can only be used for the measurement of DC values.

3. ELECTRODYNAMIC INSTRUMENTS

It is identical to the permanent magnet moving coil instrument in which permanent magnet is replaced by electromagnet. It consists the parts as shown in Figure B.15.3. Very common type of electrodynamic meter is the wattmeters and due to some practical reasons, dynamometer type ammeters and voltmeters are not in common use (generally in DC circuit). There are two coils in this type of metre in which one is fixed and other is free to move with a pointer over a graduated scale.

Two identical air coiled circular coils of many turns through which a definite fraction of the circuit current passes make the fixed coil. These are positioned parallel to each other in order to provide better uniform parallel field. Particularly, near the centre and also to allow the passage of instruments shaft is fitted in between. The fixed coil is known as current coil. The connection circuit diagram of dynamometer type wattmeter is shown in Figure B.15.4.

Moving coil is a circular air cored coil of fine wire, wound on a light frame mounted on a spindle and pivoted between jewelled bearings. This is known as pressure coil or potential coil and is connected in parallel to the load. This carries the current proportional to the voltage. The

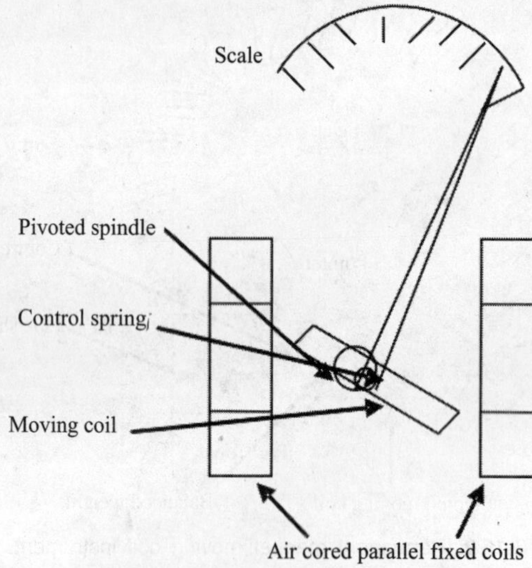

Fig. B.15.3. Dynamometer type wattmeter.

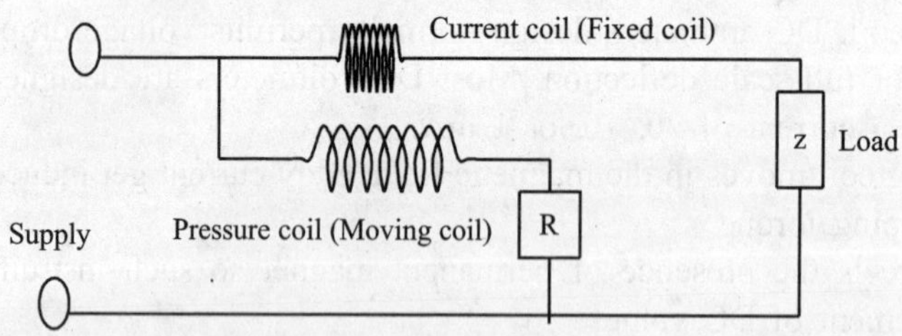

Fig. B.15.4. Dynamometer type wattmeter connection.

control springs made up of two spiral springs of phosphor bronze serve as lead-into and lead-out for the current.

A pointer attached with the spindle, moves over a calibrated and graduated scale. The scale is not even. It is non-uniform. Damping torque provided by two light valves is mounted at the bottom of the spindle. The pressure coil known as series resistance is connected across the supply through a fixed resistance. Both fixed and moving coils are air cored.

Air friction damping is mostly used in such wattmeters.

4. INDUCTION TYPES INSTRUMENTS

Induction type meters are basically integrating measuring instruments. Single-phase and three-phase energy meters come under this category for measurement of energy in domestic and industrial AC circuits.

The main parts of these meters are:

(i) Driving system

(ii) Moving system
(iii) Braking system
(iv) Registering system

Driving system of this instrument consists of silicon steel laminated core and is made up of two identical electromagnets. This is similar to that used in dynamometer type wattmeter. The load current excites one coil, known as current coil. The second coil of the electromagnet is connected across the supply and carries a current proportional to the supply voltage and is known as potential coil or pressure coil.

An aluminium disc mounted on a light alloy shaft is positioned in the air gap between the current coil and the pressure coil magnets. This forms the moving system. The shaft gets support on top and bottom and two jewel bearings supports the rotating disc.

A pinion engages the shaft with the counting and registering mechanism. The counting or registering mechanism records continuously a number proportional to the revolution made by the moving disc. In this arrangement, a train of reduction gear is driven by the pinion and drives a series of about six pointers. These pointers rotate on round dials marked with ten equal divisions.

Note: The calibration of single-phase energy meter and figure showing the parts of energy meter is described in the experiment number 20.

Conclusion: (Describe in brief, whatever you have learned.)

QUESTIONS

(To be answered in the Practical Record Book)

1. How is ammeter connected in the circuit?
2. How is wattmeter connected in the circuit?
3. How is voltmeter connected in the circuit?
4. What will happen when a permanent magnet moving coil-instrument will be connected in an AC circuit?
5. List out the differences between wattmeter and energy meter.

Experiment 16

AIM

To study direct on line starter (DOL).

THEORY

Electrical motors whether single-phase or three-phase types suffer due to malfunctioning. Malfunctioning of the motors may be due to:

(i) Overloading
(ii) Undervoltage
(iii) Fluctuations

A mechanically hold DOL starter provides only overload protection while electromagnetically hold DOL starter provides protection from fluctuation, under-voltage and overload.

The direct on line starter just connects the motor terminals to the supply and categorized for single-phase or three-phase types motors. Heavy starting current may harm the induction motors, if supply voltage is given directly. Direct on line starter is used to protect the motors from the heavy starting current.

Direct-on-line starting is a simple and cheap method and is used for up to 5 hp motors. The starting current may be as large as 10 times, the full load current and the starting torque is equal to full load torque. For the alternating current, motors higher than 5 hp rating DOL starter is not suitable because heavy starting current may damage the winding coils due to excessive heating. To avoid damage at starting, a starter is required. Connection diagram is shown schematically in the Figure B.16.1.

WORKING PRINCIPLE

Following are the main working parts of a DOL starter

(i) Connector
(ii) Thermal overload relay

(iii) Push ON button (open type)
(iv) Push OFF button (closed type)

The motor is started by connecting directly across the main supply with an association of circuit breaker or a magnetic contactor or starter. When the start button is pressed, circuit is completed through line, stop button and magnetic coil of the contactor M. Contactor 'M' has four poles, three for the motors to be connected to the three-phase supply and one for contactor coil. Pressing of the start push button is necessary because even if the three-phase supply is available, the contactor coil will not be energized. After momentarily pushing the start button in, the contactor coil gets energized and the four-pole contactor 'M' is closed. In this interval, the motor starts and runs. As the start push-button gets released, the contactor coil stays energized through the auxiliary contact 'a' in the fourth pole of the contactor 'M' and the motor keeps on running. When it is required to stop the motor intentionally by pushing the stop button in, the coil circuit gets open and the coil is de-energized which opens the contactor 'M' and the motor is isolated from the supply. These way motor stops. A thermal overload relay contact for protection against sustained overloads is provided with the DOL starter. The relay should not operate during starting period. Therefore, a combination of magnetic starter, a circuit breaker (or a

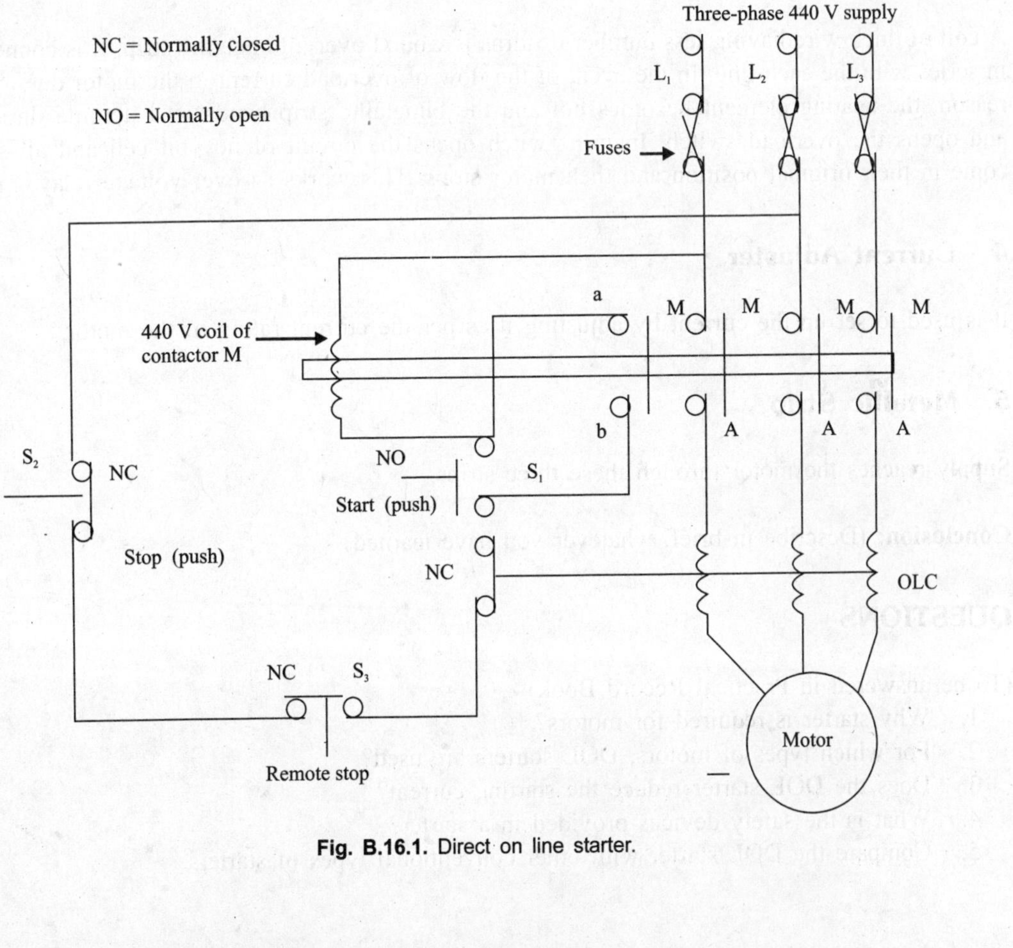

Fig. B.16.1. Direct on line starter.

switch fuse) in addition to the contactor starter (for producing starting current), is used for protection and isolating the motor from the supply mains.

FUNCTIONS OF DIFFERENT PARTS

1. Push Buttons

A red button to switch off the motor and a green button to start the motor are fixed outside the cover of the starter.

2. No Voltage Coil

The function of this coil is to keep the motor in 'ON' position and in the event of failure of supply it stops the motor. It is wound with the thin conductor having sufficient number of turns over a bobbin and operates online voltage, i.e., 400 V.

3. Overload Coil

A coil of thick wire having less number of turns is wound over a bimetallic strip. It is connected in series with the each line. In the event of the flow of overload current in the motor due to any reason, the heating element becomes hot and the bimetallic strip bends in a definite direction and opens the overload switch. In turn switch opens the circuit of no volt cell and all strips come in their original position and then motor stops. This works as over voltage relay.

4. Current Adjuster

It is used to set up the current by adjusting it as per the current rating of the motor.

5. Metallic Strip

Supply reaches the motor through these three strips.

Conclusion: (Describe in brief, whatever you have learned)

QUESTIONS

(To be answered in Practical Record Book)

1. Why starter is required for motors?
2. For which types of motors, DOL starters are used?
3. Does the DOL starter reduce the starting current?
4. What is the safety devices provided in a starter?
5. Compare the DOL starter with other conventional types of starters.

Experiment 17

AIM

To study fuses used in electrical engineering.

Note

Detailed description is given in Chapter 13.

Conclusion: (Describe in brief, whatever you have learned)

QUESTIONS

(To be answered in the Practical Record Book).

1. What is the role of a fuse in an electrical circuit?
2. Which material is used as fuse element in a re-wireable fuse?
3. Which material is used as fuse element in a HRC fuse and why?
4. List out the properties of filling powder of cartridge fuse.
5. Make comparison between:
   - (a) Re-wireable fuse and D-type cartridge fuse.
   - (b) D-type cartridge fuse and HRC fuse

Experiment 18

AIM

To study relays used in electrical engineering.

Note

Detailed description is given in Chapter 13.

Conclusion: (Describe in brief, whatever you have learned)

QUESTIONS

(To be answered in Practical Record Book)

1. Write in brief the importance of relays in electrical applications.
2. Describe in brief different types of relays.
3. What do you mean by threshold operation of the relays?
4. Describe difference between back-up protection and primary protection.
5. Make a list of the relays based upon their functions.
6. Make a list of the relays based upon their constructions.

Experiment 19

AIM

To study contactors, circuit breakers and MCBs used in electrical engineering.

Note

Detailed description is given in Chapter 13.

Conclusion: (Describe in brief, whatever you have learned)

QUESTIONS

(To be answered in Practical Record Book)

1. Why contactors are required?
2. What are the types of contactor?
3. Where MCBs are preferred?
4. Draw a cross-sectional view of MCB and explain its working.
5. How miniature circuit breakers are rated?
6. What is ELCB? List out its uses.

Experiment 20

AIM

To calibrate a given single-phase induction type energy meter.

APPARATUS REQUIRED

One single-phase induction type energy meter, One M.I. voltmeter (0-200 V), One M.I. ammeter (0-10 amps) and one stopwatch.

THEORY

Energy meters are integrating type instruments. Energy meter also known as watt-hour meter is an instrument, which measures electrical energy. These are of several types. Single-phase induction type energy meters are very commonly used to measure electrical energy consumed in domestic and commercial installations. Electrical energy is measured in units (one unit = one kilowatt-hour = 1000 watt-hour) by these energy meters. Energy meters are required to be checked and recalibrate on regular basis. The reason may be continuous friction during the disc movement and deterioration of electromagnetic property of the coil due to aging. This causes maloperation of the energy meters and gives inaccurate readings. So, it requires to be corrected by recalibrating it.

CONSTRUCTION

A single-phase induction type energy meter consists of Driving system, Moving system, Braking system, and Registering system. Each of the systems is briefly explained below:

Driving System

Driving system of energy meter consists of two silicon steel laminated electromagnets M_1 and M_2. The electromagnet M_1 is called series magnet and the electromagnet M_2 is called shunt magnet. The series magnet M_1 carries a current coil consisting of a few turns of thick wire. The load current flows through this coil. The shunt magnet M_2 carries a voltage or pressure coil consisting of many turns of thin wire. This is connected across the supply. This coil carries current proportional to the supply voltage.

Power factor compensator, the short-circuited copper bands whose position is adjustable is provided on the lower part of the central limb of the shunt magnet. The shunt magnet flux can be made to lag 90° behind the supply voltage. Copper-shading band is provided on each outer limb of the shunt magnet (FC_1 and FC_2). These bands are made to provide frictional compensation. Figure B.20.1 shows the connection diagram of single-phase energy meter.

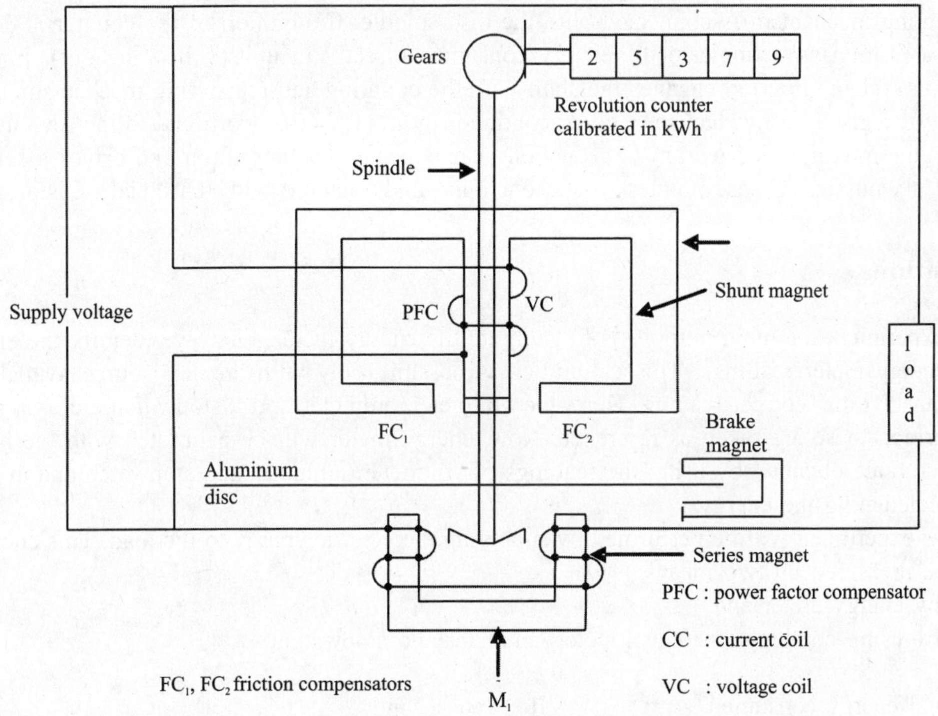

Fig. B.20.1. Single-phase energy meter.

Moving System

A thin aluminium disc mounted on a spindle (made up of light alloy shaft) is placed in the air gap between the series current coil and shunt pressure coil. It cuts the fluxes of both the magnets. Forces are produced by the fluxes of each of the electromagnets with the eddy current induced in the disc by the flux of the other magnet. A deflecting torque, which moves the shaft, is produced by these forces that acts on the disc and causes movement of the disc.

Braking System

A permanent magnet also called brake magnet placed near the edge of the aluminium disc makes braking system. Eddy current induces in the braking system when disc rotates in the field. These eddy currents react with the flux and exert torque in opposite direction of the motion of the disc. The magnitude of the braking torque is proportional to the speed of the disc.

Registering System

This system records a number, which is proportional to the number of revolutions of the disc. The counting mechanism is connected with the disc spindle. Calibration of the counter is done in such a fashion as to indicate the energy consumed directly in units (kilowatt-hour). In the registering system, a pinion engages the shaft with the counting and registering mechanism. The counting or registering mechanism records continuously a number proportional to the revolution made by the moving disc. A train of reduction gear is driven by the pinion and drives a series of about six pointers. These pointers rotate on round dials marked with ten equal divisions.

Calibration

In this experiment the main purpose is to calibrate the energy meter. First, we identify the error in the energy meter readings. This calibration is possible only with another source available, which can give the correct reading. Here, the voltmeter, ammeter and a stopwatch are assumed to be correct, these are taken as reference. Now, energy meter will be calibrated with the help of correct value obtained by using the readings of voltmeter, ammeter and stopwatch and at the last by calculating the energy.

Now, experiment will be performed by supplying the electric energy to the load. This energy would be measured by two means:

(i) by energy meter and
(ii) by using combination of voltmeter, ammeter and stopwatch.

The actual energy consumed = VIt watt second
= VIt/(3600 × 1000) kWh
= VIt/(36 × 100000) kWh

The time't' is measured in seconds for a given number of revolutions of the disc of the energy meter. A constant marked on the meters is expressed in terms of number of revolutions per kWh, i.e., 1200 revolutions per kWh. Then the energy meter records the energy. In forty revolutions it will give 40/1200 kWh. Calculation of error in the instrument is done as under.

Error = (Actual energy – Recorded energy).

Calibration curve of the energy meter is a graph of error vs load current. Calibration curve is to be plotted.

PROCEDURE

Steps

1. Connect the circuit as shown in Figure B.20.1. Also connect ammeter in series and voltmeter in parallel of the load. Apply constant AC voltage.
2. Switch 'ON' one part of the load.
3. Record the time taken for zero revolutions of the disc with the help of the stopwatch.
4. Take voltmeter and ammeter readings.
5. Repeat for more number of readings.
6. Adjust the load resistance so that a little bit high current (8-10 ampere) flows in the circuit for half an hour. Take voltmeter and. ammeter readings at regular intervals say after each 15 minutes. Note the dial reading in the beginning and at the end of half an hour. The error in energy indicated by energy meter will be the difference of the two readings. Calculate the actual value of energy consumed by using the following formula.
   $E = VI \times 15/(1000 \times 60)$ kWh
7. Calculate, % error using the formula
   $$\% \text{ Error} = 100 \times (\text{Actual energy} - \text{Recorded energy})/\text{Actual energy}$$
8. Plot the graph, error vs load current.

Observation Table

The observation taken during the experiment can be recorded as below.

S/N	V volts	I amps	t sec	Actual (VIt) Energy; kWh	Recorded Energy; kWh	%Error
1.						
2.						
3.						
4.						
5.						

PRECAUTIONS

1. All the connections should be clean and tight.
2. Check that stopwatch does not have any error.
3. The number of revolutions of the disc of the energy meter should be counted with reference to the red mark on the disc.

Conclusion: (Describe in brief, whatever you have learned.)

QUESTIONS

(To be answered in Practical Record Book)

1. What do you mean by calibration of an instrument?
2. Why calibration of energy meter is essential time to time?
3. Where is induction type meters used?
4. How many terminals an energy meter has?
5. What do you mean by the consumption of '*one unit*' electricity?

Index